中 外 物 理 学 精 品 书 系

本书出版得到“国家出版基金”资助

中 外 物 理 学 精 品 书 系

引 进 系 列 · 68

Introduction to Cosmology
2nd Edition

宇宙学导论
第二版

（影印版）

〔美〕芭芭拉·赖登（Barbara Ryden）著

著作权合同登记号　图字:01-2019-2622

图书在版编目(CIP)数据

宇宙学导论:第二版＝Introduction to Cosmology:2nd Edition:影印版:英文/(美)芭芭拉・赖登(Barbara Ryden)著.—北京:北京大学出版社,2019.10

(中外物理学精品书系)

ISBN 978-7-301-30788-5

Ⅰ.①宇…　Ⅱ.①芭…　Ⅲ.①宇宙学—英文　Ⅳ.①P159

中国版本图书馆 CIP 数据核字(2019)第 203989 号

Introduction to Cosmology, 2nd edition (ISBN-13: 9781107154834) by B. Ryden, first published by Cambridge University Press 2017.

书　　名　Introduction to Cosmology　2nd Edition（宇宙学导论 第二版）（影印版）

著作责任者　〔美〕芭芭拉・赖登（Barbara Ryden）　著

责 任 编 辑　刘　啸

标 准 书 号　ISBN 978-7-301-30788-5

出 版 发 行　北京大学出版社

地　　　址　北京市海淀区成府路 205 号　100871

网　　　址　http://www.pup.cn　　新浪微博:@北京大学出版社

电 子 信 箱　zpup@pup.cn

电　　　话　邮购部 010-62752015　发行部 010-62750672　编辑部 010-62754271

印　刷　者　北京中科印刷有限公司

经　销　者　新华书店

730 毫米×980 毫米　16 开本　17.5 印张　346 千字

2019 年 10 月第 1 版　2019 年 10 月第 1 次印刷

定　　　价　69.00 元

序　言

物理学是研究物质、能量以及它们之间相互作用的科学。她不仅是化学、生命、材料、信息、能源和环境等相关学科的基础，同时还与许多新兴学科和交叉学科的前沿紧密相关。在科技发展日新月异和国际竞争日趋激烈的今天，物理学不再囿于基础科学和技术应用研究的范畴，而是在国家发展与人类进步的历史进程中发挥着越来越关键的作用。

我们欣喜地看到，改革开放四十年来，随着中国政治、经济、科技、教育等各项事业的蓬勃发展，我国物理学取得了跨越式的进步，成长出一批具有国际影响力的学者，做出了很多为世界所瞩目的研究成果。今日的中国物理，正在经历一个历史上少有的黄金时代。

在我国物理学科快速发展的背景下，近年来物理学相关书籍也呈现百花齐放的良好态势，在知识传承、学术交流、人才培养等方面发挥着无可替代的作用。然而从另一方面看，尽管国内各出版社相继推出了一些质量很高的物理教材和图书，但系统总结物理学各门类知识和发展，深入浅出地介绍其与现代科学技术之间的渊源，并针对不同层次的读者提供有价值的学习和研究参考，仍是我国科学传播与出版领域面临的一个富有挑战性的课题。

为积极推动我国物理学研究、加快相关学科的建设与发展，特别是集中展现近年来中国物理学者的研究水平和成果，北京大学出版社在国家出版基金的支持下于 2009 年推出了"中外物理学精品书系"，并于 2018 年启动了书系的二期项目，试图对以上难题进行大胆的探索。书系编委会集结了数十位来自内地和香港顶尖高校及科研院所的知名学者。他们都是目前各领域十分活跃的知名专家，从而确保了整套丛书的权威性和前瞻性。

这套书系内容丰富、涵盖面广、可读性强，其中既有对我国物理学发展的梳理和总结，也有对国际物理学前沿的全面展示。可以说，"中外物理学精品书系"力图完整呈现近现代世界和中国物理科学发展的全貌，是一套目前国内为数不多的兼具学术价值和阅读乐趣的经典物理丛书。

“中外物理学精品书系”的另一个突出特点是，在把西方物理的精华要义“请进来”的同时，也将我国近现代物理的优秀成果“送出去”。物理学在世界范围内的重要性不言而喻。引进和翻译世界物理的经典著作和前沿动态，可以满足当前国内物理教学和科研工作的迫切需求。与此同时，我国的物理学研究数十年来取得了长足发展，一大批具有较高学术价值的著作相继问世。这套丛书首次成规模地将中国物理学者的优秀论著以英文版的形式直接推向国际相关研究的主流领域，使世界对中国物理学的过去和现状有更多、更深入的了解，不仅充分展示出中国物理学研究和积累的“硬实力”，也向世界主动传播我国科技文化领域不断创新发展的“软实力”，对全面提升中国科学教育领域的国际形象起到一定的促进作用。

习近平总书记在2018年两院院士大会开幕会上的讲话强调，“中国要强盛、要复兴，就一定要大力发展科学技术，努力成为世界主要科学中心和创新高地”。中国未来的发展在于创新，而基础研究正是一切创新的根本和源泉。我相信，在第一期的基础上，第二期“中外物理学精品书系”会努力做得更好，不仅可以使所有热爱和研究物理学的人们从中获取思想的启迪、智力的挑战和阅读的乐趣，也将进一步推动其他相关基础科学更好更快地发展，为我国的科技创新和社会进步做出应有的贡献。

“中外物理学精品书系”编委会主任
中国科学院院士，北京大学教授
王恩哥
2018年7月于燕园

Introduction to Cosmology

Second Edition

Barbara Ryden

The Ohio State University

For my husband

Contents

Preface *page* xi

1 Introduction 1

2 Fundamental Observations 6

2.1 The Night Sky is Dark 6

2.2 The Universe is Isotropic and Homogeneous 9

2.3 Redshift is Proportional to Distance 12

2.4 Different Types of Particles 18

2.5 Cosmic Microwave Background 23

Exercises 25

3 Newton versus Einstein 27

3.1 The Way of Newton 28

3.2 The Special Way of Einstein 29

3.3 The General Way of Einstein 34

3.4 Describing Curvature 37

3.5 The Robertson–Walker Metric 41

3.6 Proper Distance 43

Exercises 47

4 Cosmic Dynamics 49

4.1 Einstein's Field Equation 50

4.2 The Friedmann Equation 52

4.3 The Fluid and Acceleration Equations 58

4.4 Equations of State 60
4.5 Learning to Love Lambda 63
Exercises 67

5 Model Universes 69
5.1 Evolution of Energy Density 69
5.2 Empty Universes 74
5.3 Single-component Universes 77
5.3.1 Matter only 80
5.3.2 Radiation only 81
5.3.3 Lambda only 83
5.4 Multiple-component Universes 84
5.4.1 Matter + Curvature 86
5.4.2 Matter + Lambda 90
5.4.3 Matter + Curvature + Lambda 92
5.4.4 Radiation + Matter 95
5.5 Benchmark Model 96
Exercises 100

6 Measuring Cosmological Parameters 102
6.1 "A Search for Two Numbers" 102
6.2 Luminosity Distance 106
6.3 Angular-diameter Distance 110
6.4 Standard Candles and H_0 114
6.5 Standard Candles and Acceleration 116
Exercises 121

7 Dark Matter 123
7.1 Visible Matter 123
7.2 Dark Matter in Galaxies 128
7.3 Dark Matter in Clusters 130
7.4 Gravitational Lensing 135
7.5 What's the Matter? 139
Exercises 140

8 The Cosmic Microwave Background 142
8.1 Observing the CMB 143
8.2 Recombination and Decoupling 147
8.3 The Physics of Recombination 150
8.4 Temperature Fluctuations 157
8.5 What Causes the Fluctuations? 159
Exercises 164

9 Nucleosynthesis and the Early Universe 166
9.1 Nuclear Physics and Cosmology 167
9.2 Neutrons and Protons 169
9.3 Deuterium Synthesis 174
9.4 Beyond Deuterium 177
9.5 Baryon–Antibaryon Asymmetry 181
Exercises 183

10 Inflation and the Very Early Universe 185
10.1 The Flatness Problem 186
10.2 The Horizon Problem 187
10.3 The Monopole Problem 189
10.4 The Inflation Solution 192
10.5 The Physics of Inflation 197
Exercises 202

11 Structure Formation: Gravitational Instability 204
11.1 The Matthew Effect 206
11.2 The Jeans Length 209
11.3 Instability in an Expanding Universe 213
11.4 The Power Spectrum 217
11.5 Hot versus Cold 221
11.6 Baryon Acoustic Oscillations 226
Exercises 230

12 Structure Formation: Baryons and Photons 232

12.1 Baryonic Matter Today 233

12.2 Reionization of Hydrogen 235

12.3 The First Stars and Quasars 238

12.4 Making Galaxies 242

12.5 Making Stars 248

Exercises 254

Epilogue 256

Table of Useful Constants 258

Index 259

Preface

The first edition of this book was based on my lecture notes for an upper-level undergraduate cosmology course at The Ohio State University. The students taking the course were primarily juniors and seniors majoring in physics and astronomy. In my lectures, I assumed that my students, having triumphantly survived freshman and sophomore physics, had a basic understanding of electrodynamics, statistical mechanics, classical dynamics, and quantum physics. As far as mathematics was concerned, I assumed that, like modern major-generals, they were very good at integral and differential calculus. Readers of this book are assumed to have a similar background in physics and mathematics. In particular, no prior knowledge of general relativity is assumed; the (relatively) small amounts of general relativity needed to understand basic cosmology are introduced as needed.

The second edition that you are reading now is updated with observational and theoretical developments during the 14 years that have elapsed since the first edition. It has been improved by many comments by readers. (My thanks go to the eagle-eyed readers who caught the typographical errors that snuck into the first edition.) The second edition also contains an extended discussion of structure formation in the final two chapters. For a brief course on cosmology, the first ten chapters can stand on their own.

Unfortunately, the National Bureau of Standards has not gotten around to establishing a standard notation for cosmological equations. It seems that every cosmology book has its own notation; this book is no exception. My main motivation was to make the notation as clear as possible for the cosmological novice.

Many of the illustrations in this book were adapted from figures in published scientific papers; my thanks go to the authors of those papers for granting permission to use their figures or to replot their hard-won data. Particular thanks go to Avishai Dekel (Figure 2.2), Wendy Freedman (Figure 2.5), David Leisawitz (Figure 8.1), Alain Coc (Figure 9.4), Richard Cyburt (Figure 9.5), Rien van de Weygaert (Figure 11.1), Ashley Ross (Figure 11.6), Xiaohui Fan (Figure 12.3), and John Beacom (Figure 12.4). Extra thanks are due to Anže Slosar and José Alberto Vázquez for their assistance with Figures 6.6, 8.7, and 11.7.

Many people (too many to name individually) helped in the making of this book. I owe particular thanks to the students who took my undergraduate cosmology course at Ohio State University. Their feedback, including nonverbal feedback such as frowns and snores during lectures, greatly improved the lecture notes on which the first edition was based. The students of the graduate cosmology course at Ohio State have assisted in the development of the second edition, by field-testing the end-of-chapter problems, proposing new problems, and acting as all-around critics of the manuscript. Adam Black and Nancy Gee, at Pearson Addison Wesley, made possible the great leap from rough lecture notes to polished book. Vince Higgs and Rachel Cox, at Cambridge University Press, helped with the second great leap to a new, improved second edition. The reviewers of the text, in both its first and second editions, pointed out many omissions and suggested many improvements.

The first edition of this book was dedicated to Rick Pogge, who acted as my computer maven, graphics guru, personal chef, and general sanity check. Obviously, there was only one thing to do with such a paragon. Reader, I married him.

1

Introduction

Cosmology is the study of the universe, or cosmos, regarded as a whole. Attempting to cover the study of the entire universe in a single volume may seem like a megalomaniac's dream. The universe, after all, is richly textured, with structures on a vast range of scales; planets orbit stars, stars are collected into galaxies, galaxies are gravitationally bound into clusters, and even clusters of galaxies are found within larger superclusters. Given the complexity of the universe, the only way to condense its history into a single book is by a process of ruthless simplification. For much of this book, therefore, we will be considering the properties of an idealized, perfectly smooth, model universe. Only near the end of the book will we consider how relatively small objects, such as galaxies, clusters, and superclusters, are formed as the universe evolves. It is amusing to note in this context that the words *cosmology* and *cosmetology* come from the same Greek root: the word *kosmos*, meaning harmony or order. Just as cosmetologists try to make a human face more harmonious by smoothing over small blemishes such as pimples and wrinkles, cosmologists sometimes must smooth over small "blemishes" such as galaxies.

A science that regards entire galaxies as being small objects might seem, at first glance, very remote from the concerns of humanity. Nevertheless, cosmology deals with questions that are fundamental to the human condition. The questions that vex humanity are given in the title of a painting by Paul Gauguin (Figure 1.1): "Where do we come from? What are we? Where are we going?" Cosmology grapples with these questions by describing the past, explaining the present, and predicting the future of the universe. Cosmologists ask questions such as "What is the universe made of? Is it finite or infinite in spatial extent? Did it have a beginning some time in the past? Will it come to an end some time in the future?"

Cosmology deals with distances that are very large, objects that are very big, and timescales that are very long. Cosmologists frequently find that the standard SI units are not convenient for their purposes: the meter (m) is awkwardly

Figure 1.1 *Where Do We Come From? What Are We? Where Are We Going?* Paul Gauguin, 1897–98. [Museum of Fine Arts, Boston]

short, the kilogram (kg) is awkwardly tiny, and the second (s) is awkwardly brief. Fortunately, we can adopt the units that have been developed by astronomers for dealing with large distances, masses, and times.

One distance unit used by astronomers is the astronomical unit (AU), equal to the mean distance between the Earth and Sun; in metric units, $1\,\text{AU} = 1.50 \times 10^{11}\,\text{m}$. Although the astronomical unit is a useful length scale within the solar system, it is small compared to the distances between stars. To measure interstellar distances, it is useful to use the parsec (pc), equal to the distance at which 1 AU subtends an angle of 1 arcsecond; in metric units, $1\,\text{pc} = 3.09 \times 10^{16}\,\text{m}$. For example, we are at a distance of 1.30 pc from Proxima Centauri (a small, relatively cool star that is the Sun's nearest neighboring star); we are at a distance of 8500 pc from the center of our galaxy, the Milky Way Galaxy. Although the parsec is a useful length scale within our galaxy, it is small compared to the distances between galaxies. To measure intergalactic distances, we use the megaparsec (Mpc), equal to 10^6 pc, or 3.09×10^{22} m. For example, we are at a distance of 0.76 Mpc from M31 (otherwise known as the Andromeda galaxy) and 15 Mpc from the Virgo cluster (the nearest big cluster of galaxies).

The standard unit of mass used by astronomers is the solar mass ($\text{M}_\odot$); in metric units, the Sun's mass is $1\,\text{M}_\odot = 1.99 \times 10^{30}\,\text{kg}$. The total mass of our galaxy is not known as accurately as the mass of the Sun; in round numbers, though, it is $M_{\text{gal}} \sim 10^{12}\,\text{M}_\odot$. The Sun, incidentally, also provides the standard unit of power used in astronomy. The Sun's luminosity (that is, the rate at which it radiates away energy in the form of light) is $1\,\text{L}_\odot = 3.83 \times 10^{26}$ watts. The total luminosity of our galaxy is not known as accurately as the luminosity of the Sun; a good estimate, though, is $L_{\text{gal}} \approx 3 \times 10^{10}\,\text{L}_\odot$.

For times much longer than a second, it is convenient to use the year (yr) as a unit of time, with $1\,\text{yr} \approx 3.16 \times 10^7\,\text{s}$. In a cosmological context, a year is frequently an inconveniently short period of time, so cosmologists often use megayears (Myr), with 1 Myr $= 10^6$ yr $= 3.16 \times 10^{13}$ s. Even longer timescales call for use of gigayears (Gyr), with 1 Gyr $= 10^9$ yr $= 3.16 \times 10^{16}$ s. For example, the age of the Earth is more conveniently written as 4.57 Gyr than as 1.44×10^{17} s.

In addition to dealing with very large things, cosmology also deals with very small things. Early in its history, as we shall see, the universe was very hot and dense, and some interesting particle physics phenomena were occurring. Consequently, particle physicists have plunged into cosmology, introducing some terminology and units of their own. For instance, particle physicists tend to measure energy units in electron volts (eV) instead of joules (J). The conversion factor between electron volts and joules is $1\,\text{eV} = 1.60 \times 10^{-19}$ J. The rest energy of an electron, for instance, is $m_e c^2 = 511\,000\,\text{eV} = 0.511\,\text{MeV}$, and the rest energy of a proton is $m_p c^2 = 938.27\,\text{MeV} = 1836.1 m_e c^2$.

When you stop to think of it, you realize that the units of meters, megaparsecs, kilograms, solar masses, seconds, and gigayears could only be devised by ten-fingered Earthlings obsessed with the properties of water. An eighteen-tentacled silicon-based lifeform from a planet orbiting Betelgeuse would probably devise a different set of units. A more universal, less culturally biased system of units is the Planck system, based on the universal constants G, c, and $\hbar$. Combining the Newtonian gravitational constant, $G = 6.67 \times 10^{-11}\,\text{m}^3\,\text{kg}^{-1}\,\text{s}^{-2}$, the speed of light, $c = 3.00 \times 10^8\,\text{m}\,\text{s}^{-1}$, and the reduced Planck constant, $\hbar = h/(2\pi) = 1.05 \times 10^{-34}\,\text{J}\,\text{s} = 6.58 \times 10^{-16}\,\text{eV}\,\text{s}$, yields a unique length scale, known as the Planck length:

$$\ell_P \equiv \left(\frac{G\hbar}{c^3}\right)^{1/2} = 1.62 \times 10^{-35}\,\text{m}. \tag{1.1}$$

The same constants can be combined to yield the Planck mass,[1]

$$M_P \equiv \left(\frac{\hbar c}{G}\right)^{1/2} = 2.18 \times 10^{-8}\,\text{kg}, \tag{1.2}$$

and the Planck time,

$$t_P \equiv \left(\frac{G\hbar}{c^5}\right)^{1/2} = 5.39 \times 10^{-44}\,\text{s}. \tag{1.3}$$

Using Einstein's relation between mass and energy, we can also define the Planck energy,

$$E_P = M_P c^2 = 1.96 \times 10^9\,\text{J} = 1.22 \times 10^{28}\,\text{eV}. \tag{1.4}$$

By bringing the Boltzmann constant, $k = 8.62 \times 10^{-5}\,\text{eV}\,\text{K}^{-1}$, into the act, we can also define the Planck temperature,

$$T_P = E_P/k = 1.42 \times 10^{32}\,\text{K}. \tag{1.5}$$

[1] The Planck mass is roughly equal to the mass of a grain of sand a quarter of a millimeter across.

When distance, mass, time, and temperature are measured in the appropriate Planck units, then $c = k = \hbar = G = 1$. This is convenient for individuals who have difficulty in remembering the numerical values of physical constants. However, using Planck units can have potentially confusing side effects. For instance, many cosmology texts, after noting that $c = k = \hbar = G = 1$ when Planck units are used, then proceed to omit c, k, $\hbar$, and/or G from all equations. For instance, Einstein's celebrated equation, $E = mc^2$, becomes $E = m$. The blatant dimensional incorrectness of such an equation is jarring, but it simply means that the rest energy of an object, measured in units of the Planck energy, is equal to its mass, measured in units of the Planck mass. In this book, however, I will retain all factors of c, k, $\hbar$, and G, for the sake of clarity.

Here we will deal with distances ranging from the Planck length to 10^4 Mpc or so, a span of some 61 orders of magnitude. Dealing with such a wide range of length scales requires a stretch of the imagination, to be sure. However, cosmologists are not permitted to let their imaginations run totally unfettered. Cosmology, I emphasize strongly, is based ultimately on observation of the universe around us. Even in ancient times, cosmology was based on observations; unfortunately, those observations were frequently imperfect and incomplete. Ancient Egyptians, for instance, looked at the desert plains stretching away from the Nile valley and the blue sky overhead. Based on their observations, they developed a model of the universe in which a flat Earth (symbolized by the earth god Geb in Figure 1.2) was covered by a solid dome (symbolized by the sky goddess Nut). Underneath the sky dome, the disk of the Sun was carried from east to west by the sun god Ra. Greek cosmology was based on more precise and sophisticated observations. Ancient Greek astronomers deduced, from their observations, that the Earth and

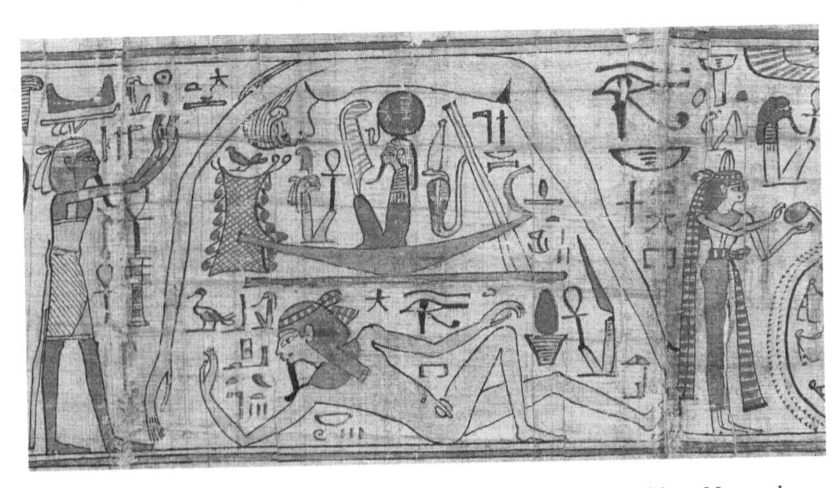

Figure 1.2 The ancient Egyptian view of the cosmos: the sky goddess Nut arches over the earth god Geb, while the sun god Ra travels between them in a reed boat. (Book of the Dead of Nespakashuty, ca. 1000 BC) [Musée du Louvre, Paris]

Moon are spherical, that the Sun is much farther from the Earth than the Moon is, and that the distance from the Earth to the stars is much greater than the Earth's diameter. Based on this knowledge, Greek cosmologists devised a "two-sphere" model of the universe, in which the spherical Earth is surrounded by a much larger celestial sphere, a spherical shell to which the stars are attached. Between the Earth and the celestial sphere, in this model, the Sun, Moon, and planets move on their complicated apparatus of epicycles and deferents.

Although cosmology is ultimately based on observation, sometimes observations temporarily lag behind theory. During periods when data are lacking, cosmologists may adopt a new model for aesthetic or philosophical reasons. For instance, when Copernicus proposed a new Sun-centered model of the universe, to replace the Earth-centered two-sphere model of the Greeks, he didn't base his model on new observational discoveries. Rather, he believed that putting the Earth in motion around the Sun resulted in a conceptually simpler, more appealing model of the universe. Direct observational evidence didn't reveal that the Earth revolves around the Sun, rather than vice versa, until the discovery of the aberration of starlight in the year 1728, nearly two centuries after the death of Copernicus. Foucault didn't demonstrate the rotation of the Earth, another prediction of the Copernican model, until 1851, over *three* centuries after the death of Copernicus. However, although observations sometimes lag behind theory in this way, every cosmological model that isn't eventually supported by observational evidence must remain pure speculation.

The current standard model for the universe is the "Hot Big Bang" model, which states that the universe has expanded from an initially hot and dense state to its current relatively cool and tenuous state, and that the expansion is still going on today. To see why cosmologists have embraced the Hot Big Bang model, let us turn, in the next chapter, to the fundamental observations on which modern cosmology is based.

2

Fundamental Observations

Some of the observations on which modern cosmology is based are highly complex, requiring elaborate apparatus and sophisticated data analysis. However, other observations are surprisingly simple. Let's start with an observation that is deceptive in its extreme simplicity.

2.1 The Night Sky is Dark

Step outside on a clear, moonless night, far from city lights, and look upward. You will see a dark sky, with roughly two thousand stars scattered across it. The fact that the night sky is dark at visible wavelengths, instead of being uniformly bright with starlight, is known as *Olbers' paradox*, after the astronomer Heinrich Olbers, who wrote a scientific paper on the subject in 1823. As it happens, Olbers was not the first person to think about Olbers' paradox. As early as 1576, Thomas Digges mentioned how strange it is that the night sky is dark, with only a few pinpoints of light to mark the location of stars.[1]

Why should it be paradoxical that the night sky is dark? Most of us simply take for granted the fact that daytime is bright and nighttime is dark. The darkness of the night sky certainly posed no problems to the ancient Egyptians or Greeks, to whom stars were lights stuck to a dome or sphere. However, the cosmological model of Copernicus required that the distance to stars be very much larger than an astronomical unit; otherwise, the parallax of the stars, as the Earth goes around on its orbit, would be large enough to see with the naked eye. Moreover, since the Copernican system no longer requires that the stars be attached to a rotating celestial sphere, the stars can be at different distances from the Sun. These

[1] The name "Olbers' paradox" is thus a prime example of what historians of science jokingly call the law of misonomy: nothing is ever named after the person who really discovers it. The law of misonomy is also known as "Stigler's law," after a statistician who admits that he (of course!) didn't discover it.

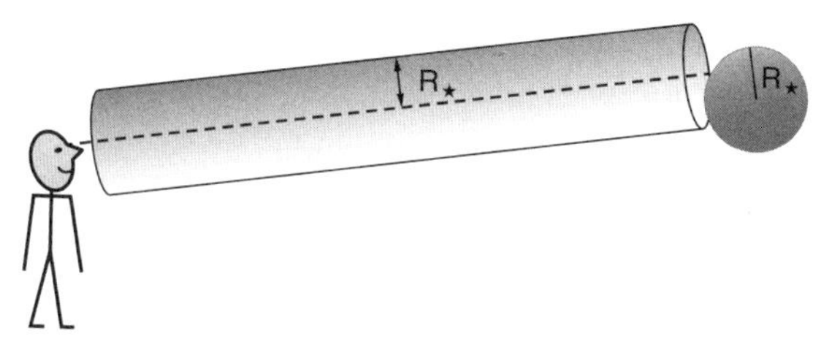

Figure 2.1 A line of sight through the universe eventually encounters an opaque star.

liberating realizations led Thomas Digges, and other post-Copernican astronomers, to embrace a model in which stars are large, opaque, glowing spheres like the Sun, scattered throughout infinite space.

Let's compute how bright we expect the night sky to be in an infinite universe. Let $n_\star$ be the number density of stars in the universe; averaged over large scales, this number is $n_\star \sim 10^9\,\mathrm{Mpc}^{-3}$. Let $R_\star$ be the typical radius of a star. Although stars have a range of sizes, from dwarfs to supergiants, we may adopt the Sun as a typical mid-sized star, with $R_\star \sim R_\odot = 7.0 \times 10^8\,\mathrm{m} = 2.3 \times 10^{-14}\,\mathrm{Mpc}$. Consider looking outward in some direction through the universe. If you draw a cylinder of radius $R_\star$ around your line of sight, as shown in Figure 2.1, then if a star's center lies within that cylinder, the opaque star will block your view of more distant objects. If the cylinder's length is λ, then its volume is $V = \lambda\pi R_\star^2$, and the average number of stars that have their centers inside the cylinder is

$$N = n_\star V = n_\star \lambda \pi R_\star^2. \tag{2.1}$$

Since it requires only one star to block your view, the typical distance you will be able to see before a star blocks your line of sight is the distance λ for which $N = 1$. From Equation (2.1), this distance is

$$\lambda = \frac{1}{n_\star \pi R_\star^2}. \tag{2.2}$$

For concreteness, if we take $n_\star \sim 10^9\,\mathrm{Mpc}^{-3}$ and $\pi R_\star^2 \sim \pi R_\odot^2 \sim 10^{-27}\,\mathrm{Mpc}^2$, then you can see a distance

$$\lambda \sim \frac{1}{(10^9\,\mathrm{Mpc}^{-3})(10^{-27}\,\mathrm{Mpc}^2)} \sim 10^{18}\,\mathrm{Mpc} \tag{2.3}$$

before your line of sight intercepts a star. This is a very large distance; but it is a *finite* distance. We therefore conclude that in an infinite universe (or one that stretches at least 10^{18} Mpc in all directions), the sky will be completely paved with stars.

What does this paving imply for the brightness of the sky? If a star of radius $R_\star$ is at a distance $r \gg R_\star$, its angular area, in steradians, will be

$$\Omega = \frac{\pi R_\star^2}{4\pi r^2} = \frac{R_\star^2}{4r^2}. \tag{2.4}$$

If the star's luminosity is $L_\star$, then its flux measured at a distance r will be

$$f = \frac{L_\star}{4\pi r^2}. \tag{2.5}$$

The surface brightness of the star, in watts per square meter of your pupil (or telescope mirror) per steradian, will then be

$$\Sigma_\star = \frac{f}{\Omega} = \frac{L_\star}{\pi R_\star^2}, \tag{2.6}$$

independent of the distance to the star. Thus, the surface brightness of a sky paved with stars will be equal to the (distance-independent) surface brightness of an individual star. We therefore conclude that in an infinite universe (or one that stretches at least 10^{18} Mpc in all directions), the entire sky, night and day, should be as dazzlingly bright as the Sun's disk.

This is utter nonsense. The surface brightness of the Sun is $\Sigma_\odot \approx 5 \times 10^{-3}$ watts m^{-2} arcsec^{-2}. By contrast, the surface brightness of the dark night sky is $\Sigma \sim 5 \times 10^{-17}$ watts m^{-2} arcsec^{-2}. Thus, my estimate of the surface brightness of the night sky ("It's the same as the Sun's") is wrong by a factor of 100 trillion.

One (or more) of the assumptions that went into my estimate of the sky brightness must be wrong. Let's scrutinize some of the assumptions. One assumption that I made is that space is transparent over distances of 10^{18} Mpc. This might not be true. Heinrich Olbers himself tried to resolve Olbers' paradox by proposing that distant stars are hidden from view by interstellar matter that absorbs starlight. This resolution does not work in the long run, because the interstellar matter is heated by starlight until it has the same temperature as the surface of a star. At that point, the interstellar matter emits as much light as it absorbs, and glows as brightly as the stars themselves.

A second assumption that I made is that the universe is infinitely large. This might not be true. If the universe extends to a maximum distance $r_{\max} \ll \lambda$, then only a fraction $F \sim r_{\max}/\lambda$ of the night sky will be covered with stars. This result will also be found if the universe is infinitely large, but is devoid of stars beyond a distance $r_{\max}$.

A third assumption, slightly more subtle than the previous ones, is that the universe is infinitely old. This might not be true. Because the speed of light is finite, when we look farther out in space, we are looking farther out in time. Thus, we see the Sun as it was 8.3 minutes ago, Proxima Centauri as it was 4.2 years ago, and M31 as it was 2.5 million years ago. If the universe has a finite age, $t_0 \ll \lambda/c$, then we are not yet able to see stars at a distance greater than $r \sim ct_0$, and only a fraction $F \sim ct_0/\lambda$ of the night sky will be covered with stars. This result will also be found if the universe is infinitely old, but has only contained stars for a finite time t_0.

A fourth assumption is that the surface brightness of a star is independent of distance, as derived in Equation 2.6. This might not be true. The assumption of constant surface brightness would have seemed totally innocuous to Olbers and other nineteenth-century astronomers, who assumed that the universe was static. However, in an expanding universe, the surface brightness of distant light sources is decreased relative to what you would see in a static universe. (In a contracting universe, the surface brightness would be increased, which would only make the problem of a bright night sky even worse.)

Thus, the infinitely large, eternally old, static universe that Thomas Digges and his successors pictured simply does not hold up to scrutiny. This is a textbook, not a suspense novel, so I'll tell you right now: the primary resolution to Olbers' paradox comes from the fact that the universe has a finite age. The stars beyond some finite distance, called the horizon distance, are invisible to us because their light hasn't had time to reach us yet. A particularly amusing bit of cosmological trivia is that the first person to hint at the correct resolution of Olbers' paradox was Edgar Allan Poe.[2] In his essay "Eureka: A Prose Poem," completed in 1848, Poe wrote, "Were the succession of stars endless, then the background of the sky would present us an [*sic*] uniform density... since there could be absolutely no point, in all that background, at which would not exist a star. The only mode, therefore, in which, under such a state of affairs, we could comprehend the voids which our telescopes find in innumerable directions, would be by supposing the distance of the invisible background so immense that no ray from it has yet been able to reach us at all."

2.2 The Universe is Isotropic and Homogeneous

What does it mean to state that the universe is isotropic and homogeneous? Saying that the universe is *isotropic* means that there are no preferred directions in the universe; it looks the same no matter which way you point your telescope. Saying that the universe is *homogeneous* means that there are no preferred locations in the universe; it looks the same no matter where you set up your telescope. Note the very important qualifier: the universe is isotropic and homogeneous *on large scales*. In this context, "large scales" means that the universe is only isotropic and homogeneous on scales of roughly 100 Mpc or more.

[2] That's right, the "Nevermore" guy. Poe was an excellent student at the University of Virginia (before he fell into debt and withdrew). He was then an excellent student at West Point (before he was court-martialed and expelled).

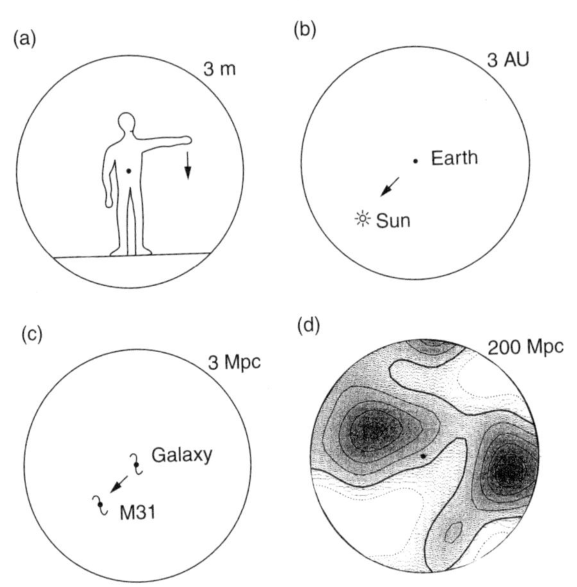

Figure 2.2 (a) A sphere 3 m in diameter, centered on your navel. (b) A sphere 3 AU in diameter, centered on your navel. (c) A sphere 3 Mpc in diameter, centered on your navel. (d) A sphere 200 Mpc in diameter, centered on your navel. Shown is the smoothed number density of galaxies. The heavy contour is drawn at the mean density; darker regions represent higher density. [Dekel *et al.* 1999, *ApJ*, **522**, 1]

The isotropy of the universe is not immediately obvious. In fact, on small scales, the universe is blatantly anisotropic. Consider, for example, a sphere 3 m in diameter, centered on your navel (Figure 2.2a). Within this sphere, there is a preferred direction; it is the direction commonly referred to as "down." It is easy to determine the vector pointing down. Just let go of a small dense object. The object doesn't hover in midair, and it doesn't move in a random direction; it falls down, toward the center of the Earth.

On significantly larger scales, the universe is still anisotropic. Consider, for example, a sphere 3 AU in diameter, centered on your navel (Figure 2.2b). Within this sphere, there is a preferred direction; it is the direction pointing toward the Sun, which is by far the most massive and most luminous object within the sphere. It is easy to determine the vector pointing toward the Sun. Just step outside on a sunny day, and point to that really bright disk of light up in the sky.

On still larger scales, the universe is *still* anisotropic. Consider, for example, a sphere 3 Mpc in diameter, centered on your navel (Figure 2.2c). This sphere contains the Local Group of galaxies, a small cluster of about a hundred galaxies. By far the most massive and most luminous galaxies in the Local Group are our own galaxy and M31, which together contribute about 86 percent of the total luminosity within the 3 Mpc sphere. Thus, within this sphere, our galaxy and

M31 define a preferred direction. It is fairly easy to determine the vector pointing from our galaxy to M31. Just step outside on a clear night when the constellation Andromeda is above the horizon, and point to the fuzzy oval in the middle of the constellation.

It isn't until you get to considerably larger scales that the universe can be considered as isotropic. Consider a sphere 200 Mpc in diameter, centered on your navel. Figure 2.2d shows a slice through such a sphere, with superclusters of galaxies indicated as dark patches. The Perseus–Pisces supercluster is on the right, the Hydra–Centaurus supercluster is on the left, and the edge of the Coma supercluster is just visible at the top of Figure 2.2d. Superclusters are typically ~ 100 Mpc along their longest dimensions, and are separated by voids (low density regions) which are typically ~ 100 Mpc across. These are the largest structures in the universe, it seems; surveys of the universe on still larger scales don't find "superduperclusters."

On small scales, the universe is obviously inhomogeneous, or lumpy, in addition to being anisotropic. For instance, a sphere 3 m in diameter, centered on your navel, will have an average density of $\sim 100\,\text{kg m}^{-3}$, in round numbers. However, the average matter density of the universe as a whole is $\rho_0 \approx 2.7 \times 10^{-27}\,\text{kg m}^{-3}$. Thus, on a scale $d \sim 3$ m, the patch of the universe surrounding you is more than 28 orders of magnitude denser than average.

On significantly larger scales, the universe is still inhomogeneous. A sphere 3 AU in diameter, centered on your navel, has an average density of $4 \times 10^{-5}\,\text{kg m}^{-3}$; that's 22 orders of magnitude denser than the average for the universe.

On still larger scales, the universe is *still* inhomogeneous. A sphere 3 Mpc in diameter, centered on your navel, will have an average density of $\sim 3 \times 10^{-26}\,\text{kg m}^{-3}$, still an order of magnitude denser than the universe as a whole. It's only when you contemplate a sphere ~ 100 Mpc in diameter that a sphere centered on your navel is not overdense compared to the universe as a whole.

Note that homogeneity does not imply isotropy. A sheet of paper printed with stripes (Figure 2.3 left) is homogeneous on scales larger than the stripe width, but it is not isotropic. The direction of the stripes provides a preferred direction by which you can orient yourself. Note also that isotropy around a single point does not imply homogeneity. A sheet of paper printed with a bullseye (Figure 2.3 right) is isotropic around the center of the bullseye, but it is not homogeneous. The rings of the bullseye look different far from the center than they look close to the center. You can tell where you are relative to the center by measuring the radius of curvature of the nearest ring.

In general, then, saying that something is homogeneous is quite different from saying it is isotropic. However, modern cosmologists have adopted the *Copernican principle*, which states "There is nothing special or privileged about our location in the universe." The Copernican principle holds true only on large scales

Figure 2.3 Left: a pattern that is anisotropic, but is homogeneous on scales larger than the stripe width. Right: a pattern that is isotropic about the origin, but is inhomogeneous.

(of 100 Mpc or more). On smaller scales, your navel obviously is in a special location. Most spheres 3 m across don't contain a sentient being; most spheres 3 AU across don't contain a star; most spheres 3 Mpc across don't contain a pair of bright galaxies. However, most spheres over 100 Mpc across do contain roughly the same pattern of superclusters and voids, statistically speaking. The universe, on scales of 100 Mpc or more, appears to be isotropic around us. Isotropy around any point in the universe, such as your navel, combined with the Copernican principle, implies isotropy around every point in the universe; and isotropy around every point in the universe *does* imply homogeneity.

The observed isotropy of the universe on scales 100 Mpc or more, combined with the assumption of the Copernican principle, leads us to state "The universe (on large scales) is homogeneous and isotropic." This statement is known as the *cosmological principle*. Although the Copernican principle forbids us to say "We're number one!", the cosmological principle permits us to say "We're second to none!"

2.3 Redshift is Proportional to Distance

When we look at a galaxy at visible wavelengths, we detect primarily the light from the stars that the galaxy contains. Thus, when we take a galaxy's spectrum at visible wavelengths, it typically contains absorption lines created in the stars' relatively cool upper atmospheres; galaxies with active galactic nuclei will also show *emission* lines from the hot gas in their nuclei. Suppose we consider a particular absorption or emission line whose wavelength, as measured in a laboratory here on Earth, is $\lambda_{\rm em}$. The wavelength we measure for the same line in a distant galaxy's observed spectrum, $\lambda_{\rm ob}$, will not, in general, be the same. We say that the galaxy has a redshift z, given by the formula

$$z \equiv \frac{\lambda_{\rm ob} - \lambda_{\rm em}}{\lambda_{\rm em}}. \tag{2.7}$$

Strictly speaking, when $z < 0$, this quantity is called a blueshift, rather than a redshift. However, the vast majority of galaxies have $z > 0$.

The fact that the light from galaxies is generally redshifted to longer wavelengths, rather than blueshifted to shorter wavelengths, was not known until the twentieth century. In 1912, Vesto Slipher at the Lowell Observatory measured the shift in wavelength of the light from the galaxy M31. He found $z = -0.001$, meaning that M31 is one of the few galaxies that exhibit a blueshift rather than a redshift. Slipher interpreted the shift in wavelength as being due to the Doppler effect. Since $|z| \ll 1$ for M31, he used the classical, nonrelativistic relation for the Doppler shift, $z = v/c$, to compute that M31 is moving toward the Earth with a speed $v = -0.001c = -300\,\mathrm{km\,s^{-1}}$.

In the year 1927, the Belgian cosmologist Georges Lemaître compiled a list of 42 galaxies whose wavelength shift had been measured, mostly by Vesto Slipher. Of these galaxies, 37 were redshifted, and only 5 were blueshifted. This is a notable excess of redshifts; by analogy, if you have a fair coin and flip it 42 times, the chance of getting "heads" 37 or more times is $P \approx 2 \times 10^{-7}$. The average radial velocity of all 42 galaxies in the sample was $v = +600\,\mathrm{km\,s^{-1}}$. Lemaître pointed out that these relatively high speeds (much higher than the average speed of stars within our galaxy) could result from an expansion of the universe. Using an estimated average distance of $r = 0.95\,\mathrm{Mpc}$ for the galaxies in his sample, he concluded that the expansion was described by the parameter $K \equiv v/r = 625\,\mathrm{km\,s^{-1}\,Mpc^{-1}}$.

Although Lemaître made an estimate of the average distance to the galaxies in his sample, finding an accurate distance to an individual galaxy was quite difficult. The astronomer Edwin Hubble invested a great deal of effort into measuring the distances to galaxies. By 1929, he had estimated distances for a sample of 20 galaxies whose value of z had been measured. Figure 2.4 shows Hubble's plot of redshift (z) versus distance (r) for these galaxies. He noted that the more distant galaxies had higher redshifts, and fitted the data with the famous linear relation now known as Hubble's law:

$$z = \frac{H_0}{c} r, \tag{2.8}$$

where H_0 is a constant (now called the Hubble constant). Interpreting the redshifts as Doppler shifts, Hubble's law takes the form

$$v = H_0 r. \tag{2.9}$$

Thus, Lemaître's expansion parameter K, if we assume $z \propto r$, can be thought of as the first measurement of the Hubble constant.

The Hubble constant H_0 can be found by dividing velocity by distance, so it is customarily written in the rather baroque units of $\mathrm{km\,s^{-1}\,Mpc^{-1}}$. When Hubble first discovered Hubble's law, he thought that the numerical value of the

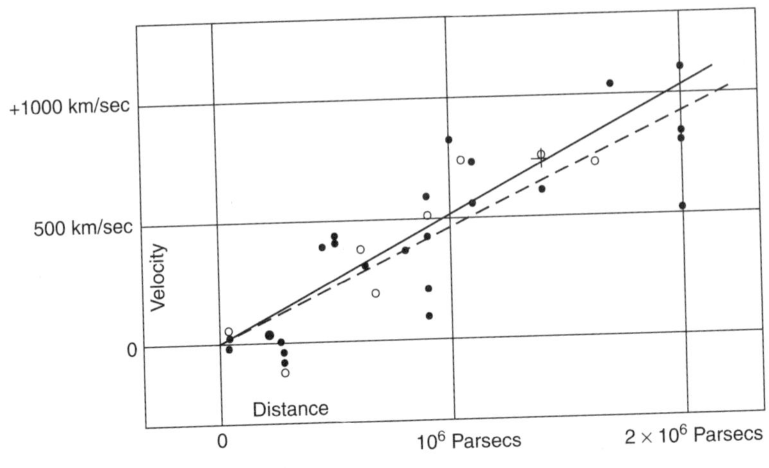

Figure 2.4 Edwin Hubble's original plot of the relation between radial velocity (assuming the formula $v = cz$) and distance. [Hubble 1929, *PNAS*, **15**, 168]

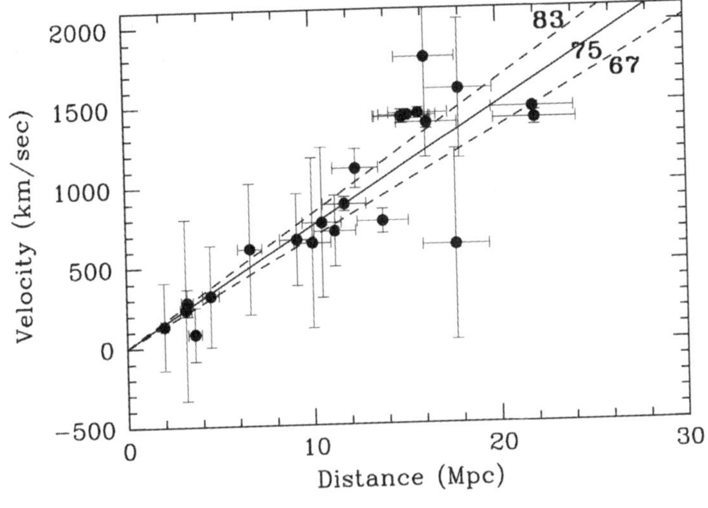

Figure 2.5 A more recent version of Hubble's plot, showing cz versus distance. In this case, the galaxy distances have been determined using Cepheid variable stars as standard candles, as described in Section 6.4. [Freedman *et al.* 2001, *ApJ*, **553**, 47]

Hubble constant was $H_0 \approx 500\,\text{km}\,\text{s}^{-1}\,\text{Mpc}^{-1}$, as shown in Figure 2.4. However, it turned out that Hubble was severely underestimating the distances to individual galaxies, just as Lemaître, who was relying on techniques pioneered by Hubble, was underestimating the average distance to nearby galaxies.

Figure 2.5 shows a more recent determination of the Hubble constant from nearby galaxies, using data obtained using the *Hubble Space Telescope*. Notice that galaxies with a radial velocity $v = cz \approx 1000\,\text{km}\,\text{s}^{-1}$, which Hubble thought were at a distance $r \approx 2\,\mathrm{Mpc}$, are now more accurately placed at a distance $r \approx 15\,\mathrm{Mpc}$. The best current estimate of the Hubble constant, combining the results from various research techniques, is

$$H_0 = 68 \pm 2\,\mathrm{km\,s^{-1}\,Mpc^{-1}}. \tag{2.10}$$

This is the value for the Hubble constant that we will use in the remainder of this book.

Cosmological innocents sometimes exclaim, when first encountering Hubble's law, "Surely it must be a violation of the Copernican principle to have all those distant galaxies moving away from *us*! It looks as if we are at a special location in the universe – the point away from which all other galaxies are fleeing." In fact, what we see here in our galaxy is exactly what you would expect to see in a universe that is undergoing homogeneous and isotropic expansion. We see distant galaxies moving away from us; but observers in any other galaxy would also see distant galaxies moving away from them.

To see on a more mathematical level what we mean by homogeneous, isotropic expansion, consider three galaxies at positions $\vec{r}_1$, $\vec{r}_2$, and $\vec{r}_3$. They define a triangle (Figure 2.6) with sides of length

$$r_{12} \equiv |\vec{r}_1 - \vec{r}_2| \tag{2.11}$$
$$r_{23} \equiv |\vec{r}_2 - \vec{r}_3| \tag{2.12}$$
$$r_{31} \equiv |\vec{r}_3 - \vec{r}_1|\,. \tag{2.13}$$

Homogeneous and uniform expansion means that the shape of the triangle is preserved as the galaxies move away from each other. Maintaining the correct relative lengths for the sides of the triangle requires an expansion law of the form

$$r_{12}(t) = a(t)r_{12}(t_0) \tag{2.14}$$
$$r_{23}(t) = a(t)r_{23}(t_0) \tag{2.15}$$
$$r_{31}(t) = a(t)r_{31}(t_0). \tag{2.16}$$

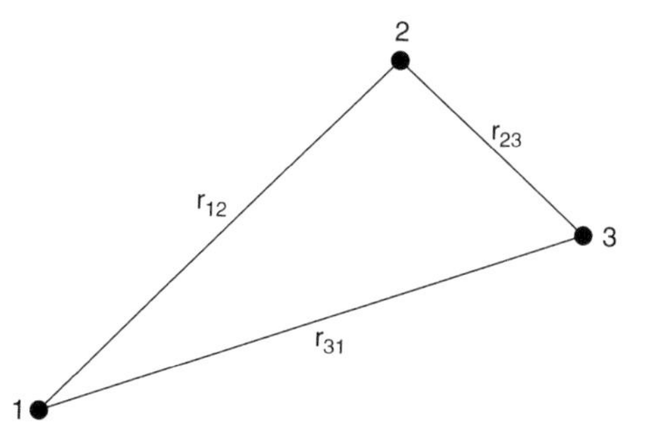

Figure 2.6 A triangle defined by three galaxies in a uniformly expanding universe.

Here the function $a(t)$ is a *scale factor*, equal to one at the present moment ($t = t_0$) and totally independent of location or direction. The scale factor $a(t)$ tells us how the expansion (or possibly contraction) of the universe depends on time. At any time t, an observer in galaxy 1 will see the other galaxies receding with a speed

$$v_{12}(t) = \frac{dr_{12}}{dt} = \dot{a}r_{12}(t_0) = \frac{\dot{a}}{a}r_{12}(t) \tag{2.17}$$

$$v_{31}(t) = \frac{dr_{31}}{dt} = \dot{a}r_{31}(t_0) = \frac{\dot{a}}{a}r_{31}(t). \tag{2.18}$$

An observer in galaxy 2 or galaxy 3 will find the same linear relation between observed recession speed and distance, with $\dot{a}/a$ playing the role of the Hubble constant. Since this argument can be applied to any trio of galaxies, it implies that in any universe where the distribution of galaxies is undergoing homogeneous, isotropic expansion, the velocity–distance relation takes the linear form $v = Hr$, with $H = \dot{a}/a$.

If galaxies are currently moving away from each other, then it implies they were closer together in the past. Consider a pair of galaxies currently separated by a distance r, with a velocity $v = H_0 r$ relative to each other. If there are no forces acting to accelerate or decelerate their relative motion, then their velocity is constant, and the time that has elapsed since they were in contact is

$$t_0 = \frac{r}{v} = \frac{r}{H_0 r} = H_0^{-1}, \tag{2.19}$$

independent of the current separation r. The time H_0^{-1} is referred to as the *Hubble time*. For $H_0 = 68 \pm 2\,\mathrm{km\,s^{-1}\,Mpc^{-1}}$, the Hubble time is $H_0^{-1} = 14.38 \pm 0.42\,\mathrm{Gyr}$. If the relative velocities of galaxies have been constant in the past, then one Hubble time ago, all the galaxies in the universe were crammed together into a small volume. Thus, the observation of galactic redshifts led naturally to a *Big Bang* model for the evolution of the universe. A Big Bang model may be broadly defined as a model in which the universe expands from an initially highly dense state to its current low-density state.

The Hubble time of $\sim 14.4\,\mathrm{Gyr}$ is comparable to the ages computed for the oldest known stars in the universe. This rough equivalence is reassuring. However, the age of the universe – that is, the time elapsed since its original highly dense state – is not necessarily exactly equal to the Hubble time. We know that gravity exists, and that galaxies contain matter. If gravity working on matter is the only force at work on large scales, then the attractive force of gravity will act to slow the expansion. In this case, the universe was expanding more rapidly in the past than it is now, and the universe is somewhat younger than H_0^{-1}. On the other hand, if the energy density of the universe is dominated by a cosmological constant (an entity we'll examine in more detail in Chapter 4), then the dominant gravitational force is repulsive, and the universe may be older than H_0^{-1}.

Just as the Hubble time provides a natural time scale for our expanding universe, the Hubble distance, $c/H_0 = 4380 \pm 130\,\text{Mpc}$, provides a natural distance scale. The age of the universe is $t_0 \sim H_0^{-1}$, with the precise age depending on the expansion history of the universe. Even if a star began shining very early in the history of the universe, the first light from that star can only have traveled a distance $d \sim ct_0 \sim c/H_0$, with the precise travel distance depending on the expansion history of the universe. The finite age of the universe thus provides the resolution for Olbers' paradox: the night sky is dark because the light from stars at a distance much greater than c/H_0 hasn't had time to reach us.

In an infinite, eternal universe, as we have seen, you could see an average distance of $\lambda \sim 10^{18}\,\text{Mpc}$ before your line of sight encountered an opaque star. In a young universe where light can travel a maximum distance $d \sim c/H_0 \sim 4000\,\text{Mpc}$, the probability that you see a star along a randomly chosen line of sight is tiny: it's of order $P \sim d/\lambda \sim 4 \times 10^{-15}$. Thus, instead of seeing a sky completely paved with stars, with surface brightness $\Sigma \sim \Sigma_\odot \sim 5 \times 10^{-3}\,\text{watts m}^{-2}\,\text{arcsec}^{-2}$, you see a sky severely underpaved with stars, with an average surface brightness[3] $\Sigma \sim P\Sigma_\odot \sim 2 \times 10^{-17}\,\text{watts m}^{-2}\,\text{arcsec}^{-2}$. For the night sky to be completely paved with stars, the universe would have to be over 100 trillion times older than it is; *and* you'd have to keep the stars shining during all that time.

Hubble's law occurs naturally in a Big Bang model for the universe, in which homogeneous and isotropic expansion causes the density of the universe to decrease steadily from its initial high value. In a Big Bang model, the properties of the universe evolve with time; the average density decreases, the mean distance between galaxies increases, and so forth. However, Hubble's law can also be explained by a *Steady State* model. The Steady State model was first proposed in the 1940s by Hermann Bondi, Thomas Gold, and Fred Hoyle, who were proponents of the *perfect cosmological principle*, which states that not only are there no privileged locations in space, there are no privileged moments in time. Thus, a Steady State universe is one in which the global properties of the universe, such as the mean density ρ_0 and the Hubble constant H_0, remain constant with time.

In a Steady State universe, the velocity–distance relation

$$\frac{dr}{dt} = H_0 r \tag{2.20}$$

can be easily integrated, since H_0 is constant with time, to yield an exponential law:

$$r(t) \propto e^{H_0 t}. \tag{2.21}$$

Note that $r \to 0$ only in the limit $t \to -\infty$; a Steady State universe is infinitely old. If there existed an instant in time at which the universe started expanding

[3] This crude back-of-envelope calculation doesn't exactly match the observed surface brightness of the night sky, but it's surprisingly close.

(as in a Big Bang model), that would be a special moment, in violation of the assumed "perfect cosmological principle." The volume of a spherical region of space, in a Steady State model, increases exponentially with time:

$$V = \frac{4\pi}{3} r^3 \propto e^{3H_0 t}. \tag{2.22}$$

However, if the universe is in a steady state, the density of the sphere must remain constant. To have a constant density of matter within a growing volume, matter must be continuously created at a rate

$$\dot{M}_{\rm ss} = \rho_0 \dot{V} = \rho_0 3 H_0 V. \tag{2.23}$$

If our own universe, with matter density $\rho_0 \approx 2.7 \times 10^{-27}\,\mathrm{kg\,m^{-3}}$, happened to be a Steady State universe, then matter would have to be created at a rate

$$\frac{\dot{M}_{\rm ss}}{V} = 3 H_0 \rho_0 \approx 5.6 \times 10^{-28}\,\mathrm{kg\,m^{-3}\,Gyr^{-1}}. \tag{2.24}$$

This corresponds to creating roughly one hydrogen atom per cubic kilometer per year.

During the 1950s and 1960s, the Big Bang and Steady State models battled for supremacy. Critics of the Steady State model pointed out that the continuous creation of matter violates mass-energy conservation. Supporters of the Steady State model pointed out that the continuous creation of matter is no more absurd than the instantaneous creation of the entire universe in a single "Big Bang" billions of years ago.[4] The Steady State model finally fell out of favor when observational evidence increasingly indicated that the perfect cosmological principle is not true. The properties of the universe *do*, in fact, change with time. The discovery of the cosmic microwave background, discussed in Section 2.5, is commonly regarded as the observation that decisively tipped the scales in favor of the Big Bang model.

2.4 Different Types of Particles

It doesn't take a brilliant observer to confirm that the universe contains a variety of different things: shoes, ships, sealing wax, cabbages, kings, galaxies, and what have you. From a cosmologist's viewpoint, though, cabbages and kings are nearly indistinguishable – the main difference between them is that the mean mass per king is greater than the mean mass per cabbage. From a cosmological viewpoint, the most significant difference between the different components of the universe is that they are made of different elementary particles. The properties of the most cosmologically important particles are summarized in Table 2.1.

[4] The name "Big Bang" was actually coined by Fred Hoyle, a supporter of the Steady State model.

Table 2.1 Elementary particle properties.

Particle	*Symbol*	*Rest energy (MeV)*	*Charge*
Proton	p	938.27	+1
Neutron	n	939.57	0
Electron	e^-	0.5110	−1
Neutrino	ν_e, ν_μ, ν_τ	$< 3 \times 10^{-7}$	0
Photon	γ	0	0
Dark matter	?	?	0

The material objects that surround us in our everyday life are made up of *protons*, *neutrons*, and *electrons*.[5] Protons and neutrons are examples of *baryons*, where a baryon is defined as a particle made of three quarks. A proton (p) contains two "up" quarks, each with an electrical charge of $+2/3$, and a "down" quark, with charge $-1/3$. A neutron (n) contains one "up" quark and two "down" quarks. Thus a proton has a net positive charge of $+1$, while a neutron is electrically neutral. Protons and neutrons also differ in their mass – or equivalently, in their rest energies. The proton mass is $m_pc^2 = 938.27\,\text{MeV}$, while the neutron mass is $m_nc^2 = 939.57\,\text{MeV}$, about 0.1% greater. Free neutrons are unstable, decaying into protons with a decay time of $\tau_n = 880\,\text{s}$, about a quarter of an hour. By contrast, protons are extremely stable; the lower limit on the decay time of the proton is $\tau_p > 10^{24}H_0^{-1}$. Neutrons can be preserved against decay by binding them into an atomic nucleus with one or more protons.

Electrons (e^-) are examples of *leptons*, a class of elementary particles that are not made of quarks. The mass of an electron is much smaller than that of a neutron or proton; the rest energy of an electron is $m_ec^2 = 0.511\,\text{MeV}$. An electron has an electric charge equal in magnitude to that of a proton, but opposite in sign. On large scales, the universe is electrically neutral; the number of electrons is equal to the number of protons. Since protons outmass electrons by a factor of 1836 to 1, the mass density of electrons is only a small perturbation to the mass density of protons and neutrons. For this reason, the component of the universe made up of ions, atoms, and molecules is generally referred to as *baryonic matter*, since only the baryons (protons and neutrons) contribute significantly to the mass density. Protons and neutrons are 800-pound gorillas; electrons are only 7-ounce bushbabies.

About three-fourths of the baryonic matter in the universe is currently in the form of ordinary hydrogen, the simplest of all elements. In addition, when we look at the remainder of the baryonic matter, it is primarily in the form of helium, the next simplest element. When astronomers look at a wide range of astronomical objects – stars and interstellar gas clouds, for instance – they find a minimum

[5] For that matter, we ourselves are made of protons, neutrons, and electrons.

helium mass fraction of 24%. The baryonic component of the universe can be described, to lowest order, as a mix of three parts hydrogen to one part helium, with only minor contamination by heavier elements.

Another type of lepton, in addition to the electron, is the *neutrino* (ν). The most poetic summary of the properties of the neutrino was made by John Updike, in his poem "Cosmic Gall":[6]

Neutrinos, they are very small.
They have no charge and have no mass
And do not interact at all.
The earth is just a silly ball
To them, through which they simply pass,
Like dustmaids down a drafty hall
Or photons through a sheet of glass.

In truth, Updike was using a bit of poetic license here. It is definitely true that neutrinos have no electric charge. However, it is not true that neutrinos "do not interact at all"; they actually are able to interact with other particles via the weak nuclear force. The weak nuclear force, though, is very weak indeed; a typical neutrino emitted by the Sun would have to pass through a few parsecs of solid lead before having a 50 percent chance of interacting with a lead atom. Since neutrinos pass through neutrino detectors with the same facility with which they pass through the Earth, detecting neutrinos from astronomical sources is difficult.

There are three types, or "flavors," of neutrinos: electron neutrinos (ν_e), muon neutrinos (ν_μ), and tau neutrinos (ν_τ). What Updike didn't know in 1960, when he wrote his poem, is that each flavor of neutrino has a small mass. In addition to there being three flavor states of neutrino, (ν_e, ν_μ, ν_τ), there are also three *mass* states of neutrino, (ν_1, ν_2, ν_3), with masses m_1, m_2, and m_3. Each of the three flavor states is a quantum superposition of the three different mass states. The presence of three neutrino mass states, at least two of which have a non-zero mass, is known indirectly from the search for neutrino oscillations. An *oscillation* is the transmutation of one flavor of neutrino into another. For instance, an electron neutrino produced by a fusion reaction in the core of the Sun will be converted into some combination of an electron neutrino, a muon neutrino, and a tau neutrino as it moves away from the Sun. These oscillations can only occur, according to the laws of quantum mechanics, if the different mass states have masses that differ from each other. The oscillations of electron neutrinos from the Sun are explained if the two first mass states have $(m_2^2 - m_1^2)c^4 \approx 7.5 \times 10^{-5}\,\text{eV}^2$. The oscillations of muon neutrinos created by cosmic rays striking the Earth's

[6] From TELEPHONE POLES AND OTHER POEMS, by John Updike, ©1958, 1959, 1960, 1961, 1962, 1963 by John Updike. Used by permission of Alfred A. Knopf, an imprint of the Knopf Doubleday Publishing Group, a division of Penguin Random House LLC.

upper atmosphere are most easily explained if $(m_3^2 - m_2^2)c^4 \approx 2.4 \times 10^{-3}\,\text{eV}^2$. Unfortunately, knowing the differences of the squares of the masses doesn't tell us the values of the masses themselves. Given $m_1 \geq 0$, there is a lower limit on the sum of the neutrino masses:

$$(m_1 + m_2 + m_3)c^2 = [m(\nu_e) + m(\nu_\mu) + m(\nu_\tau)]c^2 \geq 0.057\,\text{eV}. \qquad (2.25)$$

The best upper limit on the sum of the neutrino masses is given by observations of the large scale structure of the universe, as we shall see in Chapter 11. This upper limit is

$$(m_1 + m_2 + m_3)c^2 = [m(\nu_e) + m(\nu_\mu) + m(\nu_\tau)]c^2 \leq 0.3\,\text{eV}\,. \qquad (2.26)$$

In any case, although John Updike was not strictly correct about neutrinos being massless, they are constrained to be very much lower in mass than electrons.

A particle which is known to be massless is the *photon*. Electromagnetic radiation can be thought of either as a wave or as a stream of particles, called photons. Light, when regarded as a wave, is characterized by its frequency f or its wavelength $\lambda = c/f$. When light is regarded as a stream of photons, each photon is characterized by its energy, $E_\gamma = hf$, where $h = 2\pi\hbar$ is the Planck constant. Photons of a wide range of energy, from radio to gamma rays, pervade the universe. Unlike neutrinos, photons interact readily with electrons, protons, and neutrons. For instance, photons can ionize an atom by kicking an electron out of its orbit, a process known as *photoionization*. Higher-energy photons can break an atomic nucleus apart, a process known as *photodissociation*.

Photons, in general, are easily created. One way to make photons is to take a dense, opaque object – such as the filament of an incandescent lightbulb – and heat it up. If an object is opaque, then the protons, neutrons, electrons, and photons that it contains frequently interact, and attain thermal equilibrium; that is, they all have the same temperature T. The density of photons in the object, as a function of photon energy, will depend only on T. It doesn't matter whether the system is a tungsten filament, an ingot of steel, or a sphere of ionized hydrogen and helium. The energy density of photons in the frequency range $f \to f + df$ is given by the *blackbody* function

$$\varepsilon(f)df = \frac{8\pi h}{c^3}\frac{f^3\,df}{\exp(hf/kT) - 1}, \qquad (2.27)$$

illustrated in Figure 2.7.

The peak in the blackbody function occurs at $hf_{\text{peak}} \approx 2.82kT$. Integrated over all frequencies, Equation 2.27 yields a total energy density for blackbody radiation of

$$\varepsilon_\gamma = \alpha T^4, \qquad (2.28)$$

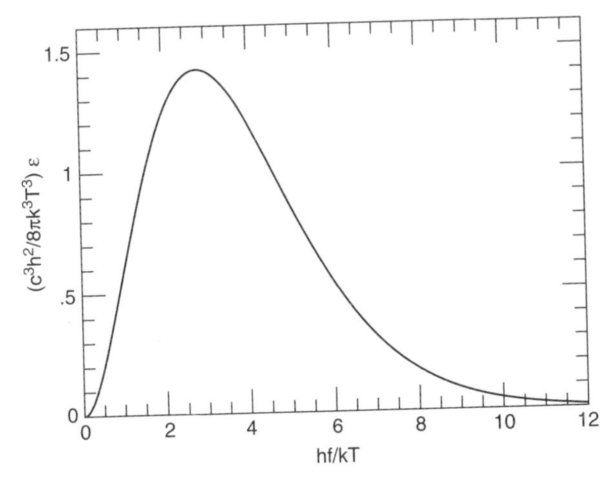

Figure 2.7 The energy density of blackbody radiation, expressed as a function of frequency f.

where

$$\alpha = \frac{\pi^2}{15}\frac{k^4}{\hbar^3 c^3} = 7.566 \times 10^{-16}\,\mathrm{J\,m^{-3}\,K^{-4}}. \tag{2.29}$$

Since the energy of a photon is $E_\gamma = hf$, the number density of photons in the frequency range $f \to f + df$ is, from Equation 2.27,

$$n(f)df = \frac{\varepsilon(f)df}{hf} = \frac{8\pi}{c^3}\frac{f^2\,df}{\exp(hf/kT) - 1}. \tag{2.30}$$

Integrated over all frequencies, the number density of photons in blackbody radiation is

$$n_\gamma = \beta T^3, \tag{2.31}$$

where

$$\beta = \frac{2.4041}{\pi^2}\frac{k^3}{\hbar^3 c^3} = 2.029 \times 10^7\,\mathrm{m^{-3}\,K^{-3}}. \tag{2.32}$$

Division of Equation 2.28 by Equation 2.31 yields a mean photon energy $E_{\rm mean} = hf_{\rm mean} \approx 2.70kT$, close to the peak in the spectrum. You have a temperature $T_{\rm you} = 310\,\mathrm{K}$, assuming you are not running a fever, and you radiate an approximate blackbody spectrum, with a mean photon energy $E_{\rm mean} \approx 0.072\,\mathrm{eV}$, corresponding to a wavelength $\lambda \approx 1.7 \times 10^{-5}\,\mathrm{m} \approx 17\,000\,\mathrm{nm}$, in the mid-infrared. By contrast, the Sun produces an approximate blackbody spectrum with a temperature $T_\odot \approx 5800\,\mathrm{K}$. This implies a mean photon energy $E_{\rm mean} \approx 1.3\,\mathrm{eV}$, corresponding to $\lambda \approx 9.0 \times 10^{-7}\,\mathrm{m} \approx 900\,\mathrm{nm}$, in the near infrared. Note, however, that although the mean photon energy in a blackbody spectrum is $2.70kT$, Figure 2.7 shows us that there is a long exponential tail to higher photon energies. A large fraction of the Sun's output is at wavelengths of $400 \to 700\,\text{nm}$, which our eyes are equipped to detect.

A more mysterious component of the universe is *dark matter*. When observational astronomers refer to dark matter, they often mean any massive component of the universe that is too dim to be detected readily using current technology. Theoretical astrophysicists often use a more stringent definition of dark matter than do observers, defining dark matter as any massive component of the universe which doesn't emit, absorb, or scatter light at all.[7] If neutrinos have mass, for instance, as the neutrino oscillation results indicate, they qualify as dark matter. In some extensions to the Standard Model of particle physics, there exist massive particles that interact, like neutrinos, only through the weak nuclear force and through gravity. These particles, which have not yet been detected in the laboratory, are generically referred to as weakly interacting massive particles, or WIMPs.

In this book, we will generally adopt the broader definition of dark matter as something which is too dim for us to see, even with our best available technology. Detecting dark matter is, naturally, difficult. The standard method of detecting dark matter is by measuring its gravitational effect on luminous matter, just as the planet Neptune was first detected by its gravitational effect on the planet Uranus. Although Neptune no longer qualifies as dark matter, observations of the motions of stars within galaxies and of galaxies within clusters indicate that a significant amount of dark matter is in the universe. Exactly how much there is, and what it's made of, is a topic of great interest to cosmologists.

2.5 Cosmic Microwave Background

The discovery of the cosmic microwave background (CMB) by Arno Penzias and Robert Wilson in 1965 has entered cosmological folklore. Using a microwave antenna at Bell Labs, they found an isotropic background of microwave radiation. More recently, space-based experiments have revealed that the cosmic microwave background is exquisitely well fitted by a blackbody spectrum (Equation 2.27) with a temperature

$$T_0 = 2.7255 \pm 0.0006\,\text{K}. \tag{2.33}$$

The energy density of the CMB is, from Equation 2.28,

$$\varepsilon_\gamma = 4.175 \times 10^{-14}\,\text{J}\,\text{m}^{-3} = 0.2606\,\text{MeV}\,\text{m}^{-3}. \tag{2.34}$$

[7] Using this definition, an alternate name for dark matter might be "transparent matter" or "invisible matter." However, the name "dark matter" is the commonly adopted term.

The number density of CMB photons is, from Equation 2.31,

$$n_\gamma = 4.107 \times 10^8 \,\mathrm{m}^{-3}. \tag{2.35}$$

Thus, there are about 411 CMB photons per cubic centimeter of the universe at the present day. The mean energy of CMB photons, however, is quite low, only

$$E_{\rm mean} = 6.344 \times 10^{-4} \,\mathrm{eV}. \tag{2.36}$$

This is too low in energy to photoionize an atom, much less photodissociate a nucleus. The mean CMB photon energy corresponds to a wavelength of 2 millimeters, in the microwave region of the electromagnetic spectrum – hence the name "cosmic *microwave* background."

The existence of the CMB is a very important cosmological clue. In particular, it is the clue that caused the Big Bang model for the universe to be favored over the Steady State model. In a Steady State universe, the existence of blackbody radiation at 2.7255 K is not easily explained. In a Big Bang universe, however, a cosmic background radiation arises naturally if the universe was initially very hot as well as very dense. If mass is conserved in an expanding universe, then in the past the universe was denser than it is now. Assume that the early dense universe was very hot ($T \gg 10^4$ K, or $kT \gg 1$ eV). At such high temperatures, the baryonic matter in the universe was completely ionized, and the free electrons rendered the universe opaque. A dense, hot, opaque body, as described in Section 2.4, produces blackbody radiation. So, the early hot dense universe was full of photons, banging off the electrons like balls in a pinball machine, with a spectrum typical of a blackbody (Equation 2.27). However, as the universe expanded, it cooled. Eventually, the temperature became sufficiently low that ions and electrons combined to form neutral atoms. When the universe no longer contained a significant number of free electrons, the blackbody photons started streaming freely through the universe, without further scattering off free electrons.

The blackbody radiation that fills the universe today can be explained as a relic of the time when the universe was sufficiently hot and dense to be opaque. However, at the time the universe became transparent, its temperature was 2970 K. The temperature of the CMB today is 2.7255 K, a factor of 1090 lower. The drop in temperature of the blackbody radiation is a direct consequence of the expansion of the universe. Consider a region of volume V that expands at the same rate as the universe, so that $V \propto a(t)^3$. The blackbody radiation in the volume can be thought of as a photon gas with energy density $\varepsilon_\gamma = \alpha T^4$. Moreover, since the photons in the volume have momentum as well as energy, the photon gas has a pressure; the pressure of a photon gas is $P_\gamma = \varepsilon_\gamma/3$. The photon gas within our imaginary box follows the laws of thermodynamics; in particular, the boxful of photons obeys the first law

$$dQ = dE + PdV, \tag{2.37}$$

where dQ is the amount of heat flowing into or out of the photon gas in the volume V, dE is the change in the internal energy, P is the pressure, and dV is the change in volume of the box. Since, in a homogeneous universe, there is no net flow of heat (everything is the same temperature, after all), $dQ = 0$. Thus, the first law of thermodynamics, applied to an expanding homogeneous universe, is

$$\frac{dE}{dt} = -P(t)\frac{dV}{dt}. \tag{2.38}$$

Since, for the photons of the CMB, $E = \varepsilon_\gamma V = \alpha T^4 V$ and $P = P_\gamma = \alpha T^4/3$, Equation 2.38 can be rewritten in the form

$$\alpha\left(4T^3\frac{dT}{dt}V + T^4\frac{dV}{dt}\right) = -\frac{1}{3}\alpha T^4\frac{dV}{dt}, \tag{2.39}$$

or

$$\frac{1}{T}\frac{dT}{dt} = -\frac{1}{3V}\frac{dV}{dt}. \tag{2.40}$$

However, since $V \propto a(t)^3$ as the box expands, this means that the rate of change of the photons' temperature is related to the rate of expansion of the universe by the relation

$$\frac{d}{dt}(\ln T) = -\frac{d}{dt}(\ln a). \tag{2.41}$$

This implies the simple relation $T(t) \propto a(t)^{-1}$; the temperature of the cosmic background radiation has dropped by a factor of 1090 since the universe became transparent, because the scale factor $a(t)$ has increased by a factor of 1090 since then. What we now see as a cosmic microwave background was once, at the time the universe became transparent, a cosmic *near infrared* background, with a temperature comparable to that of a relatively cool star like Proxima Centauri.

The evidence cited so far can all be explained within the framework of a *Hot Big Bang* model, in which the universe was originally very hot and very dense, and since then has been expanding and cooling. The remainder of this book will be devoted to working out the details of the Hot Big Bang model that best fits the universe in which we live.

Exercises

2.1 Assume you are a perfect blackbody at a temperature of $T = 310\,\text{K}$. What is the rate, in watts, at which you radiate energy? (For the purposes of this problem, you may assume you are spherical.)

2.2 Since you are made mostly of water, you are very efficient at absorbing microwave photons. If you were in intergalactic space, how many CMB photons would you absorb per second? (The assumption that you are spherical will be useful.) What is the rate, in watts, at which you would absorb radiative energy from the CMB?

2.3 Suppose that intergalactic space pirates toss you out the airlock of your spacecraft without a spacesuit. Combining the results of the two previous questions, at what rate would your temperature change? (Assume your heat capacity is that of pure water, $C = 4200\,\mathrm{J\,kg^{-1}\,K^{-1}}$.) Would you be most worried about overheating, freezing, or asphyxiating?

2.4 A hypothesis once used to explain the Hubble relation is the "tired light hypothesis." The tired light hypothesis states that the universe is not expanding, but that photons simply lose energy as they move through space (by some unexplained means), with the energy loss per unit distance being given by the law

$$\frac{dE}{dr} = -kE, \tag{2.42}$$

where k is a constant. Show that this hypothesis gives a distance–redshift relation that is linear in the limit $z \ll 1$. What must the value of k be in order to yield a Hubble constant of $H_0 = 68\,\mathrm{km\,s^{-1}\,Mpc^{-1}}$?

2.5 Consider blackbody radiation at a temperature T. Show that for an energy threshold $E_0 \gg kT$, the fraction of the blackbody photons that have energy $hf > E_0$ is

$$\frac{n(hf > E_0)}{n_\gamma} \approx 0.42 \left(\frac{E_0}{kT}\right)^2 \exp\left(-\frac{E_0}{kT}\right). \tag{2.43}$$

The cosmic background radiation is currently called the "cosmic *microwave* background." However, photons with $\lambda < 1\,\mathrm{mm}$ actually lie in the *far infrared* range of the electromagnetic spectrum. It's time for truth in advertising: what fraction of the photons in today's "cosmic microwave background" are actually far infrared photons?

2.6 Show that for an energy threshold $E_0 \ll kT$, the fraction of blackbody photons that have energy $hf < E_0$ is

$$\frac{n(hf < E_0)}{n_\gamma} \approx 0.21 \left(\frac{E_0}{kT}\right)^2. \tag{2.44}$$

Microwave (and far infrared) photons with a wavelength $\lambda < 3\,\mathrm{cm}$ are strongly absorbed by H_2O and O_2 molecules. What fraction of the photons in today's cosmic microwave background have $\lambda > 3\,\mathrm{cm}$, and thus are capable of passing through the Earth's atmosphere and being detected on the ground?

3

Newton versus Einstein

On cosmological scales (that is, on scales greater than 100 Mpc or so), the dominant force determining the evolution of the universe is gravity. The weak and strong nuclear forces are short-range forces; the weak force is effective only on scales of $\ell_w \sim 10^{-18}$ m or less, and the strong force on scales of $\ell_s \sim 10^{-15}$ m or less. Both gravity and electromagnetism are long-range forces. On small scales, gravity is negligibly small compared to electromagnetic forces; for instance, the electrostatic repulsion between a pair of protons is larger by a factor $\sim 10^{36}$ than the gravitational attraction between them. However, on large scales, the universe is electrically neutral, so there are no electrostatic forces on large scales. Moreover, intergalactic magnetic fields are sufficiently small that magnetic forces are also negligibly tiny on cosmological scales.

In referring to gravity as a force, we are implicitly adopting a Newtonian viewpoint. In physics, the two useful ways of looking at gravity are the Newtonian (classical) viewpoint and the Einsteinian (general relativistic) viewpoint. In Isaac Newton's view, as formulated by his laws of motion and law of gravity, gravity is a force that causes massive bodies to be accelerated. By contrast, in Einstein's view, gravity is a manifestation of the curvature of spacetime. Although Newton's view and Einstein's view are conceptually very different, in most contexts they yield the same predictions. The Newtonian predictions differ significantly from the predictions of general relativity only in the limit of deep potential minima (to use Newtonian language) or strong spatial curvature (to use general relativistic language). In these limits, general relativity yields the correct result.

In the limit of shallow potential minima and weak spatial curvature, it is permissible to switch back and forth between a Newtonian and a general relativistic viewpoint, adopting whichever one is more convenient. I will frequently adopt the Newtonian view of gravity in this book because, in many contexts, it is mathematically simpler and conceptually more familiar. The question of *why* it is possible to switch back and forth between the two very different viewpoints of Newton and Einstein is an intriguing one, and deserves closer investigation.

3.1 The Way of Newton

In Newton's view of the universe, space is unchanging and Euclidean. In Euclidean space, all the axioms and theorems of plane geometry, as codified by Euclid in the third century BC, hold true. (Euclidean space is also referred to as "flat" space. In this context, "flat" doesn't mean two-dimensional, like a piece of paper; you can have three-dimensional flat spaces as well as two-dimensional flat spaces.) In Euclidean space, the shortest distance between two points is a straight line, the angles at the vertices of a triangle sum to π radians, the circumference of a circle is 2π times its radius, and so on, through all the other axioms and theorems you learned in high school geometry. In Newton's view, moreover, an object with no net force acting on it moves in a straight line at constant speed. However, when we look at objects in the Solar System such as planets, moons, comets, and asteroids, we find that they move on curved lines, with constantly changing speed. Why is this? Newton would tell us, "Their velocities are changing because there is a force acting on them; the force called *gravity*."

Newton devised a formula for computing the gravitational force between two objects. Every object in the universe, said Newton, has a property that we may call the "gravitational mass." Let the gravitational masses of two objects be M_g and m_g, and let the distance between their centers be r. The gravitational force acting between the two objects (assuming they are both spherical) is

$$F = -\frac{GM_g m_g}{r^2}. \tag{3.1}$$

The negative sign in the above equation indicates that gravity, in the Newtonian view, is always an attractive force, tending to draw two bodies closer together.

What is the acceleration that results from this gravitational force? Newton had something to say about that as well. Every object in the universe, said Newton, has a property that we may call the "inertial mass." Let the inertial mass of an object be m_i. Newton's second law of motion says that force and acceleration are related by the equation

$$F = m_i a. \tag{3.2}$$

In Equations 3.1 and 3.2 we have distinguished, through the use of different subscripts, between the gravitational mass m_g and the inertial mass m_i. One of the fundamental principles of physics is that the gravitational mass and the inertial mass of an object are identical:

$$m_g = m_i. \tag{3.3}$$

When you stop to think about it, this equality is a remarkable fact. The property of an object that determines how strongly it is pulled on by the force of gravity is equal to the property that determines its resistance to acceleration by *any* force, not just the force of gravity. The equality of gravitational mass and inertial mass is called the *equivalence principle*.

If the equivalence principle did not hold, then the gravitational acceleration of an object toward a mass M_g would be (combining Equations 3.1 and 3.2)

$$a = -\frac{GM_g}{r^2}\left(\frac{m_g}{m_i}\right), \tag{3.4}$$

with the ratio m_g/m_i varying from object to object. However, when Galileo dropped objects from towers and slid objects down inclined planes, he found that the acceleration (barring the effects of air resistance and friction) was always the same, regardless of the mass and composition of the object. The magnitude of the gravitational acceleration close to the Earth's surface is $a = GM_{\text{Earth}}/r^2_{\text{Earth}} = 9.8\,\text{m}\,\text{s}^{-2}$. Modern tests of the equivalence principle, which are basically more sensitive versions of Galileo's experiments, reveal that the inertial and gravitational masses are the same to within one part in 10^{13}. For the rest of this book, therefore, we'll just use the symbol m for mass, where $m = m_i = m_g$.

The equivalence principle implies that at every point $\vec{r}$ in the universe there is a unique gravitational acceleration $\vec{a}(\vec{r})$. It is useful to compute this acceleration in terms of a gravitational potential $\Phi(\vec{r})$. If the mass density of the universe is $\rho(\vec{r})$, then the gravitational potential is given by Poisson's equation:

$$\nabla^2\Phi = 4\pi G\rho. \tag{3.5}$$

If we start with a known density distribution ρ and want to find the associated potential Φ, it is more useful to use Poisson's equation in its integral form:

$$\Phi(\vec{r}) = -G\int \frac{\rho(\vec{x})}{|\vec{x}-\vec{r}|}d^3x. \tag{3.6}$$

The gravitational acceleration is then $\vec{a} = -\vec{\nabla}\Phi$.

3.2 The Special Way of Einstein

After the publication of Newton's *Principia Mathematica* in 1687, the immense power of Newtonian physics became apparent to Newton's contemporaries. As Alexander Pope wrote shortly after Newton's death:

Nature and Nature's law lay hid in night.
God said *Let Newton be!* and all was light.

Two centuries later, however, the poet John Collings Squire was able to write:

It did not last: the Devil howling *Ho!*
Let Einstein be! restored the status quo.

In popular culture, Newton's laws were regarded as rational and comprehensible; Einstein's theories were regarded as esoteric and incomprehensible. In fact, the theory of special relativity (as first published by Einstein in 1905) is mathematically rather simple. It's only when we turn to general relativity (as published by Einstein in 1915) that the mathematics becomes more complicated. Let's start, as a warmup exercise, by considering special relativity.

Special relativity deals with the *special* case in which gravity is not present. In the absence of gravity, space is Euclidean, just as in Newtonian theory. Suppose we place a particle of mass m in three-dimensional Euclidean space. It is straightforward to measure the particle's coordinates (x, y, z) relative to a set of cartesian coordinate axes, which provide a *reference frame* for measuring positions, velocities, and accelerations. The reference frame is *inertial* if the motion of a particle, with speed $v \ll c$ relative to the reference frame, obeys Newton's second law of motion,

$$\frac{d^2\vec{r}}{dt^2} = \frac{1}{m}\vec{F}, \tag{3.7}$$

when the acceleration is measured relative to the reference frame. A rotating reference frame, for example, is *not* an inertial frame, since the equation of motion in a rotating frame contains a Coriolis term and a centrifugal term. Whether or not a reference frame is inertial can be determined empirically. Take a particle, apply a known force to it, and measure whether its acceleration is equal to that predicted by Newton's second law. (The necessary caution is that your test is limited by the precision and accuracy with which you can measure accelerations. Newton, after all, devised his second law after performing experiments in which accelerations were measured relative to a frame of reference attached to the rotating Earth. The resulting Coriolis and centrifugal terms, however, were too small for Newton to measure.)

Suppose you've taken out your accelerometer and have satisfied yourself that your cartesian reference frame is inertial. Now consider a second reference frame, moving relative to the first at a constant speed v in the $+x$ direction, as shown in Figure 3.1. If the first reference frame (let's call it the "unprimed" frame) is

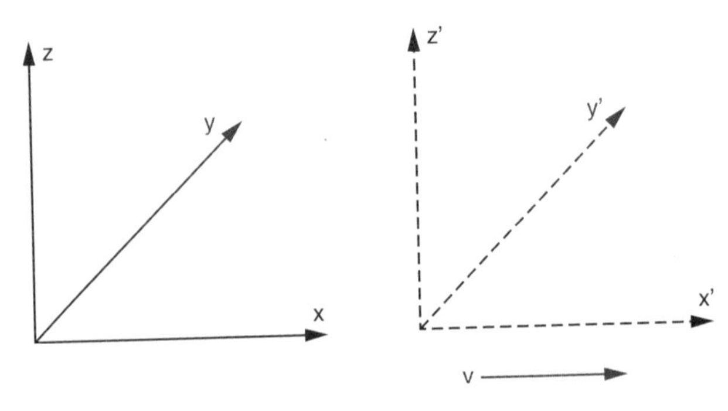

Figure 3.1 A pair of inertial reference frames (unprimed and primed), moving at a constant relative velocity $\vec{v}$.

inertial, the second reference frame (the "primed" frame) is inertial as well, as long as the relative velocity $\vec{v}$ of the two frames is constant.

In Newtonian physics, time is independent of the reference frame in which it is measured. As Newton himself put it, "Absolute, true, and mathematical time, of itself, and from its own nature, flows equably without relation to anything external." If the origins of the unprimed and primed reference frames coincide at some time $t = t' = 0$, then at some other time $t = t' \neq 0$, the coordinates in the two frames are related, in Newtonian physics, by the *Galilean transformation*:

$$\begin{aligned} x' &= x - vt \\ y' &= y \\ z' &= z \\ t' &= t. \end{aligned} \tag{3.8}$$

The Galilean transformation implies that a particle that has a velocity $\vec{u}$ measured relative to the unprimed frame has a velocity $\vec{u}' = \vec{u} - v\hat{e}_x$ relative to the primed frame.

Newtonian physics and the Galilean transformation were seriously questioned by Einstein at the beginning of the 20th century. Einstein's *first postulate of special relativity* is:

1st: The equations describing the basic laws of physics are the same in all inertial frames of reference.

Einstein's first postulate, on its surface, doesn't seem very radical. It's just an extension of what Galileo said in the 17th century, even before the birth of Newton. Galileo pointed out that if you were below decks in a sailing ship with no portholes, there would be no experiment you could conduct that would enable you to tell whether you were anchored on a placid sea or sailing along at a constant velocity. Einstein's key realization, though, was that Maxwell's equations, as well as Newton's laws of motion, are unchanged in a switch between inertial reference frames. Maxwell's equations, which describe the behavior of electric and magnetic fields, imply the existence of electromagnetic waves traveling through a vacuum at speed c. If Maxwell's equations are identical in all inertial frames of reference, as Einstein assumed, then electromagnetic waves must travel with the identical speed c in all inertial frames. This realization led Einstein to what is sometimes called the *second postulate of special relativity*:

2nd: The speed of light in a vacuum has the same value c in all inertial frames of reference.

The constancy of the speed of light had been demonstrated by Michelson and Morley as early as 1887 (although it is unclear whether Einstein was aware of their results when he published the theory of special relativity in 1905).

Let's return to the unprimed and primed frames of reference shown in Figure 3.1. At the instant when the origins of the two frames coincide, we synchronize the clocks associated with the frames, so that $t = t' = 0$. We celebrate the synchronization by having a lamp located at the joint origin emit a brief flash of light. If space is empty, then a spherical shell of light expands outward with speed c, regardless of the frame in which it is observed. At a later time $t > 0$, the equation giving the size of the shell in the unprimed frame is

$$c^2t^2 = x^2 + y^2 + z^2. \tag{3.9}$$

At the corresponding time t' in the primed frame,

$$c^2(t')^2 = (x')^2 + (y')^2 + (z')^2. \tag{3.10}$$

Equations 3.9 and 3.10 are incompatible with the Galilean transformation, as you can verify by substitution from Equation 3.8.

Equations 3.9 and 3.10 are, however, compatible with the *Lorentz transformation*:[1]

$$\begin{aligned} x' &= \gamma(x - vt) \\ y' &= y \\ z' &= z \\ t' &= \gamma(t - vx/c^2), \end{aligned} \tag{3.11}$$

where γ is the *Lorentz factor*,

$$\gamma \equiv \frac{1}{\sqrt{1 - v^2/c^2}}. \tag{3.12}$$

In special relativity, the Lorentz transformation is the correct way to convert between coordinates in two inertial frames of reference.

To see how the Lorentz transformation disrupts Newtonian ideas about space and time, consider two events. In the unprimed frame, event 1 occurs at time t_1 at location (x_1, y_1, z_1); event 2 occurs at time t_2 at location (x_2, y_2, z_2). Since space is Euclidean in special relativity, we can easily compute the spatial distance between the two events in the unprimed frame,

$$(\Delta\ell)^2 = (x_1 - x_2)^2 + (y_1 - y_2)^2 + (z_1 - z_2)^2. \tag{3.13}$$

The time elapsed between the two events in the unprimed frame is

$$\Delta t = t_1 - t_2. \tag{3.14}$$

We can use the Lorentz transformation to compute the spatial distance between the two events measured in the primed frame,

[1] The Lorentz transformation was first published by Joseph Larmor in 1897; Hendrik Lorentz didn't independently find the Lorentz transformation until 1899. (The law of misonomy strikes again.)

$$
\begin{aligned}
(\Delta\ell')^2 &= (x_1' - x_2')^2 + (y_1' - y_2')^2 + (z_1' - z_2')^2 \\
&= \gamma^2 [x_1 - x_2 - v(t_1 - t_2)]^2 + (y_1 - y_2)^2 + (z_1 - z_2)^2.
\end{aligned}
\tag{3.15}
$$

The time elapsed between the two events in the primed frame is

$$
\Delta t' = t_1' - t_2' = \gamma \left[t_1 - t_2 - \frac{v}{c^2}(x_1 - x_2) \right]. \tag{3.16}
$$

Observers in the primed and unprimed frames will measure different spatial distances between the two events. They will also measure different time intervals between the two events; under some circumstances, they will even disagree on which event occurred first. Contrary to Newton's thinking, special relativity tells us that there is no "absolute time." Observers in different reference frames will measure time differently.

Although observers in different inertial reference frames will disagree on the spatial distance between two events, and also on the time interval between the events, there is something that they will agree on: the *spacetime* separation between the events. In the unprimed frame, the spacetime separation between event 1 and event 2 is

$$
(\Delta s)^2 = -c^2(t_1 - t_2)^2 + (x_1 - x_2)^2 + (y_1 - y_2)^2 + (z_1 - z_2)^2 \tag{3.17}
$$

or

$$
(\Delta s)^2 = -c^2(\Delta t)^2 + (\Delta\ell)^2. \tag{3.18}
$$

Notice the choice of signs in this relation: two events have a spacetime separation $\Delta s = 0$ if the light travel time between their spatial locations, $\Delta\ell/c$, is equal to the time that elapses between the events, $|\Delta t|$.

The spacetime separation in the primed frame is

$$
(\Delta s')^2 = -c^2(\Delta t')^2 + (\Delta\ell')^2, \tag{3.19}
$$

where $\Delta\ell'$ is given by Equation 3.15 and $\Delta t'$ is given by Equation 3.16. Making the substitutions into Equation 3.19, we find

$$
\begin{aligned}
(\Delta s')^2 = &-\gamma^2 \left[c(t_1 - t_2)^2 - \frac{v}{c}(x_1 - x_2)^2 \right]^2 \\
&+ \gamma^2 [x_1 - x_2 - v(t_1 - t_2)]^2 + (y_1 - y_2)^2 + (z_1 - z_2)^2.
\end{aligned}
\tag{3.20}
$$

A little algebraic simplification reveals that

$$
(\Delta s')^2 = -c^2(t_1 - t_2)^2 + (x_1 - x_2)^2 + (y_1 - y_2)^2 + (z_1 - z_2)^2, \tag{3.21}
$$

and therefore, comparing Equations 3.17 and 3.21, that $(\Delta s)^2 = (\Delta s')^2$.

Using the Galilean transformation, the separation in *time* between two events is the same in all inertial frames of reference. Using the Lorentz transformation, the separation in *spacetime* is the same in all inertial frames. In Newtonian physics, it makes sense to think about space and time as two separate entities;

however, in special relativity, it is more useful to think about a four-dimensional spacetime, with the four-dimensional separation Δs between two events being given by Equation 3.17.

3.3 The General Way of Einstein

The theory of special relativity has limited usefulness, since it deals only with the case in which gravity is non-existent. It took Einstein a decade, from 1905 to 1915, to generalize his theory. To see how Einstein was inspired by the equivalence principle to devise his theory of general relativity, let's begin with a thought experiment. Suppose you wake up one morning to find that you have been sealed up (bed and all) within an opaque, soundproof, hermetically sealed box. "Oh no!" you say. "This is what I've always feared would happen. I've been abducted by space aliens who are taking me away to their home planet." Startled by this realization, you drop your teddy bear. Observing the bear, you find that it falls toward the floor of the box with an acceleration $a = 9.8\,\mathrm{m\,s^{-2}}$. "Whew!" you say, with some relief. "At least I am still on the Earth's surface; they haven't taken me away in their spaceship yet." At that moment, a window in the side of the box opens to reveal (much to your horror) that you are inside an alien spaceship that is being accelerated at $a = 9.8\,\mathrm{m\,s^{-2}}$ by a rocket engine.

When you drop a teddy bear, or any other object, within a sealed box (Figure 3.2), the equivalence principle permits two possible interpretations, with no way of distinguishing between them:

(1) The box is static, or moving with a constant velocity, and the bear is being accelerated downward by a gravitational force.

(2) The bear is static, or moving at a constant velocity, and the box is being accelerated upward by a non-gravitational force.

The behavior of the bear in each case is identical. In each case, a big bear falls at the same rate as a little bear; in each case, a bear stuffed with cotton falls at the same rate as a bear stuffed with lead; and in each case, a sentient anglophone bear would say, "Oh, bother. I'm weightless," during the interval before it collides with the floor of the box.

Einstein's insight, starting from the equivalence principle, led him to the theory of general relativity. To understand Einstein's thought processes, imagine yourself back in the sealed box, being accelerated through interplanetary space at $9.8\,\mathrm{m\,s^{-2}}$. You grab the flashlight that you keep on the bedside table and shine a beam of light perpendicular to the acceleration vector (Figure 3.3). Since the box is accelerating upward, the path of the light beam will appear to you to be bent downward, as the floor of the box rushes up to meet the photons. However, thanks to the equivalence principle, we can replace the accelerated box with a

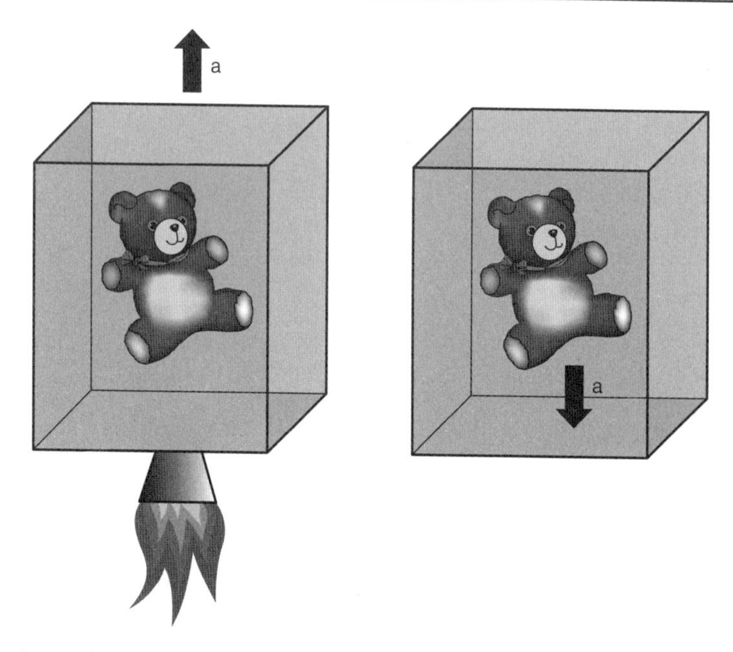

Figure 3.2 Equivalence principle (teddy bear version). The behavior of a bear in an accelerated box (left) is identical to that of a bear being accelerated by gravity (right).

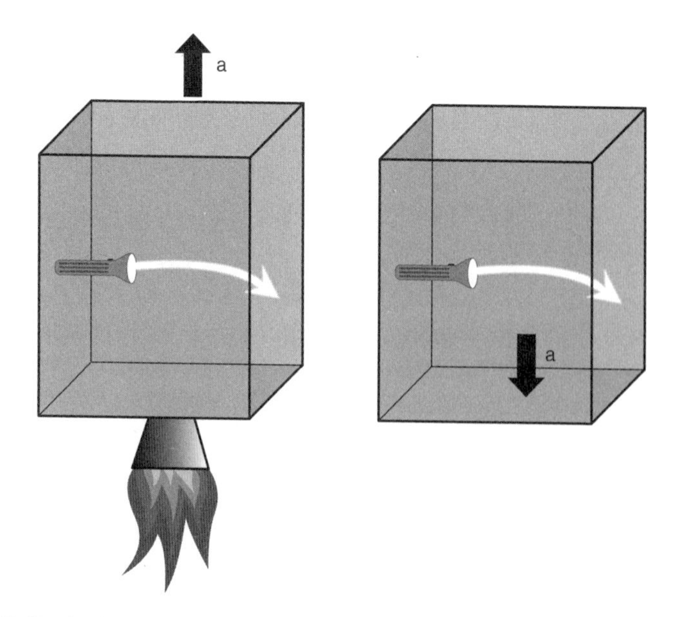

Figure 3.3 Equivalence principle (photon version). The path followed by a light beam in an accelerated box (left) is identical to the path followed by a light beam being accelerated by gravity (right). [The deflection shown is greatly exaggerated for the sake of visualization. The actual deflection will be $\sim 2 \times 10^{-14}$ m if the box is 2 meters across.]

stationary box experiencing a constant gravitational acceleration. Since there's no way to distinguish between these two cases, we are led to the conclusion that the paths of photons will be curved downward in the presence of a gravitational field. Gravity affects photons, Einstein concluded, even though they have no mass. Contemplating the curved path of the light beam, Einstein had one more insight. One of the fundamental principles of optics is *Fermat's principle*, which states that light travels between two points along a path that minimizes the travel time required. (More generally, Fermat's principle requires that the travel time be an extremum – either a minimum or a maximum. In most situations, however, the path taken by light minimizes the travel time rather than maximizing it.) In a vacuum, where the speed of light is constant, this translates into the requirement that light takes the shortest path between two points. In Euclidean, or flat, space, the shortest path between two points is a straight line. However, in the presence of gravity, the path taken by light is not a straight line. Thus, Einstein concluded, space is *not* Euclidean.

The presence of mass, in Einstein's view, causes space to be curved. In fact, in the fully developed theory of general relativity, mass and energy (which Newton thought of as two separate entities) are interchangeable, via the famous equation $E = mc^2$. Moreover, space and time (which Newton thought of as two separate entities) form a four-dimensional spacetime. A more accurate summary of Einstein's viewpoint, therefore, is that the presence of mass-energy causes spacetime to be curved. We now have a third way of thinking about the motion of the teddy bear in the box:

(3) No forces are acting on the bear; it is simply following a *geodesic* in curved spacetime.

If you take two points in an N-dimensional space or spacetime, a geodesic is defined as the locally shortest path between them.

We now have two ways of describing how gravity works.

The Way of Newton:

Mass tells gravity how to exert a force ($F = -GMm/r^2$),
Force tells mass how to accelerate ($F = ma$).

The (General) Way of Einstein:

Mass-energy tells spacetime how to curve,
Curved spacetime tells mass-energy how to move.[2]

Einstein's description of gravity gives a natural explanation for the equivalence principle. In the Newtonian description of gravity, the equality of the gravitational mass and the inertial mass is a remarkable coincidence. However, in Einstein's theory of general relativity, curvature is a property of spacetime itself.

2 This pocket summary of general relativity was coined by the physicist John Wheeler, who also popularized the term "black hole."

It then follows automatically that the gravitational acceleration of an object should be independent of mass and composition – it's just following a geodesic, which is dictated by the geometry of spacetime.

3.4 Describing Curvature

In developing his theory of general relativity, Einstein faced multiple challenges. Ultimately, he wanted a mathematical formula (called a *field equation*) that relates the curvature of spacetime to its mass-energy density, similar to the way in which Poisson's equation relates the gravitational potential of space to its mass density. En route to this ultimate goal, however, Einstein needed a way of mathematically describing curvature. Since picturing the curvature of a four-dimensional spacetime is difficult, let's start by considering ways of describing the curvature of two-dimensional spaces, and then extend what we have learned to higher dimensions.

The simplest of two-dimensional spaces is a plane, as illustrated in Figure 3.4, for which Euclidean geometry holds true. On a plane, a geodesic is a straight line. If a triangle is constructed on a plane by connecting three points with geodesics, the angles at its vertices (α, β, and γ in Figure 3.4) obey the relation

$$\alpha + \beta + \gamma = \pi, \tag{3.22}$$

where angles are measured in radians. On a plane, we can set up a cartesian coordinate system, and assign to every point a coordinate (x, y). On a plane, the Pythagorean theorem holds, so the distance $d\ell$ between points (x, y) and $(x + dx, y + dy)$ is given by the relation[3]

$$d\ell^2 = dx^2 + dy^2. \tag{3.23}$$

Stating that Equation 3.23 holds true everywhere in two-dimensional space is equivalent to saying that the space is a plane. Of course, other coordinate systems can be used in place of cartesian coordinates. For instance, in a polar coordinate system, the distance between points (r, θ) and $(r + dr, \theta + d\theta)$ is

$$d\ell^2 = dr^2 + r^2 d\theta^2. \tag{3.24}$$

Although Equations 3.23 and 3.24 are different in appearance, they both represent the same flat geometry, as you can verify by making the simple coordinate substitution $x = r\cos\theta$, $y = r\sin\theta$.

Now consider another simple two-dimensional space, the surface of a sphere (Figure 3.5). On the surface of a sphere, a geodesic is a portion of a great circle; that is, a circle whose center corresponds to the center of the sphere. If a triangle is

[3] Starting with this equation, I adopt the convention, commonly used among relativists, that $d\ell^2 = (d\ell)^2$, and not $d(\ell^2)$. Omitting the parentheses makes the equations less cluttered.

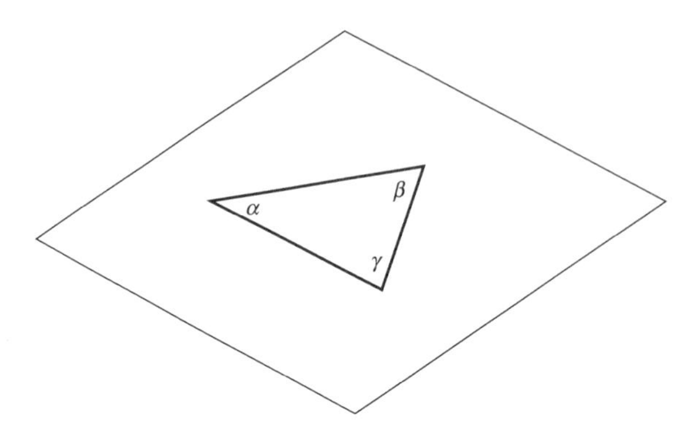

Figure 3.4 A Euclidean, or flat, two-dimensional space.

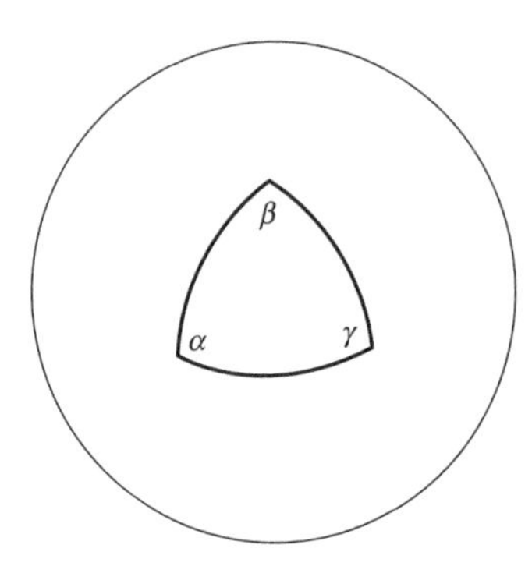

Figure 3.5 A positively curved two-dimensional space.

constructed on the surface of a sphere by connecting three points with geodesics, the angles at its vertices (α, β, and γ) obey the relation

$$\alpha + \beta + \gamma = \pi + A/R^2, \tag{3.25}$$

where A is the area of the triangle, and R is the radius of the sphere. All spaces in which $\alpha + \beta + \gamma > \pi$ are called positively curved spaces. The surface of a sphere is a special variety of positively curved space; it has curvature that is both homogeneous and isotropic. That is, no matter where you draw a triangle on the surface of a sphere, or how you orient it, it must always satisfy Equation 3.25, with the radius R being the same everywhere and in all directions. For brevity, we can describe a space where the curvature is homogeneous and isotropic as having "uniform curvature." Thus, the surface of a sphere can be described as a two-dimensional space with uniform positive curvature.

On the surface of a sphere, we can set up a polar coordinate system by picking a pair of antipodal points to be the "north pole" and "south pole" and by picking a geodesic from the north to the south pole to be the "prime meridian." If r is the distance from the north pole, and θ is the azimuthal angle measured relative to the

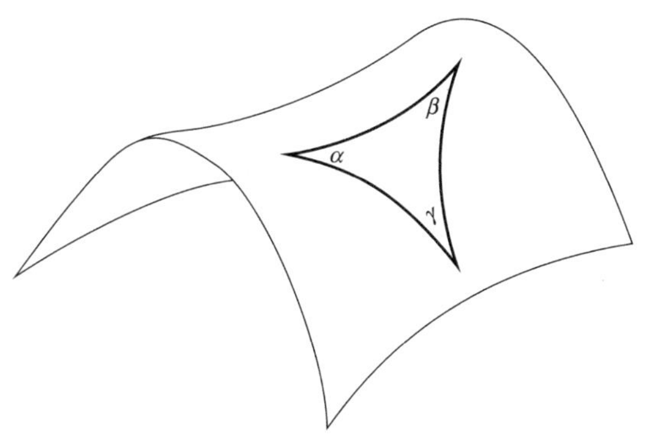

Figure 3.6 A negatively curved two-dimensional space.

prime meridian, then the distance $d\ell$ between a point (r, θ) and another nearby point $(r + dr, \theta + d\theta)$ is given by the relation

$$d\ell^2 = dr^2 + R^2 \sin^2(r/R) d\theta^2. \quad (3.26)$$

Note that the surface of a sphere has a finite area, equal to $4\pi R^2$, and a maximum possible distance between points. (In a non-Euclidean space, the distance between two points is defined as the length of the geodesic connecting them.) The distance between antipodal points, at the maximum possible separation, is $\ell_{\rm max} = \pi R$. By contrast, a plane has infinite area, and has no upper limit on the possible distance between points.[4]

In addition to flat spaces and positively curved spaces, there exist negatively curved spaces. An example of a negatively curved two-dimensional space is the hyperboloid, or saddle shape, shown in Figure 3.6. For illustrative purposes, it would be useful to show you a surface of uniform negative curvature, just as the surface of a sphere has uniform positive curvature. Unfortunately, the mathematician David Hilbert proved that a two-dimensional surface of uniform negative curvature cannot be constructed in a three-dimensional Euclidean space. The saddle shape illustrated in Figure 3.6 has uniform curvature only in the central region, near the "seat" of the saddle.

Despite the difficulties in visualizing a surface of uniform negative curvature, its properties can be written down easily. Consider a two-dimensional surface of uniform negative curvature, with radius of curvature R. If a triangle is constructed on this surface by connecting three points with geodesics, the angles at its vertices (α, β, and γ) obey the relation

$$\alpha + \beta + \gamma = \pi - A/R^2, \quad (3.27)$$

where A is the area of the triangle.

[4] Since the Syndicate of Cambridge University Press objected to producing a book of infinite size, Figure 3.4 actually shows only a portion of a plane.

On a surface of uniform negative curvature, we can set up a polar coordinate system by choosing some point as the pole, and some geodesic leading away from the pole as the prime meridian. If r is the distance from the pole, and θ is the azimuthal angle measured relative to the prime meridian, then the distance $d\ell$ between a point (r, θ) and a nearby point $(r + dr, \theta + d\theta)$ is given by

$$d\ell^2 = dr^2 + R^2 \sinh^2(r/R) d\theta^2. \tag{3.28}$$

A surface of uniform negative curvature has infinite area, and has no upper limit on the possible distance between points.

Relations like those presented in Equations 3.24, 3.26, and 3.28, which give the distance $d\ell$ between two nearby points in space, are known as *metrics*. In general, curvature is a local property. A rubber tablecloth can be badly rumpled at one end of the table and smooth at the other end; a bagel (or other toroidal object) is negatively curved on part of its surface and positively curved on other portions.[5] However, if you want a two-dimensional space to be homogeneous and isotropic, only three possibilities can fit the bill: the space can be uniformly *flat*; it can have uniform *positive* curvature; or it can have uniform *negative* curvature. Thus, if a two-dimensional space has curvature that is homogeneous and isotropic, its geometry can be specified by two quantities, κ, and R. The number κ, called the *curvature constant*, is $\kappa = 0$ for a flat space, $\kappa = +1$ for a positively curved space, and $\kappa = -1$ for a negatively curved space. If the space is curved, then the quantity R, which has dimensions of length, is the radius of curvature.

The results for two-dimensional space can be extended straightforwardly to three dimensions. A three-dimensional space, if its curvature is homogeneous and isotropic, must be flat, or have uniform positive curvature, or have uniform negative curvature. If a three-dimensional space is flat ($\kappa = 0$), it has the metric

$$d\ell^2 = dx^2 + dy^2 + dz^2, \tag{3.29}$$

expressed in cartesian coordinates, or

$$d\ell^2 = dr^2 + r^2[d\theta^2 + \sin^2\theta d\phi^2], \tag{3.30}$$

expressed in spherical coordinates.

If a three-dimensional space has uniform positive curvature ($\kappa = +1$), its metric is

$$d\ell^2 = dr^2 + R^2 \sin^2(r/R)[d\theta^2 + \sin^2\theta d\phi^2]. \tag{3.31}$$

A positively curved three-dimensional space has finite volume, just as a positively curved two-dimensional space has finite area. The point at $r = \pi R$ is the antipodal point to the origin, just as the south pole is the antipodal point to the north pole

[5] You can test this assertion, if you like, by drawing triangles on a bagel.

on the surface of a sphere. By traveling a distance $C = 2\pi R$, it is possible to "circumnavigate" a space of uniform positive curvature.

Finally, if a three-dimensional space has uniform negative curvature ($\kappa = -1$), its metric is

$$d\ell^2 = dr^2 + R^2 \sinh^2(r/R)[d\theta^2 + \sin^2\theta d\phi^2]. \tag{3.32}$$

Like flat space, negatively curved space has infinite volume.

The three possible metrics for a homogeneous, isotropic, three-dimensional space can be written more compactly in the form

$$d\ell^2 = dr^2 + S_\kappa(r)^2 d\Omega^2, \tag{3.33}$$

where

$$d\Omega^2 \equiv d\theta^2 + \sin^2\theta d\phi^2 \tag{3.34}$$

and

$$S_\kappa(r) = \begin{cases} R\sin(r/R) & (\kappa = +1) \\ r & (\kappa = 0) \\ R\sinh(r/R) & (\kappa = -1). \end{cases} \tag{3.35}$$

In the limit $r \ll R$, $S_\kappa \approx r$, regardless of the value of κ. When space is flat, or negatively curved, S_κ increases monotonically with r, with $S_\kappa \to \infty$ as $r \to \infty$. By contrast, when space is positively curved, S_κ increases to a maximum of $S_{\rm max} = R$ at $r/R = \pi/2$, then decreases again to 0 at $r/R = \pi$, the antipodal point to the origin.

The coordinate system (r, θ, ϕ) is not the only possible system. For instance, if we switch the radial coordinate from r to $x \equiv S_\kappa(r)$, the metric for a homogeneous, isotropic, three-dimensional space can be written in the form

$$d\ell^2 = \frac{dx^2}{1 - \kappa x^2/R^2} + x^2 d\Omega^2. \tag{3.36}$$

Although the metrics written in Equations 3.33 and 3.36 appear different on the page, they represent the same homogeneous, isotropic spaces. They merely have a different functional form because of the different choice of radial coordinates.

3.5 The Robertson–Walker Metric

So far, we've considered the metrics for simple two-dimensional and three-dimensional spaces. However, relativity teaches us that space and time together constitute a four-dimensional spacetime. Just as we can compute the distance between two points in space using the appropriate metric for that space, so we can compute the four-dimensional separation between two events in spacetime.

Consider two events, one occurring at the spacetime location (t, r, θ, ϕ), and another occurring at the spacetime location $(t + dt, r + dr, \theta + d\theta, \phi + d\phi)$. According to the laws of special relativity, the spacetime separation between these two events is

$$ds^2 = -c^2 dt^2 + dr^2 + r^2 d\Omega^2. \tag{3.37}$$

The metric given in Equation 3.37 is called the *Minkowski metric*, and the spacetime that it describes is called Minkowski spacetime. Note, from comparison with Equation 3.33, that the spatial component of Minkowski spacetime is Euclidean, or flat.

A photon's path through spacetime is a four-dimensional geodesic – and not just any geodesic, mind you, but a special variety called a *null geodesic*. A null geodesic is one for which, along every infinitesimal segment of the photon's path, $ds = 0$. In Minkowski spacetime, then, a photon's trajectory obeys the relation

$$ds^2 = 0 = -c^2 dt^2 + dr^2 + r^2 d\Omega^2. \tag{3.38}$$

If the photon is moving along a radial path, toward or away from the origin, this means, since θ and ϕ are constant,

$$c^2 dt^2 = dr^2, \tag{3.39}$$

or

$$\frac{dr}{dt} = \pm c. \tag{3.40}$$

The Minkowski metric of Equation 3.37 applies only within the context of special relativity. With no gravity present, Minkowski spacetime is flat and static. When gravity is added, however, the permissible spacetimes are more interesting. In the 1930s, the physicists Howard Robertson and Arthur Walker asked, "What form can the metric of spacetime assume if the universe is spatially homogeneous and isotropic at all time, and if distances are allowed to expand or contract as a function of time?" The metric they derived (independently of each other) is called the *Robertson–Walker metric*.[6] It can be written in the form

$$ds^2 = -c^2 dt^2 + a(t)^2 \left[dr^2 + S_\kappa(r)^2 d\Omega^2\right], \tag{3.41}$$

where the function $S_\kappa(r)$ is given by Equation 3.35, with $R = R_0$. The spatial component of the Robertson–Walker metric consists of the spatial metric for a uniformly curved space of radius R_0 (compare Equation 3.33), scaled by the square of the scale factor $a(t)$. The scale factor, first introduced in Section 2.3,

[6] The Robertson–Walker metric is also called the Friedmann–Robertson–Walker (FRW) metric or the Friedmann–Lemaître–Robertson–Walker (FLRW) metric, depending on which subset of pioneering cosmologists you want to acknowledge.

describes how distances in a homogeneous, isotropic universe expand or contract with time.

The time variable t in the Robertson–Walker metric is the cosmological proper time, called the *cosmic time* for short, and is the time measured by an observer who sees the universe expanding uniformly around him or her. The spatial variables (r, θ, ϕ) are called the *comoving coordinates* of a point in space; if the expansion of the universe is perfectly homogeneous and isotropic, then the comoving coordinates of any point remain constant with time.

The assumption of homogeneity and isotropy is an extremely powerful one. If the universe is perfectly homogeneous and isotropic, then everything we need to know about its geometry is contained within the scale factor $a(t)$, the curvature constant κ (which can be $\kappa = +1$, 0, or -1), and, if $\kappa \neq 0$, the present-day radius of curvature R_0. Much of modern cosmology is devoted in one way or another to finding the values of $a(t)$, κ, and R_0. The assumption of spatial homogeneity and isotropy is so powerful that it was adopted by cosmologists such as Einstein, Friedmann, Lemaître, Robertson, and Walker long before the available observational evidence gave support for such an assumption.[7]

The Robertson–Walker metric is an approximation that holds good only on large scales; on smaller scales, the universe is lumpy, and hence does not expand uniformly. Small, dense lumps, such as humans, teddy bears, and interstellar dust grains, are held together by electromagnetic forces, and hence do not expand. Larger lumps, as long as they are sufficiently dense, are held together by their own gravity, and hence do not expand. Examples of such gravitationally bound systems are galaxies (such as the Milky Way Galaxy in which we live) and clusters of galaxies (such as the Local Group in which we live). It's only on scales larger than $\sim 100\,\text{Mpc}$ that the expansion of the universe can be treated as the ideal, homogeneous, isotropic expansion described by the single scale factor $a(t)$.

3.6 Proper Distance

Consider a galaxy far away from us – sufficiently far away that we may ignore the small scale perturbations of spacetime and adopt the Robertson–Walker metric. One question we may ask is, "Exactly how far away is this galaxy?" In an expanding universe, the distance between two objects is increasing with time. Thus, if we want to assign a spatial distance d between two objects, we must specify the time t at which the distance is the correct one. Suppose that you are at the origin, and that the galaxy that you are observing is at a comoving coordinate position (r, θ, ϕ), as illustrated in Figure 3.7. The proper distance $d_p(t)$ between two points

[7] If homogeneity and isotropy did not exist, as Voltaire might have said, it would be necessary to invent them – at least if your desire is to have a simple, analytically tractable form for the metric of spacetime.

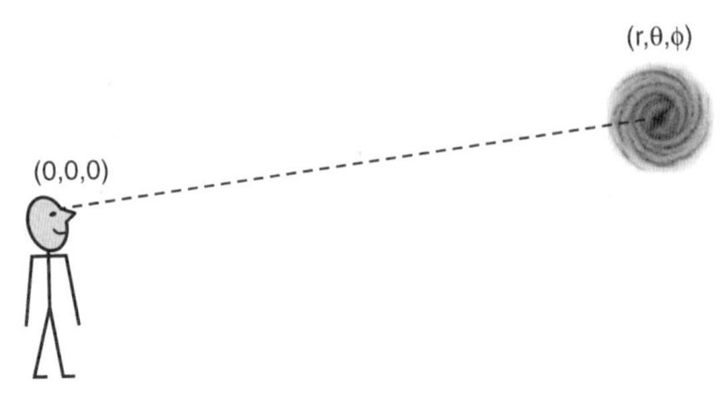

Figure 3.7 An observer at the origin observes a galaxy at coordinate position (r, θ, ϕ). A photon emitted by the galaxy at cosmic time t_e reaches the observer at cosmic time t_0.

is equal to the length of the spatial geodesic between them when the scale factor is fixed at the value $a(t)$. The proper distance between the observer and galaxy in Figure 3.7 can be found using the Robertson–Walker metric at a fixed time t:

$$ds^2 = a(t)^2[dr^2 + S_\kappa(r)^2 d\Omega^2]. \tag{3.42}$$

Along the spatial geodesic between the observer and galaxy, the angle (θ, ϕ) is constant, and thus

$$ds = a(t)dr. \tag{3.43}$$

The proper distance d_p is found by integrating over the radial comoving coordinate r:

$$d_p(t) = a(t) \int_0^r dr = a(t)r. \tag{3.44}$$

Because the proper distance has the form $d_p(t) = a(t)r$, with the comoving coordinate r constant with time, the rate of change for the proper distance between us and a distant galaxy is

$$\dot{d}_p = \dot{a}r = \frac{\dot{a}}{a} d_p. \tag{3.45}$$

Thus, at the current time $(t = t_0)$, there is a linear relation between the proper distance to a galaxy and its recession speed:

$$v_p(t_0) = H_0 d_p(t_0), \tag{3.46}$$

where

$$v_p(t_0) \equiv \dot{d}_p(t_0) \tag{3.47}$$

and

$$H_0 = \left(\frac{\dot{a}}{a}\right)_{t=t_0}. \tag{3.48}$$

In a sense, this is just a repetition of what was demonstrated in Section 2.3; if the distance between points is proportional to $a(t)$, there will be a linear relation between the relative velocity of two points and the distance between them. Now, however, we are interpreting the change in distance between widely separated galaxies as being associated with the expansion of space. As the distance between galaxies increases, the radius of curvature of the universe, $R(t) = a(t)R_0$, increases at the same rate.

Some cosmology books contain a statement like "As space expands, it drags galaxies away from each other." Statements of this sort are somewhat misleading because they make galaxies appear to be entirely passive. On the other hand, a statement like "As galaxies move apart, they drag space along with them" would be equally misleading because it makes space appear to be entirely passive. As the theory of general relativity points out, spacetime and mass-energy are intimately linked. Yes, the curvature of spacetime does tell mass-energy how to move, but then it's mass-energy which tells spacetime how to curve.

The linear velocity–distance relation given in Equation 3.46 implies that points separated by a proper distance greater than the Hubble distance,

$$d_H(t_0) \equiv c/H_0, \tag{3.49}$$

will have

$$v_p = \dot{d}_p > c. \tag{3.50}$$

Using the observationally determined value of $H_0 = 68 \pm 2\,\mathrm{km\,s^{-1}\,Mpc^{-1}}$, the current value of the Hubble distance in our universe is

$$d_H(t_0) = c/H_0 = 4380 \pm 130\,\mathrm{Mpc}. \tag{3.51}$$

Thus, galaxies farther than $\sim$ 4400 megaparsecs from us are currently moving away from us at speeds greater than that of light. Cosmological innocents sometimes exclaim, "Gosh! Doesn't this violate the law that massive objects can't travel faster than the speed of light?" Actually, it doesn't. The speed limit that states that massive objects must travel with $v < c$ relative to each other is one of the results of special relativity, and refers to the relative motion of objects within a static space. In the context of general relativity, there is no objection to having two points moving away from each other at superluminal speed due to the expansion of space.

When we observe a distant galaxy, we know its angular position very well, but not its distance. That is, we can point in its direction, but we don't know its current proper distance $d_p(t_0)$. We can, however, measure the redshift z of the light we receive from the galaxy. Although the redshift doesn't tell us the proper distance to the galaxy, it does tell us what the scale factor a was at the time the light from that galaxy was emitted. To see the link between a and z, consider the galaxy illustrated in Figure 3.7. Light that was emitted by the galaxy at a time t_e

is observed by us at a time t_0. During its travel from the distant galaxy to us, the light traveled along a null geodesic, with $ds = 0$. The null geodesic has θ and ϕ constant.[8] Thus, along the light's null geodesic,

$$c^2 dt^2 = a(t)^2 dr^2. \tag{3.52}$$

Rearranging this relation, we find

$$c\frac{dt}{a(t)} = dr. \tag{3.53}$$

In Equation 3.53, the left-hand side is a function only of t, and the right-hand side is independent of t. Suppose the distant galaxy emits light with a wavelength λ_e, as measured by an observer in the emitting galaxy. Fix your attention on a single wave crest of the emitted light. The wave crest is emitted at a time t_e and observed at a time t_0, such that

$$c\int_{t_e}^{t_0} \frac{dt}{a(t)} = \int_0^r dr = r. \tag{3.54}$$

The next wave crest of light is emitted at a time $t_e + \lambda_e/c$, and is observed at a time $t_0 + \lambda_0/c$, where, in general, $\lambda_0 \neq \lambda_e$. For the second wave crest,

$$c\int_{t_e+\lambda_e/c}^{t_0+\lambda_0/c} \frac{dt}{a(t)} = \int_0^r dr = r. \tag{3.55}$$

Comparing Equations 3.54 and 3.55, we find that

$$\int_{t_e}^{t_0} \frac{dt}{a(t)} = \int_{t_e+\lambda_e/c}^{t_0+\lambda_0/c} \frac{dt}{a(t)}. \tag{3.56}$$

That is, the integral of $dt/a(t)$ between the time of emission and the time of observation is the same for every wave crest in the emitted light. If we subtract the integral

$$\int_{t_e+\lambda_e/c}^{t_0} \frac{dt}{a(t)} \tag{3.57}$$

from each side of Equation 3.56, we find the relation

$$\int_{t_e}^{t_e+\lambda_e/c} \frac{dt}{a(t)} = \int_{t_0}^{t_0+\lambda_0/c} \frac{dt}{a(t)}. \tag{3.58}$$

That is, the integral of $dt/a(t)$ between the emission of successive wave crests is equal to the integral of $dt/a(t)$ between the observation of the same two wave crests. This relation becomes still simpler when we realize that during the time between the emission or observation of two wave crests, the universe doesn't have time to expand by a significant amount. The time scale for expansion of the

[8] In a homogeneous, isotropic universe there's no reason for the light to swerve to one side or the other.

universe is the Hubble time, $H_0^{-1} \approx 14\,\text{Gyr}$. The time between wave crests, for visible light, is $\lambda/c \approx 2 \times 10^{-15}\,\text{s} \approx 10^{-32} H_0^{-1}$. Thus, $a(t)$ is effectively constant in the integrals of Equation 3.58. We may then write

$$\frac{1}{a(t_e)} \int_{t_e}^{t_e+\lambda_e/c} dt = \frac{1}{a(t_0)} \int_{t_0}^{t_0+\lambda_0/c} dt, \tag{3.59}$$

or

$$\frac{\lambda_e}{a(t_e)} = \frac{\lambda_0}{a(t_0)}. \tag{3.60}$$

Using the definition of redshift, $z = (\lambda_0 - \lambda_e)/\lambda_e$, we find that the redshift of light from a distant object is related to the expansion factor at the time it was emitted via the equation

$$1 + z = \frac{a(t_0)}{a(t_e)} = \frac{1}{a(t_e)}. \tag{3.61}$$

Here, we have used the usual convention that $a(t_0) = 1$.

Thus, if we observe a galaxy with a redshift $z = 2$, we are observing it as it was when the universe had a scale factor $a(t_e) = 1/3$. The redshift we observe for a distant object depends only on the relative scale factors at the time of emission and the time of observation. It doesn't depend on how the transition between $a(t_e)$ and $a(t_0)$ was made. It doesn't matter if the expansion was gradual or abrupt; it doesn't matter if the transition was monotonic or oscillatory. All that matters is the scale factors at the time of emission and the time of observation.

Exercises

3.1 What evidence can you provide to support the assertion that the universe is electrically neutral on large scales?

3.2 Suppose you are a two-dimensional being, living on the surface of a sphere with radius R. An object of width $d\ell \ll R$ is at a distance r from you (remember, all distances are measured on the surface of the sphere). What angular width $d\theta$ will you measure for the object? Explain the behavior of $d\theta$ as $r \to \pi R$.

3.3 Suppose you are *still* a two-dimensional being, living on the same sphere of radius R. Show that if you draw a circle of radius r, the circle's circumference will be

$$C = 2\pi R \sin(r/R). \tag{3.62}$$

Idealize the Earth as a perfect sphere of radius $R = 6371\,\text{km}$. If you could measure distances with an error of ± 1 meter, how large a circle would you have to draw on the Earth's surface to convince yourself that the Earth is spherical rather than flat?

3.4 Consider an equilateral triangle, with sides of length L, drawn on a two-dimensional surface of uniform curvature. Can you draw an equilateral triangle of arbitrarily large area A on a surface with $\kappa = +1$ and radius of curvature R? If not, what is the maximum possible value of A? Can you draw an equilateral triangle of arbitrarily large area A on a surface with $\kappa = 0$? If not, what is the maximum possible value of A? Can you draw an equilateral triangle of arbitrarily large area A on a surface with $\kappa = -1$ and radius of curvature R? If not, what is the maximum possible value of A?

3.5 By making the substitutions $x = r \sin\theta \cos\phi$, $y = r \sin\theta \sin\phi$, and $z = r \cos\theta$, demonstrate that Equations 3.29 and 3.30 represent the same metric.

4

Cosmic Dynamics

The idea that the universe could be curved, or non-Euclidean, long predates Einstein's theory of general relativity. As early as 1829, half a century before Einstein's birth, Nikolai Ivanovich Lobachevski, one of the founders of non-Euclidean geometry, proposed observational tests to demonstrate whether the universe was curved. In principle, measuring the curvature of the universe is simple; in practice, it is much more difficult. In principle, we could determine the curvature by drawing a really, really big triangle, and measuring the angles α, β, and γ at the vertices. Equations 3.22, 3.25, and 3.27 generalize to the equation

$$\alpha + \beta + \gamma = \pi + \frac{\kappa A}{R_0^2}, \tag{4.1}$$

where A is the area of the triangle. Therefore, if $\alpha + \beta + \gamma > \pi$ radians, the universe is positively curved, and if $\alpha + \beta + \gamma < \pi$ radians, the universe is negatively curved. If, in addition, we measure the area of the triangle, we can determine the radius of curvature R_0. Unfortunately for this elegant geometric plan, the area of the biggest triangle we can draw is much smaller than R_0^2, and the deviation of $\alpha + \beta + \gamma$ from π radians would be too small to measure.

We can conclude from geometric arguments that if the universe is curved, it can't have a radius of curvature R_0 that is significantly smaller than the current Hubble distance, $c/H_0 \approx 4380\,\text{Mpc}$. To see why, consider a galaxy of diameter D that is at a distance r from the Earth. In a flat universe, in the limit $D \ll r$, we can use the small angle formula to compute the observed angular size α of the galaxy:

$$\alpha = \frac{D}{r}. \tag{4.2}$$

In a positively curved universe, the angular size is

$$\alpha_+ = \frac{D}{R_0 \sin(r/R_0)}. \tag{4.3}$$

When $r < \pi R_0$, then $\alpha_+ > D/r$, and the galaxy appears larger in size than it would in a flat universe. That is, in a positively curved universe, the curvature acts as a magnifying lens. Notice that the angular size α_+ blows up at $r = \pi R_0$; physically, this means that when a galaxy is at a distance corresponding to half the circumference of the universe, it fills the entire sky. No such bloated, highly-magnified galaxies are seen, even though we can see galaxies at distances as great as $r \sim c/H_0$. Thus, we conclude that if the universe is positively curved, it must have $\pi R_0 > c/H_0$.

In a negatively curved universe, the observed angular size of the galaxy is

$$\alpha_- = \frac{D}{R_0 \sinh(r/R_0)} < \frac{D}{r}. \tag{4.4}$$

At a distance $r \gg R_0$, we can use the approximation $\sinh x \approx e^x/2$ to find

$$\alpha_- \approx \frac{2D}{R_0} \exp\left(-\frac{r}{R_0}\right). \tag{4.5}$$

In a negatively curved universe, galaxies at a distance much greater than the radius of curvature R_0 appear exponentially tiny in angle. However, in our universe, galaxies are seen to be resolved in angle out to distances $r \sim c/H_0$. Thus, we conclude that if the universe is negatively curved, it must have $R_0 > c/H_0$.

4.1 Einstein's Field Equation

In the 19th century, mathematicians such as Lobachevski were able to conceive of curved space. However, it wasn't until Einstein published his theory of general relativity in 1915 that anyone related the curvature of space (and time) to the physical content of the universe. The key equation of general relativity, which gives the mathematical relation between spacetime curvature and the energy density and pressure of the universe, is the field equation.

Einstein's field equation plays a role in general relativity that is analogous to the role played by Poisson's equation in Newtonian dynamics. Poisson's equation,

$$\nabla^2 \Phi = 4\pi G \rho, \tag{4.6}$$

tells you how to compute the gravitational potential Φ, given the mass density ρ of the material filling the universe. By taking the gradient of Φ, you determine the acceleration, and can then compute the trajectory of objects moving freely through space under the influence of gravity. Analogously, you can use Einstein's field equation to compute the curvature of spacetime, given the energy density ε, pressure P, and other properties of the material filling the universe. You can then compute the trajectory of a freely moving object by finding the appropriate geodesic in the curved spacetime.

Einstein's field equation looks simple when it is written down:

$$G_{\mu\nu} = \frac{8\pi G}{c^4} T_{\mu\nu}. \tag{4.7}$$

The simplicity is deceptive, since the compact notation hides a great deal of necessary detail. The quantity $G_{\mu\nu}$ on the left-hand side of Equation 4.7 is the *Einstein tensor*, which is a 4×4 tensor that describes the curvature of spacetime at every location (t, x, y, z). It is a symmetric tensor, with $G_{\mu\nu} = G_{\nu\mu}$, so it has ten independent components. On the right-hand side of Equation 4.7, the quantity $T_{\mu\nu}$ is the *stress-energy tensor*, sometimes called the energy-momentum tensor; like the Einstein tensor, it is a 4×4 symmetric tensor.

The deceptively simple field equation, $G_{\mu\nu} = (8\pi G/c^4)T_{\mu\nu}$, is actually a set of ten nonlinear second-order differential equations. Even without making the herculean effort to solve them exactly to find the curvature everywhere (and everywhen) in spacetime, we can make general statements about the properties of the solution. Since the ten differential equations are second order, this means that spacetime can have nonzero curvature even in spacetime neighborhoods where the stress-energy tensor $T_{\mu\nu}$ is zero. (Analogously, Poisson's equation is a second-order differential equation, and gravitational acceleration can be nonzero even in spatial neighborhoods where the mass density is zero.) Another property of second-order differential equations involving space and time is that they can yield propagating wave solutions, in which disturbances propagate through space as a function of time. Just as a time-varying electric dipole creates electromagnetic waves, a time-varying mass-energy quadrupole creates gravitational waves.[1]

In the most general case, the stress-energy tensor $T_{\mu\nu}$ can be very complicated, and difficult to calculate. However, things become much simpler if the universe is filled with a homogeneous and isotropic perfect gas. In that case, an observer who sees the universe expanding uniformly around her will measure an energy density $\varepsilon(t)$ and a pressure $P(t)$ for the ideal gas that are a function only of cosmic time; the observer will not measure any bulk velocity $\vec{u}$ for the gas, since that would break the isotropy. In this idealized case (which fortunately is a good approximation for our purposes), the stress-energy tensor $T_{\mu\nu}$ depends only on $\varepsilon(t)$ and $P(t)$. The metric describing the curvature of spacetime, in this case, is the homogeneous and isotropic Robertson–Walker metric (Equation 3.41):

$$ds^2 = -c^2 dt^2 + a(t)^2 \left[dr^2 + S_\kappa(r)^2 d\Omega^2\right], \tag{4.8}$$

[1] Einstein predicted the existence of gravitational waves in 1916. He then "un-predicted" them in 1936, when he erroneously thought they were a byproduct of the approximations he had made in 1916. In the 1950s, however, physicists "re-predicted" the existence of gravitational waves, which were finally detected by the Laser Interferometry Gravitational-wave Observatory on 14 Sept. 2015, just in time for the centenary of Einstein's prediction.

where

$$S_\kappa(r) = \begin{cases} R_0 \sin(r/R_0) & (\kappa = +1) \\ r & (\kappa = 0) \\ R_0 \sinh(r/R_0) & (\kappa = -1). \end{cases} \quad (4.9)$$

Our remaining goal is to find how $a(t)$, κ, and R_0, the parameters that describe curvature, are linked to $\varepsilon(t)$ and $P(t)$, the parameters that describe the contents of the universe.

4.2 The Friedmann Equation

The equation that links together $a(t)$, κ, R_0, and $\varepsilon(t)$ is known as the *Friedmann equation*, after Alexander Friedmann, the Russian physicist who first derived the equation in 1922.[2] Friedmann actually started his scientific career as a meteorologist. Later, however, he taught himself general relativity, and used Einstein's field equation to describe how a spatially homogeneous and isotropic universe expands or contracts as a function of time. Friedmann published his first results, implying expanding or contracting space, five years before Lemaître interpreted the observed galaxy redshifts in terms of an expanding universe, and seven years before Hubble published Hubble's law.

Friedmann derived his eponymous equation starting from Einstein's field equation, using the full power of general relativity. Even without bringing relativity into play, some (though not all) of the aspects of the Friedmann equation can be understood with the use of purely Newtonian dynamics. To see how the expansion or contraction of the universe can be viewed from a Newtonian viewpoint, I will first derive the nonrelativistic equivalent of the Friedmann equation, starting from Newton's law of gravity and second law of motion. Then I will state (without proof) the modifications that must be made to find the more correct, general relativistic form of the Friedmann equation.

To begin, consider a homogeneous sphere of matter, with total mass M_s constant with time (Figure 4.1). The sphere is expanding or contracting isotropically, so that its radius $R_s(t)$ is increasing or decreasing with time. Place a test mass, of infinitesimal mass m, at the surface of the sphere. The gravitational force F experienced by the test mass will be, from Newton's law of gravity,

$$F = -\frac{GM_s m}{R_s(t)^2}. \quad (4.10)$$

[2] Using the Library of Congress transliteration system for Cyrillic, his name would be "Aleksandr Fridman." However, in the German scientific journals where he published his main results, he alternated between the spellings "Friedman" and "Friedmann" for his last name. The two-n spelling is more popular among historians of science.

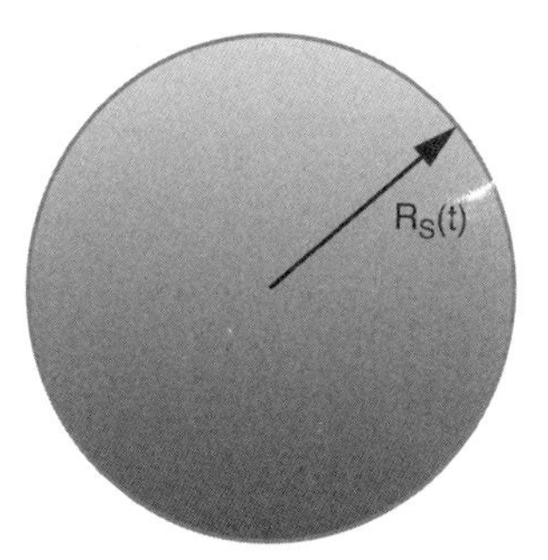

Figure 4.1 A sphere of radius $R_s(t)$ and mass M_s, expanding or contracting under its own gravity.

The gravitational acceleration at the surface of the sphere will then be, from Newton's second law of motion,

$$\frac{d^2R_s}{dt^2} = -\frac{GM_s}{R_s(t)^2}. \tag{4.11}$$

Multiply each side of the equation by dR_s/dt and integrate to find

$$\frac{1}{2}\left(\frac{dR_s}{dt}\right)^2 = \frac{GM_s}{R_s(t)} + U, \tag{4.12}$$

where U is a constant of integration. Equation 4.12 simply states that the sum of the kinetic energy per unit mass,

$$\epsilon_{\text{kin}} = \frac{1}{2}\left(\frac{dR_s}{dt}\right)^2, \tag{4.13}$$

and the gravitational *potential* energy per unit mass,

$$\epsilon_{\text{pot}} = -\frac{GM_s}{R_s(t)}, \tag{4.14}$$

is constant for a bit of matter at the surface of a sphere, as the sphere expands or contracts under its own gravitational influence.

Since the mass of the sphere is constant, we may write

$$M_s = \frac{4\pi}{3}\rho(t)R_s(t)^3. \tag{4.15}$$

Since the expansion or contraction is isotropic about the sphere's center, we may write the radius $R_s(t)$ in the form

$$R_s(t) = a(t)r_s, \tag{4.16}$$

where $a(t)$ is the scale factor and r_s is the comoving radius of the sphere. In terms of $\rho(t)$ and $a(t)$, the energy conservation Equation 4.12 can be rewritten in the form

$$\frac{1}{2}r_s^2\dot{a}^2 = \frac{4\pi}{3}Gr_s^2\rho(t)a(t)^2 + U. \quad (4.17)$$

Dividing each side of Equation 4.17 by $r_s^2a^2/2$ yields the equation

$$\left(\frac{\dot{a}}{a}\right)^2 = \frac{8\pi G}{3}\rho(t) + \frac{2U}{r_s^2}\frac{1}{a(t)^2}. \quad (4.18)$$

Equation 4.18 gives the Friedmann equation in its Newtonian form.

Note that the time derivative of the scale factor only enters into Equation 4.18 as $\dot{a}^2$; a contracting sphere ($\dot{a} < 0$) is simply the time reversal of an expanding sphere ($\dot{a} > 0$). Let's concentrate on the case of an expanding sphere, analogous to the expanding universe in which we find ourselves. The future of the expanding sphere falls into one of three classes, depending on the sign of U. First, consider the case $U > 0$. In this case, the right-hand side of Equation 4.18 is always positive. Therefore, $\dot{a}^2$ is always positive, and the expansion of the sphere never stops. Second, consider the case $U < 0$. In this case, the right-hand side of Equation 4.18 starts out positive. However, at a maximum scale factor

$$a_{\max} = -\frac{GM_s}{Ur_s}, \quad (4.19)$$

the right-hand side will equal zero, and expansion will stop. Since $\ddot{a}$ will still be negative, the sphere will then contract. Third, and finally, consider the case $U = 0$. This is the boundary case in which $\dot{a} \to 0$ as $t \to \infty$ and $\rho \to 0$.

The three possible fates of an expanding sphere in a Newtonian universe are analogous to the three possible fates of a ball thrown upward from the surface of the Earth. First, the ball can be thrown upward with a speed greater than the escape speed; in this case, the ball continues to go upward forever. Second, the ball can be thrown upward with a speed less than the escape speed; in this case, the ball reaches a maximum altitude, then falls back down. Third, and finally, the ball can be thrown upward with a speed exactly equal to the escape speed; in this case, the speed of the ball approaches zero as $t \to \infty$.

The Friedmann equation in its Newtonian form (Equation 4.18) is useful in picturing how isotropically expanding objects behave under the influence of their self-gravity. However, its application to the real universe must be regarded with considerable skepticism. First of all, a spherical volume of finite radius R_s cannot represent a homogeneous, isotropic universe. In a finite spherical volume, there exists a special location (the center of the sphere), violating the assumption of homogeneity, and at any point there exists a special direction (the direction pointing toward the center), violating the assumption of isotropy. What if we instead regard the sphere of radius R_s as being carved out of an infinite, homogeneous, isotropic universe? In that case, Newtonian dynamics tell us that the gravitational acceleration inside a hollow spherically symmetric shell is equal to zero. We divide up the region outside the sphere into concentric shells, and thus conclude that the test mass m at R_s experiences no net acceleration from matter at $R > R_s$. Unfortunately, a Newtonian argument of this sort assumes that space is perfectly Euclidean, an assumption that we can't necessarily make in the real universe. A derivation of the correct Friedmann equation has to begin, as Friedmann himself began, with Einstein's field equation.

The correct form of the Friedmann equation, including all general relativistic effects, is

$$\left(\frac{\dot{a}}{a}\right)^2 = \frac{8\pi G}{3c^2}\varepsilon(t) - \frac{\kappa c^2}{R_0^2}\frac{1}{a(t)^2}. \tag{4.20}$$

Note the changes made in going from the Newtonian form of the Friedmann equation (Equation 4.18) to the correct relativistic form (Equation 4.20). The first change is that the mass density ρ has been replaced by an energy density ε divided by the square of the speed of light. One of Einstein's insights was that in determining the gravitational influence of a particle, the important quantity was not its mass m but its energy,

$$E = (m^2c^4 + p^2c^2)^{1/2}. \tag{4.21}$$

Here p is the momentum of the particle, as measured by an observer at the particle's location who sees the universe expanding isotropically. Any motion that a particle has, in addition to the motion associated with the expansion or contraction of the universe, is called the particle's *peculiar* motion.[3] If a massive particle is nonrelativistic, with a peculiar velocity $v \ll c$, then its peculiar momentum will be $p \approx mv$ and its energy will be

$$E_{\rm nonrel} \approx mc^2(1 + v^2/c^2)^{1/2} \approx mc^2 + \frac{1}{2}mv^2. \tag{4.22}$$

Thus, if the universe contained only massive, slowly moving particles, then the energy density ε would be nearly equal to ρc^2, with only a small correction for the kinetic energy $mv^2/2$ of the particles. However, photons and other massless particles also have an energy,

$$E_{\rm rel} = pc = hf, \tag{4.23}$$

[3] The adjective "peculiar" comes from the Latin word *peculium*, meaning "private property." The peculiar motion of a particle is thus the motion that belongs to the particle alone, and not to the global expansion or contraction of the universe.

which contributes to the energy density ε. Not only do photons respond to the curvature of spacetime, they also contribute to it.

The second change that must be made in going from the Newtonian form of the Friedmann equation to the correct relativistic form is making the substitution

$$\frac{2U}{r_s^2} = -\frac{\kappa c^2}{R_0^2}. \tag{4.24}$$

The Newtonian case with $U < 0$ corresponds to the relativistic case with positive curvature ($\kappa = +1$); conversely, $U > 0$ corresponds to negative curvature ($\kappa = -1$). The Newtonian special case with $U = 0$ corresponds to the relativistic special case where the space is perfectly flat ($\kappa = 0$). Although I have not given the derivation of the Friedmann equation in the general relativistic case, it makes sense that the curvature, given by κ and R_0, the expansion rate, given by $a(t)$, and the energy density ε should be bound up together in the same equation. After all, in Einstein's view, the energy density of the universe determines both the curvature of space and the overall dynamics of the expansion.

The Friedmann equation is a Very Important Equation in cosmology.[4] However, if we want to apply the Friedmann equation to the real universe, we must have some way of tying it to observable properties. For instance, the Friedmann equation can be tied to the Hubble constant, H_0. Remember, in a universe whose expansion (or contraction) is described by a scale factor $a(t)$, there's a linear relation between recession speed v and proper distance d:

$$v(t) = H(t)d(t), \tag{4.25}$$

where $H(t) \equiv \dot{a}/a$. Thus, the Friedmann equation can be rewritten in the form

$$H(t)^2 = \frac{8\pi G}{3c^2}\varepsilon(t) - \frac{\kappa c^2}{R_0^2 a(t)^2}. \tag{4.26}$$

At the present moment,

$$H_0 = H(t_0) = \left(\frac{\dot{a}}{a}\right)_{t=t_0} = 68 \pm 2\,\text{km}\,\text{s}^{-1}\,\text{Mpc}^{-1}. \tag{4.27}$$

(As an aside, the time-varying function $H(t)$ is generally known as the "Hubble parameter," while H_0, the value of $H(t)$ at the present day, is known as the "Hubble constant.") The Friedmann equation evaluated at the present moment is

$$H_0^2 = \frac{8\pi G}{3c^2}\varepsilon_0 - \frac{\kappa c^2}{R_0^2}, \tag{4.28}$$

using the convention that a subscript "0" indicates the value of a time-varying quantity evaluated at the present. Thus, the Friedmann equation gives a relation

[4] You should consider writing it in reverse on your forehead so that you can see it every morning in the mirror when you comb your hair.

between the Hubble parameter H_0, which tells us the current rate of expansion, ε_0, which tells us the current energy density, and κ/R_0^2, which tells us the current curvature.

In a spatially flat universe ($\kappa = 0$), the Friedmann equation takes a particularly simple form:

$$H(t)^2 = \frac{8\pi G}{3c^2}\varepsilon(t). \tag{4.29}$$

Thus, for a given value of the Hubble parameter, there is a *critical density*,

$$\varepsilon_c(t) \equiv \frac{3c^2}{8\pi G}H(t)^2. \tag{4.30}$$

If the energy density $\varepsilon(t)$ is greater than this value, the universe is positively curved ($\kappa = +1$). If $\varepsilon(t)$ is less than this value, the universe is negatively curved ($\kappa = -1$). Knowing the Hubble constant to within about 3%, we can compute the current value of the critical density to within about 6%:

$$\varepsilon_{c,0} = \frac{3c^2}{8\pi G}H_0^2 = (7.8 \pm 0.5) \times 10^{-10}\,\mathrm{J\,m^{-3}} = 4870 \pm 290\,\mathrm{MeV\,m^{-3}}. \tag{4.31}$$

The critical density is frequently written as the equivalent mass density,

$$\begin{aligned}\rho_{c,0} \equiv \frac{\varepsilon_{c,0}}{c^2} &= (8.7 \pm 0.5) \times 10^{-27}\,\mathrm{kg\,m^{-3}} \\ &= (1.28 \pm 0.08) \times 10^{11}\,\mathrm{M_\odot\,Mpc^{-3}}.\end{aligned} \tag{4.32}$$

The current critical density is roughly equivalent to a density of one proton per 200 liters. This is definitely not a large density, by terrestrial standards; a 200-liter drum filled with liquid water, for instance, contains $\sim 10^{29}$ protons and neutrons. The critical density is not even a large density by the standards of interstellar space within our galaxy; even the hottest, most tenuous regions of the interstellar medium have a few protons per liter. However, keep in mind that most of the volume of the universe consists of intergalactic voids, where the density is extraordinarily low. When averaged over scales of 100 Mpc or more, the mean energy density of the universe, as it turns out, is very close to the critical density.

When discussing the curvature of the universe, it is useful to talk about the energy density in terms of the dimensionless *density parameter*

$$\Omega(t) \equiv \frac{\varepsilon(t)}{\varepsilon_c(t)}. \tag{4.33}$$

The current value of the density parameter, determined from a combination of observational data, lies in the range $0.995 < \Omega_0 < 1.005$. Written in terms of the density parameter Ω, the Friedmann equation becomes

$$1 - \Omega(t) = -\frac{\kappa c^2}{R_0^2 a(t)^2 H(t)^2}. \tag{4.34}$$

Note that, since the right-hand side of Equation 4.34 cannot change sign as the universe expands, neither can the left-hand side. If $\Omega < 1$ at any time, it remains less than one for all time; similarly, if $\Omega > 1$ at any time, it remains greater than one for all time, and if $\Omega = 1$ at any time, $\Omega = 1$ at all times. At the present moment, the relation between curvature, density, and expansion rate can be written in the form

$$1 - \Omega_0 = -\frac{\kappa c^2}{R_0^2 H_0^2}, \tag{4.35}$$

or

$$\frac{\kappa}{R_0^2} = \frac{H_0^2}{c^2}(\Omega_0 - 1). \tag{4.36}$$

If you know Ω_0, you know the sign of the curvature (κ). If, in addition, you know the Hubble distance, c/H_0, you can compute the radius of curvature (R_0).

4.3 The Fluid and Acceleration Equations

Although the Friedmann equation is indeed important, it cannot, all by itself, tell us how the scale factor $a(t)$ evolves with time. Even if we had accurate boundary conditions (precise values for ε_0 and H_0, for instance), it still remains a single equation in two unknowns, $a(t)$ and $\varepsilon(t)$.

We need another equation involving a and ε if we are to solve for a and ε as functions of time. The Friedmann equation, in the Newtonian approximation, is a statement of energy conservation; in particular, it says that the sum of the gravitational potential energy and the kinetic energy of expansion is constant. Energy conservation is a generally useful concept in Newtonian physics, so let's look at another manifestation of the same concept – the first law of thermodynamics:

$$dQ = dE + PdV, \tag{4.37}$$

where dQ is the heat flow into or out of a volume, dE is the change in internal energy, P is the pressure, and dV is the change in volume. This equation was applied in Section 2.5 to a comoving volume filled with photons, but it applies equally well to a comoving volume filled with any sort of fluid. If the universe is perfectly homogeneous, then for any volume $dQ = 0$; that is, there is no bulk flow of heat.

Processes for which $dQ = 0$ are known as *adiabatic* processes. The term "adiabatic" comes from the Greek word *adiabatos*, meaning "not to be passed through," referring to the fact that heat does not pass through the boundary of the volume.[5] Saying that the expansion of the universe is adiabatic is also a statement about entropy. The change in entropy dS within a region is given by the relation $dS = dQ/T$; thus, an adiabatic process is one in which entropy is not increased. A homogeneous, isotropic expansion of the universe does not increase the universe's entropy.

Since $dQ = 0$ for a comoving volume as the universe expands, the first law of thermodynamics, as applied to the expanding universe, reduces to the form

$$\dot{E} + P\dot{V} = 0. \tag{4.38}$$

For concreteness, consider a sphere of comoving radius r_s expanding along with the universal expansion, so that its proper radius is $R_s(t) = a(t)r_s$. The volume of the sphere is

$$V(t) = \frac{4\pi}{3} r_s^3 a(t)^3, \tag{4.39}$$

so the rate of change of the sphere's volume is

$$\dot{V} = \frac{4\pi}{3} r_s^3 (3a^2\dot{a}) = V\left(3\frac{\dot{a}}{a}\right). \tag{4.40}$$

The internal energy of the sphere is

$$E(t) = V(t)\varepsilon(t), \tag{4.41}$$

so the rate of change of the sphere's internal energy is

$$\dot{E} = V\dot{\varepsilon} + \dot{V}\varepsilon = V\left(\dot{\varepsilon} + 3\frac{\dot{a}}{a}\varepsilon\right). \tag{4.42}$$

Combining Equations 4.38, 4.40, and 4.42, we find that the first law of thermodynamics in an expanding (or contracting) universe takes the form

$$V\left(\dot{\varepsilon} + 3\frac{\dot{a}}{a}\varepsilon + 3\frac{\dot{a}}{a}P\right) = 0, \tag{4.43}$$

or

$$\dot{\varepsilon} + 3\frac{\dot{a}}{a}(\varepsilon + P) = 0. \tag{4.44}$$

This equation is called the *fluid equation*, and is the second of the key equations describing the expansion of the universe.[6] Unlike the Friedmann equation, whose relativistic form is different from its Newtonian form, the fluid equation is unchanged by the switch from Newtonian physics to general relativity.

[5] The word "adiabatic" is thus etymologically related to "diabetes," a word that refers to the quick passage of liquids through the human body, creating the increased thirst and frequent urination that are symptomatic of untreated diabetes. If you do not have diabetes, you could, I suppose, refer to yourself as "a-diabetic."

[6] Write it on your forehead just underneath the Friedmann equation.

The Friedmann equation and fluid equation can be combined into an acceleration equation that tells how the universe speeds up or slows down with time. The Friedmann equation (Equation 4.20), multiplied by a^2, takes the form

$$\dot{a}^2 = \frac{8\pi G}{3c^2}\varepsilon a^2 - \frac{\kappa c^2}{R_0^2}. \tag{4.45}$$

Taking the time derivative yields

$$2\dot{a}\ddot{a} = \frac{8\pi G}{3c^2}(\dot{\varepsilon}a^2 + 2\varepsilon a\dot{a}). \tag{4.46}$$

Dividing by $2\dot{a}a$ tells us

$$\frac{\ddot{a}}{a} = \frac{4\pi G}{3c^2}\left(\dot{\varepsilon}\frac{a}{\dot{a}} + 2\varepsilon\right). \tag{4.47}$$

Using the fluid equation (Equation 4.44), we may make the substitution

$$\dot{\varepsilon}\frac{a}{\dot{a}} = -3(\varepsilon + P) \tag{4.48}$$

to find the usual form of the *acceleration equation*,

$$\frac{\ddot{a}}{a} = -\frac{4\pi G}{3c^2}(\varepsilon + 3P). \tag{4.49}$$

If the energy density ε is positive, then it provides a negative acceleration – that is, it decreases the value of $\dot{a}$ and reduces the relative velocity of any two points in the universe. The acceleration equation also includes the pressure P associated with the material filling the universe.[7]

A gas made of ordinary baryonic matter has a positive pressure P, resulting from the random thermal motions of the molecules, atoms, or ions of which the gas is made. A gas of photons also has a positive pressure, as does a gas of neutrinos or WIMPs. The positive pressure associated with these components of the universe will cause the expansion to slow down. Suppose, though, that the universe had a component with $\varepsilon > 0$ and

$$P < -\frac{1}{3}\varepsilon. \tag{4.50}$$

Inspection of the acceleration equation (Equation 4.49) shows us that such a component will cause the expansion of the universe to speed up rather than slow down.

4.4 Equations of State

To recap, we now have three key equations that describe how the universe expands. There's the Friedmann equation,

[7] Although we think of ε as an energy per unit volume and P as a force per unit area, they both have the same dimensionality: $1\,\mathrm{J\,m^{-3}} = 1\,\mathrm{N\,m^{-2}} = 1\,\mathrm{kg\,m^{-1}\,s^{-2}}$.

$$\left(\frac{\dot{a}}{a}\right)^2 = \frac{8\pi G}{3c^2}\varepsilon - \frac{\kappa c^2}{R_0^2 a^2}, \tag{4.51}$$

the fluid equation,

$$\dot{\varepsilon} + 3\frac{\dot{a}}{a}(\varepsilon + P) = 0, \tag{4.52}$$

and the acceleration equation,

$$\frac{\ddot{a}}{a} = -\frac{4\pi G}{3c^2}(\varepsilon + 3P). \tag{4.53}$$

Of these three equations, only two are independent, because Equation 4.53, as we've just seen, can be derived from Equations 4.51 and 4.52. Thus, we have a system of two independent equations in three unknowns – the functions $a(t)$, $\varepsilon(t)$, and $P(t)$. To solve for the scale factor, energy density, and pressure as a function of cosmic time, we need another equation. What we need is an *equation of state*; that is, a mathematical relation between the pressure and energy density of the stuff that fills up the universe. If only we had a relation of the form

$$P = P(\varepsilon), \tag{4.54}$$

life would be complete – or at least, our set of equations would be complete. We could then, given the appropriate boundary conditions, solve them to find how the universe expanded in the past, and how it will expand (or contract) in the future.

In general, equations of state can be dauntingly complicated. Condensed matter physicists frequently deal with substances in which the pressure is a complicated nonlinear function of the density. Fortunately, cosmology usually deals with dilute gases, for which the equation of state is simple. For substances of cosmological importance, the equation of state can be written in a simple linear form:

$$P = w\varepsilon, \tag{4.55}$$

where w is a dimensionless number.

Consider, for instance, a low-density gas of nonrelativistic massive particles. Nonrelativistic, in this case, means that the random thermal motions of the gas particles have peculiar velocities which are tiny compared to the speed of light. Such a nonrelativistic gas obeys the perfect gas law,

$$P = \frac{\rho}{\mu}kT, \tag{4.56}$$

where μ is the mean mass of the gas particles. The energy density ε of a nonrelativistic gas is almost entirely contributed by the mass of the gas particles: $\varepsilon \approx \rho c^2$. Thus, in terms of ε, the perfect gas law is

$$P \approx \frac{kT}{\mu c^2}\varepsilon. \tag{4.57}$$

For a nonrelativistic gas, the temperature T and the root mean square thermal velocity $\langle v^2 \rangle$ are associated by the relation

$$3kT = \mu \langle v^2 \rangle. \tag{4.58}$$

Thus, the equation of state for a nonrelativistic gas can be written in the form

$$P_{\rm nonrel} = w\varepsilon_{\rm nonrel}, \tag{4.59}$$

where

$$w \approx \frac{\langle v^2 \rangle}{3c^2} \ll 1. \tag{4.60}$$

Most of the gases we encounter in everyday life are nonrelativistic. For instance, in air at room temperature, nitrogen molecules are slow-poking along with a root mean square velocity of $\sim 500\,{\rm m\,s^{-1}}$, yielding $w \sim 10^{-12}$. Even in astronomical contexts, gases are mainly nonrelativistic at the present moment. Within a gas of ionized hydrogen, for instance, the electrons are nonrelativistic as long as $T \ll 6 \times 10^9$ K; the protons are nonrelativistic when $T \ll 10^{13}$ K.

A gas of photons, or other massless particles, is guaranteed to be relativistic. Although photons have no mass, they have momentum, and hence exert pressure. The equation of state of photons, or of any other relativistic gas, is

$$P_{\rm rel} = \frac{1}{3}\varepsilon_{\rm rel}. \tag{4.61}$$

(This relation has already been used in Section 2.5, to compute how the cosmic microwave background cools as the universe expands.) A gas of highly relativistic massive particles (with $\langle v^2 \rangle \sim c^2$) will also have $w = 1/3$; a gas of mildly relativistic particles (with $0 < \langle v^2 \rangle < c^2$) will have $0 < w < 1/3$.

Some values of w are of particular interest. For instance, the case $w = 0$ is of interest, because we know that our universe contains nonrelativistic matter. The case $w = 1/3$ is of interest, because we know that our universe contains photons. For simplicity, we will refer to the component of the universe that consists of nonrelativistic particles (and hence has $w \approx 0$) as "matter," and the component that consists of photons and other relativistic particles (and hence has $w = 1/3$) as "radiation." The case $w < -1/3$ is of interest, because a component with $w < -1/3$ provides a positive acceleration ($\ddot{a} > 0$ in Equation 4.53). A component of the universe with $w < -1/3$ is referred to generically as "dark energy" (a phrase coined by the cosmologist Michael Turner). One form of dark energy is of special interest; observational evidence, which we'll review in future chapters, indicates that our universe may contain a *cosmological constant*. A cosmological constant may be defined simply as a component of the universe that has $w = -1$, and hence has $P = -\varepsilon$. The cosmological constant, also designated by the Greek letter Λ, has had a controversial history. To learn why cosmologists have had such a long-standing love/hate affair with the cosmological constant Λ, it is necessary to make a brief historical review.

4.5 Learning to Love Lambda

The cosmological constant Λ was first introduced by Albert Einstein. After publishing his first paper on general relativity in 1915, Einstein, naturally enough, wanted to apply his field equation to the real universe. He looked around, and noted that the universe contains both radiation and matter. Since Einstein, along with every other earthling of his time, was unaware of the existence of the cosmic microwave background, he thought that most of the radiation in the universe was in the form of starlight. He also noted, quite correctly, that the energy density of starlight in our galaxy is much less than the rest energy density of the stars. Thus, Einstein concluded that the primary contribution to the energy density of the universe was from nonrelativistic matter, and that he could safely make the approximation that we live in a pressureless universe.

So far, Einstein was on the right track. However, in 1915, astronomers were unaware of the existence of the expansion of the universe. In fact, it was by no means settled that galaxies besides our own actually existed. After all, the sky is full of faint fuzzy patches of light. It took some time to sort out that some of the faint fuzzy patches are glowing clouds of gas within our galaxy and that some of them are galaxies in their own right, far beyond our own galaxy. Thus, when Einstein asked, "Is the universe expanding or contracting?" he looked, not at the motions of galaxies, but at the motions of stars within our galaxy. Einstein noted that some stars are moving toward us and that others are moving away from us, with no evidence that the galaxy is expanding or contracting.

The incomplete evidence available to Einstein led him to the belief that the universe is static – neither expanding nor contracting – and that it has a positive energy density but negligible pressure. Einstein then had to ask the question, "Can a universe filled with nonrelativistic matter, and nothing else, be static?" The answer to this question is "No!" A universe containing nothing but matter must, in general, be either expanding or contracting. The reason why this is true can be illustrated in a Newtonian context. If the mass density of the universe is ρ, then the gravitational potential Φ is given by Poisson's equation:

$$\nabla^2\Phi = 4\pi G\rho. \tag{4.62}$$

The gravitational acceleration $\vec{a}$ at any point in space is then found by taking the gradient of the potential:

$$\vec{a} = -\vec{\nabla}\Phi. \tag{4.63}$$

In a permanently static universe, $\vec{a}$ must vanish everywhere, implying the potential Φ must be constant in space. However, if Φ is constant, then (from Equation 4.62)

$$\rho = \frac{1}{4\pi G}\nabla^2\Phi = 0. \tag{4.64}$$

The only permissible static universe, in this analysis, is a totally empty universe. If you create a matter-filled universe that is initially static, then gravity will cause it to contract. If you create a matter-filled universe that is initially expanding, then it will either expand forever (if the Newtonian energy U is greater than or equal to zero) or reach a maximum radius and then collapse (if $U < 0$). Trying to make a matter-filled universe that doesn't expand or collapse is like throwing a ball into the air and expecting it to hover there.

How did Einstein surmount this problem? How did he reconcile the fact that the universe contains matter with his desire for a static universe? Basically, he added a fudge factor to the equations. In Newtonian terms, what he did was analogous to rewriting Poisson's equation in the form

$$\nabla^2\Phi + \Lambda = 4\pi G\rho. \tag{4.65}$$

The new term, symbolized by the Greek letter Λ, came to be known as the *cosmological constant*. Introducing Λ into Poisson's equation allows the universe to be static if you set $\Lambda = 4\pi G\rho$.

In general relativistic terms, what Einstein did was to add an additional term, involving Λ, to his field equation. If the Friedmann equation is re-derived from Einstein's field equation, with the Λ term added, it becomes

$$\left(\frac{\dot{a}}{a}\right)^2 = \frac{8\pi G}{3c^2}\varepsilon - \frac{\kappa c^2}{R_0^2 a^2} + \frac{\Lambda}{3}. \tag{4.66}$$

The fluid equation is unaffected by the presence of a Λ term, so it still has the form

$$\dot{\varepsilon} + 3\frac{\dot{a}}{a}(\varepsilon + P) = 0. \tag{4.67}$$

With the Λ term present, the acceleration equation becomes

$$\frac{\ddot{a}}{a} = -\frac{4\pi G}{3c^2}(\varepsilon + 3P) + \frac{\Lambda}{3}. \tag{4.68}$$

A look at the Friedmann equation (Equation 4.66) tells us that adding the Λ term is equivalent to adding a new component to the universe with energy density

$$\varepsilon_\Lambda \equiv \frac{c^2}{8\pi G}\Lambda. \tag{4.69}$$

If Λ remains constant with time, then so does its associated energy density ε_Λ. The fluid equation (Equation 4.67) tells us that to have ε_Λ constant with time, the Λ term must have an associated pressure

$$P_\Lambda = -\varepsilon_\Lambda = -\frac{c^2}{8\pi G}\Lambda. \tag{4.70}$$

Thus, we can think of the cosmological constant as a component of the universe, which has a constant density ε_Λ and a constant pressure $P_\Lambda = -\varepsilon_\Lambda$.

By introducing a Λ term into his equations, Einstein got the static model universe he wanted. For the universe to remain static, both $\dot{a}$ and $\ddot{a}$ must be equal to zero. If $\ddot{a} = 0$, then in a universe with matter density ρ and cosmological constant Λ, the acceleration equation (Equation 4.68) reduces to

$$0 = -\frac{4\pi G}{3}\rho + \frac{\Lambda}{3}. \tag{4.71}$$

Thus, Einstein had to set $\Lambda = 4\pi G\rho$ in order to produce a static universe, just as in the Newtonian case. If $\dot{a} = 0$, the Friedmann equation (Equation 4.66) reduces to

$$0 = \frac{8\pi G}{3}\rho - \frac{\kappa c^2}{R_0^2} + \frac{\Lambda}{3} = 4\pi G\rho - \frac{\kappa c^2}{R_0^2}. \tag{4.72}$$

Einstein's static model therefore had to be positively curved ($\kappa = +1$), with a radius of curvature

$$R_0 = \frac{c}{2(\pi G\rho)^{1/2}} = \frac{c}{\Lambda^{1/2}}. \tag{4.73}$$

Although Einstein published the details of his static, positively curved, matter-filled model in 1917, he was dissatisfied with the model. He believed that the cosmological constant was "gravely detrimental to the formal beauty of the theory." In addition to its aesthetic shortcomings, the model had a practical defect; it was unstable. Although Einstein's static model was in equilibrium, with the repulsive force of Λ balancing the attractive force of ρ, it was an unstable equilibrium. Consider expanding Einstein's universe just a tiny bit. The energy density of Λ remains unchanged, but the energy density of matter drops. Thus, the repulsive force is greater than the attractive force, and the universe expands further. This causes the matter density to drop further, which causes the expansion to accelerate, which causes the matter density to drop further, and so forth. Expanding Einstein's static universe triggers runaway expansion; similarly, compressing it causes a runaway collapse.

Einstein was willing, even eager, to dispose of the "ugly" cosmological constant in his equations. Hubble's 1929 paper on the redshift–distance relation gave Einstein the necessary excuse for tossing Λ onto the rubbish heap.[8]

[8] According to the physicist George Gamow, writing his memoirs much later, Einstein "remarked that the introduction of the cosmological term [Λ] was the biggest blunder of his life."

Ironically, however, the same paper that caused Einstein to abandon the cosmological constant caused other scientists to embrace it. In his initial analysis, Hubble underestimated the distance to galaxies, and hence overestimated the Hubble constant. Hubble's initial value of $H_0 = 500\,\text{km}\,\text{s}^{-1}\,\text{Mpc}^{-1}$ leads to a Hubble time $H_0^{-1} = 2\,\text{Gyr}$. However, by the year 1929, the technique of radiometric dating, as pioneered by the geologist Arthur Holmes, was indicating that the Earth was $\sim 3\,\text{Gyr}$ old. How could cosmologists reconcile a short Hubble time with an old Earth? Some cosmologists pointed out that one way to increase the age of the universe for a given value of H_0^{-1} was to introduce a cosmological constant. If the value of Λ is large enough to make $\ddot{a} > 0$, then $\dot{a}$ was smaller in the past than it is now, and consequently the universe is older than H_0^{-1}.

If Λ has a value greater than $4\pi G\rho_0$, then the expansion of the universe is accelerating, and the universe can be arbitrarily old for a given value of H_0^{-1}. Since 1917, the cosmological constant has gone in and out of fashion, like sideburns or short skirts. It has been particularly fashionable during periods when the favored value of the Hubble time H_0^{-1} has been embarrassingly short compared to the estimated ages of astronomical objects. Currently, the cosmological constant is popular, thanks to observations (discussed in Section 6.5) indicating that the expansion of the universe has a positive acceleration.

A question that has been asked since the time of Einstein – and one which we've assiduously dodged until this moment – is "What is the physical cause of the cosmological constant?" In order to give Λ a physical meaning, we need to identify some component of the universe whose energy density ε_Λ remains constant as the universe expands or contracts. Currently, a leading candidate for this component is the *vacuum energy*.

In classical physics, the idea of a vacuum having energy is nonsense. A vacuum, from the classical viewpoint, contains nothing; and as King Lear would say, "Nothing can come of nothing." In quantum physics, however, a vacuum is not a sterile void. The Heisenberg uncertainty principle permits particle–antiparticle pairs to spontaneously appear and then annihilate in an otherwise empty vacuum. The total energy ΔE and the lifetime Δt of these pairs of virtual particles must satisfy the relation[9]

$$\Delta E \Delta t \leq \hbar. \tag{4.74}$$

Just as there's an energy density associated with the real particles in the universe, there is an energy density ε_{vac} associated with the virtual particle–antiparticle pairs. The vacuum density ε_{vac} is a quantum phenomenon that doesn't give a hoot

[9] The usual analogy that's made is to an embezzling bank teller who takes money from the till but who always replaces it before the auditor comes around. Naturally, the more money a teller is entrusted with, the more frequently the auditor makes random checks.

about the expansion of the universe and is independent of time as the universe expands or contracts.

Unfortunately, computing the numerical value of ε_{vac} is an exercise in quantum field theory that has not yet been successfully completed. It has been suggested that the natural value for the vacuum energy density is the Planck energy density,

$$\varepsilon_{\text{vac}} \sim \frac{E_P}{\ell_P^3} \qquad (???). \tag{4.75}$$

As we've seen in Chapter 1, the Planck energy is large by particle physics standards ($E_P = 1.22 \times 10^{28}$ eV $=$ 540 kilowatt-hours), while the Planck length is small by anybody's standards ($\ell_P = 1.62 \times 10^{-35}$ m). This gives an energy density

$$\varepsilon_{\text{vac}} \sim 3 \times 10^{132}\,\text{eV m}^{-3} \qquad (!\,!\,!). \tag{4.76}$$

This is 123 orders of magnitude larger than the current critical density for our universe, and represents a spectacularly bad match between theory and observations. Obviously, we don't know much yet about the energy density of the vacuum! This is a situation where astronomers can help particle physicists, by deducing the value of ε_Λ from observations of the expansion of the universe. By looking at the universe at extremely large scales, we are indirectly examining the structure of the vacuum on extremely small scales.

Exercises

4.1 Suppose the energy density of the cosmological constant is equal to the present critical density $\varepsilon_\Lambda = \varepsilon_{c,0} = 4870\,\text{MeV m}^{-3}$. What is the total energy of the cosmological constant within a sphere 1 AU in radius? What is the rest energy of the Sun ($E_\odot = M_\odot c^2$)? Comparing these two numbers, do you expect the cosmological constant to have a significant effect on the motion of planets within the solar system?

4.2 Consider Einstein's static universe, in which the attractive force of the matter density ρ is exactly balanced by the repulsive force of the cosmological constant, $\Lambda = 4\pi G\rho$. Suppose that some of the matter is converted into radiation (by stars, for instance). Will the universe start to expand or contract? Explain your answer.

4.3 If $\rho = 2.7 \times 10^{-27}\,\text{kg m}^{-3}$, what is the radius of curvature R_0 of Einstein's static universe? How long would it take a photon to circumnavigate such a universe?

4.4 Suppose that the universe were full of regulation baseballs, each of mass $m_{\text{bb}} = 0.145$ kg and radius $r_{\text{bb}} = 0.0369$ m. If the baseballs were distributed uniformly throughout the universe, what number density of baseballs would be required to make the density equal to the critical density? (Assume non-relativistic baseballs.) Given this density of baseballs, how far would you be able to see, on average, before your line of sight intersected a baseball? In fact, we can see galaxies at a distance $\sim c/H_0 \sim 4000\,\text{Mpc}$; does the transparency of the universe on this length scale place useful limits on the number density of intergalactic baseballs? (Note to readers outside North America or Japan: feel free to substitute regulation cricket balls, with $m_{\text{cr}} = 0.160\,\text{kg}$ and $r_{\text{cr}} = 0.0360\,\text{m}$.)

4.5 The principle of wave-particle duality tells us that a particle with momentum p has an associated de Broglie wavelength of $\lambda = h/p$; this wavelength increases as $\lambda \propto a$ as the universe expands. The total energy density of a gas of particles can be written as $\varepsilon = nE$, where n is the number density of particles, and E is the energy per particle. For simplicity, let's assume that all the gas particles have the same mass m and momentum p. The energy per particle is then simply

$$E = (m^2c^4 + p^2c^2)^{1/2} = (m^2c^4 + h^2c^2/\lambda^2)^{1/2} . \qquad (4.77)$$

Compute the equation-of-state parameter w for this gas as a function of the scale factor a. Show that $w = 1/3$ in the highly relativistic limit ($a \to 0$, $p \to \infty$) and that $w = 0$ in the highly nonrelativistic limit ($a \to \infty, p \to 0$).

5

Model Universes

In a spatially homogeneous and isotropic universe, the relation between the energy density $\varepsilon(t)$, the pressure $P(t)$, and the scale factor $a(t)$ is given by the Friedmann equation,

$$\left(\frac{\dot{a}}{a}\right)^2 = \frac{8\pi G}{3c^2}\varepsilon - \frac{\kappa c^2}{R_0^2 a^2}, \tag{5.1}$$

the fluid equation,

$$\dot{\varepsilon} + 3\frac{\dot{a}}{a}(\varepsilon + P) = 0, \tag{5.2}$$

and the equation of state,

$$P = w\varepsilon. \tag{5.3}$$

In principle, given the appropriate boundary conditions, we can solve Equations 5.1, 5.2, and 5.3 to yield $\varepsilon(t)$, $P(t)$, and $a(t)$ for all times, past and future. In reality, the evolution of our universe is complicated by the fact that it contains different components with different equations of state. Let's start by seeing how the energy density ε of the different components changes as the universe expands.

5.1 Evolution of Energy Density

The universe contains nonrelativistic matter and radiation – that's a conclusion as firm as the earth under your feet and as plain as daylight. Thus, the universe contains components with both $w = 0$ and $w = 1/3$. It contains dark energy that is consistent with being a cosmological constant ($w = -1$). Moreover, the possibility exists that it may contain still more exotic components, with different values of w. Fortunately for the cause of simplicity, the energy density and pressure for the different components of the universe are additive. Suppose that the universe contains N different components, with the ith component having an energy density ε_i

and an equation-of-state parameter w_i. We may then write the total energy density ε as the sum of the energy density of the different components:

$$\varepsilon = \sum_i \varepsilon_i. \tag{5.4}$$

The total pressure P is the sum of the pressures of the different components:

$$P = \sum_i w_i \varepsilon_i. \tag{5.5}$$

Because the energy densities and pressures add in this way, the fluid equation must hold for each component separately, as long as there is no interaction between the different components. If this is so, then the component with equation-of-state parameter w_i obeys the equation

$$\dot{\varepsilon}_i + 3\frac{\dot{a}}{a}(\varepsilon_i + P_i) = 0 \tag{5.6}$$

or

$$\dot{\varepsilon}_i + 3\frac{\dot{a}}{a}(1 + w_i)\varepsilon_i = 0. \tag{5.7}$$

Equation 5.7 can be rearranged to yield

$$\frac{d\varepsilon_i}{\varepsilon_i} = -3(1 + w_i)\frac{da}{a}. \tag{5.8}$$

If we assume that w_i is constant, then

$$\varepsilon_i(a) = \varepsilon_{i,0} a^{-3(1+w_i)}. \tag{5.9}$$

Note that Equation 5.9 is derived solely from the fluid equation and the equation of state; the Friedmann equation doesn't enter into it.

From Equation 5.9, we conclude that the energy density ε_m associated with nonrelativistic matter decreases as the universe expands with the dependence

$$\varepsilon_m(a) = \varepsilon_{m,0}/a^3. \tag{5.10}$$

The energy density in radiation, ε_r, drops at the steeper rate

$$\varepsilon_r(a) = \varepsilon_{r,0}/a^4. \tag{5.11}$$

Why this difference between matter and radiation? We may write the energy density of either component in the form $\varepsilon = nE$, where n is the number density of particles and E is the mean energy per particle. For both relativistic and nonrelativistic particles, the number density has the dependence $n \propto a^{-3}$ as the universe expands, assuming that particles are neither created nor destroyed.

The energy of nonrelativistic particles, shown in the top panel of Figure 5.1, is contributed solely by their rest mass ($E = mc^2$) and remains constant as the universe expands. Thus, for nonrelativistic matter, $\varepsilon_m = nE = n(mc^2) \propto a^{-3}$.

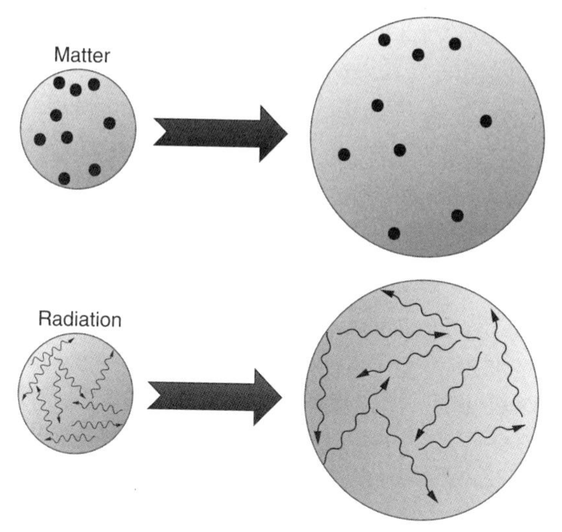

Figure 5.1 The dilution of nonrelativistic particles ("matter") and relativistic particles ("radiation") as the universe expands.

The energy of photons or other massless particles, shown in the bottom panel of Figure 5.1, has the dependence $E = hc/\lambda \propto a^{-1}$, since their wavelength λ expands along with the expansion of the universe. Thus, for photons and other massless particles, $\varepsilon_r = nE = n(hc/\lambda) \propto a^{-3}a^{-1} \propto a^{-4}$.

Although we've explained why photons have an energy density $\varepsilon_r \propto a^{-4}$, the explanation required the assumption that photons are neither created nor destroyed. This assumption is wrong: photons are always being created by luminous objects and absorbed by opaque objects.[1] However, it turns out that the energy density of the cosmic microwave background is larger than the energy density of all the photons emitted by all the stars in the history of the universe. To see why this is true, remember, from Section 2.5, that the present energy density of the CMB, which has a temperature $T_0 = 2.7255\,\mathrm{K}$, is

$$\varepsilon_{\mathrm{CMB},0} = \alpha T_0^4 = 4.175 \times 10^{-14}\,\mathrm{J\,m^{-3}} = 0.2606\,\mathrm{MeV\,m^{-3}}. \tag{5.12}$$

Expressed as a fraction of the critical density, the CMB has a density parameter

$$\Omega_{\mathrm{CMB},0} = \frac{\varepsilon_{\mathrm{CMB},0}}{\varepsilon_{c,0}} = \frac{0.2606\,\mathrm{MeV\,m^{-3}}}{4870\,\mathrm{MeV\,m^{-3}}} = 5.35 \times 10^{-5}. \tag{5.13}$$

Although the energy density of the CMB is small compared to the critical density, it is large compared to the energy density of starlight. Galaxy surveys tell us that the present luminosity density of galaxies is

$$\Psi \approx 1.7 \times 10^8\,\mathrm{L_\odot\,Mpc^{-3}} \approx 2.2 \times 10^{-33}\,\mathrm{watts\,m^{-3}}. \tag{5.14}$$

[1] The Sun, for instance, is emitting 10^{45} photons every second, and thus acts as a glaring example of photon non-conservation.

(By terrestrial standards, the universe is not a well-lit place; this luminosity density is equivalent to a single 30-watt bulb within a sphere 1 AU in radius.) As a *very* rough estimate, let's assume that galaxies have been emitting light at this rate for the entire age of the universe, $t_0 \approx H_0^{-1} \approx 4.5 \times 10^{17}$ s. This gives an energy density in starlight of

$$\begin{aligned}\varepsilon_{\text{starlight},0} &\sim \Psi t_0 \sim (2.2 \times 10^{-33}\,\text{J}\,\text{s}^{-1}\,\text{m}^{-3})(4.5 \times 10^{17}\,\text{s}) \\ &\sim 10^{-15}\,\text{J}\,\text{m}^{-3} \sim 0.006\,\text{MeV}\,\text{m}^{-3}.\end{aligned} \tag{5.15}$$

Thus, we expect the average energy density of starlight to be just a few percent of the energy density of the CMB. In fact, the estimate given above is a very rough one indeed. Measurements of background radiation from ultraviolet to infrared, including both direct starlight and starlight absorbed and reradiated by dust, yield the larger value $\varepsilon_{\text{starlight}}/\varepsilon_{\text{CMB}} \approx 0.1$. In the past, however, the ratio of starlight density to CMB density was smaller than it is today. For most purposes, it is an acceptable approximation to ignore non-CMB photons when computing the mean energy density of photons in the universe.

The cosmic microwave background, remember, is a relic of the time when the universe was hot and dense enough to be opaque to photons. If we extrapolate further back, we reach a time when the universe was hot and dense enough to be opaque to neutrinos. As a consequence, there should be a cosmic *neutrino* background today, analogous to the cosmic microwave background. The energy density in neutrinos should be comparable to, but not exactly equal to, the energy density in photons. A detailed calculation indicates that the energy density of each neutrino flavor should be

$$\varepsilon = \frac{7}{8}\left(\frac{4}{11}\right)^{4/3} \varepsilon_{\text{CMB}} = 0.227\,\varepsilon_{\text{CMB}}. \tag{5.16}$$

(The above result assumes that the neutrinos are relativistic, or, equivalently, that their energy is much greater than their rest energy $m_\nu c^2$.) The density parameter of the cosmic neutrino background, taking into account all three flavors of neutrino, should then be $\Omega_\nu = 0.681\Omega_{\text{CMB}}$, as long as all neutrino flavors are relativistic. The mean energy per neutrino will be comparable to, but not exactly equal to, the mean energy per photon:

$$E_\nu \approx \frac{5 \times 10^{-4}\,\text{eV}}{a}, \tag{5.17}$$

as long as $E_\nu > m_\nu c^2$. When the mean energy of a particular neutrino species drops to $\sim m_\nu c^2$, then it makes the transition from being "radiation" to being "matter."

If all neutrino species were effectively massless today, with $m_\nu c^2 \ll 5 \times 10^{-4}$ eV, then the present density parameter in radiation would be

$$\Omega_{r,0} = \Omega_{\text{CMB},0} + \Omega_{\nu,0} = 5.35 \times 10^{-5} + 3.65 \times 10^{-5} = 9.00 \times 10^{-5}. \tag{5.18}$$

We know the energy density of the cosmic microwave background with high precision. We can calculate theoretically what the energy density of the cosmic neutrino background should be. The total energy density of nonrelativistic matter, and that of dark energy, is not quite as well determined. The available evidence favors a universe in which the density parameter for matter is currently $\Omega_{m,0} \approx 0.31$, while the density parameter for the cosmological constant is currently $\Omega_{\Lambda,0} \approx 0.69$. Thus, when we want to employ a model that matches the observed properties of the real universe, we will use what I call the "Benchmark Model"; this model has $\Omega_{r,0} = 9.0 \times 10^{-5}$ in radiation, $\Omega_{m,0} = 0.31$ in nonrelativistic matter, and $\Omega_{\Lambda,0} = 1 - \Omega_{r,0} - \Omega_{m,0} \approx 0.69$ in a cosmological constant.[2]

In the Benchmark Model, at the present moment, the ratio of the energy density in Λ to the energy density in matter is

$$\frac{\varepsilon_{\Lambda,0}}{\varepsilon_{m,0}} = \frac{\Omega_{\Lambda,0}}{\Omega_{m,0}} \approx \frac{0.69}{0.31} \approx 2.23. \tag{5.19}$$

In the language of cosmologists, the cosmological constant is "dominant" over matter today in the Benchmark Model. In the past, however, when the scale factor was smaller, the ratio of densities was

$$\frac{\varepsilon_{\Lambda}(a)}{\varepsilon_m(a)} = \frac{\varepsilon_{\Lambda,0}}{\varepsilon_{m,0}/a^3} = \frac{\Omega_{\Lambda,0}}{\Omega_{m,0}} a^3. \tag{5.20}$$

If the universe has been expanding from an initial very dense state, at some moment in the past, the energy density of matter and Λ must have been equal. This moment of matter–Λ equality occurred when the scale factor was

$$a_{m\Lambda} = \left(\frac{\Omega_{m,0}}{\Omega_{\Lambda,0}}\right)^{1/3} \approx \left(\frac{0.31}{0.69}\right)^{1/3} \approx 0.766. \tag{5.21}$$

Similarly, the ratio of the energy density in matter to the energy density in radiation is currently

$$\frac{\varepsilon_{m,0}}{\varepsilon_{r,0}} = \frac{\Omega_{m,0}}{\Omega_{r,0}} \approx \frac{0.31}{9.0 \times 10^{-5}} \approx 3400, \tag{5.22}$$

if all three neutrino flavors in the cosmic neutrino background are assumed to be relativistic today. (It's even larger if some or all of the neutrino flavors are massive enough to be nonrelativistic today.) Thus, matter is now strongly dominant over radiation. However, in the past, the ratio of matter density to energy density was

$$\frac{\varepsilon_m(a)}{\varepsilon_r(a)} = \frac{\varepsilon_{m,0}}{\varepsilon_{r,0}} a. \tag{5.23}$$

Thus, the moment of radiation–matter equality took place when the scale factor was

[2] Note that the Benchmark Model is defined to be spatially flat.

$$a_{rm} = \frac{\varepsilon_{m,0}}{\varepsilon_{r,0}} \approx \frac{1}{3400} \approx 2.9 \times 10^{-4}. \tag{5.24}$$

Note that as long as a neutrino's mass is $m_\nu c^2 \ll (3400)(5 \times 10^{-4}\,\mathrm{eV}) \sim 2\,\mathrm{eV}$, then it was relativistic at a scale factor $a = 1/3400$, and hence would have been "radiation" then even if it is "matter" today.

To generalize, if the universe contains different components with different values of w, Equation 5.9 tells us that in the limit $a \to 0$, the component with the largest value of w is dominant. If the universe expands forever, then as $a \to \infty$, the component with the smallest value of w is dominant. The evidence indicates we live in a universe where radiation ($w = \frac{1}{3}$) was dominant during the early stages, followed by a period when matter ($w = 0$) was dominant, followed by a period when the cosmological constant ($w = -1$) is dominant.

In a continuously expanding universe, the scale factor a is a monotonically increasing function of t. Thus, in a continuously expanding universe, the scale factor a can be used as a surrogate for the cosmic time t. We can refer, for instance, to the moment when $a = 0.766$ with the assurance that we are referring to a unique moment in the history of the universe. In addition, because of the simple relation between scale factor and redshift, $1+z = 1/a$, cosmologists often use redshift as a surrogate for time. For example, they make statements such as, "Matter–lambda equality took place at a redshift $z_{m\Lambda} \approx 0.31$." That is, light that was emitted at the time of matter–lambda equality is observed by us with its wavelength stretched by a factor $1 + z_{m\Lambda} \approx 1.31$.

One reason why cosmologists use scale factor or redshift as a surrogate for time is that the conversion from a to t is not simple to calculate in a multiple-component universe like our own. In a universe with many components, the Friedmann equation can be written in the form

$$\dot{a}^2 = \frac{8\pi G}{3c^2} \sum_i \varepsilon_{i,0} a^{-1-3w_i} - \frac{\kappa c^2}{R_0^2}. \tag{5.25}$$

Each term on the right-hand side of Equation 5.25 has a different dependence on scale factor; radiation contributes a term $\propto a^{-2}$, matter contributes a term $\propto a^{-1}$, curvature contributes a term independent of a, and the cosmological constant Λ contributes a term $\propto a^2$. Solving Equation 5.25 for a multiple-component model like the Benchmark Model does not yield a simple analytic form for $a(t)$. However, looking at single-component universes, in which there is only one term on the right-hand side of Equation 5.25, yields useful insight into the physics of an expanding universe.

5.2 Empty Universes

A particularly simple universe is one that is empty – no radiation, no matter, no cosmological constant, no contribution to ε of any sort. For this universe, the Friedmann equation takes the form (compare to Equation 5.25)

$$\dot{a}^2 = -\frac{\kappa c^2}{R_0^2}. \tag{5.26}$$

One solution to this equation has $\dot{a} = 0$ and $\kappa = 0$. An empty, static, spatially flat universe is a permissible solution to the Friedmann equation. This is the universe whose geometry is described by the Minkowski metric of Equation 3.37, and in which all the transformations of special relativity hold true.

However, Equation 5.26 tells us that it is also possible to have an empty universe with $\kappa = -1$. (Positively curved empty universes are forbidden, since that would require an imaginary value of $\dot{a}$ in Equation 5.26.) A negatively curved empty universe must be expanding or contracting, with

$$\dot{a} = \pm \frac{c}{R_0}. \tag{5.27}$$

In an expanding empty universe, integration of this relation yields a scale factor of the form[3]

$$a(t) = \frac{t}{t_0}, \tag{5.28}$$

where $t_0 = R_0/c$. In Newtonian terms, if there's no gravitational force at work, then the relative velocity of any two points is constant, and the scale factor a simply increases linearly with time in an empty universe.

The scale factor in an empty, expanding universe is shown as the dashed line in Figure 5.2. Note that in an empty universe, $t_0 = H_0^{-1}$; with nothing to speed or slow the expansion, the age of the universe is exactly equal to the Hubble time.

An empty, expanding universe might seem nothing more than a mathematical curiosity.[4] However, if a universe has a density ε that is very small compared to the critical density ε_c (that is, if $\Omega \ll 1$), then the linear scale factor of Equation 5.28 is a good approximation to the true scale factor. Imagine you are in an expanding universe with a negligibly small value for the density parameter Ω, so that you can reasonably approximate it as an empty, negatively curved universe, with $t_0 = H_0^{-1} = R_0/c$. You observe a distant light source, such as a galaxy, which has a redshift z. The light you observe now, at $t = t_0$, was emitted at an earlier time, $t = t_e$. In an empty expanding universe,

$$1 + z = \frac{1}{a(t_e)} = \frac{t_0}{t_e}, \tag{5.29}$$

so it is easy to compute the time when the light you observe from the source was emitted:

$$t_e = \frac{t_0}{1+z} = \frac{H_0^{-1}}{1+z}. \tag{5.30}$$

[3] Such an empty, negatively curved, expanding universe is sometimes called a *Milne universe*, after the cosmologist E. A. Milne, who pioneered its study in the 1930s.

[4] If a universe contains nothing, there will be no observers in it to detect the expansion.

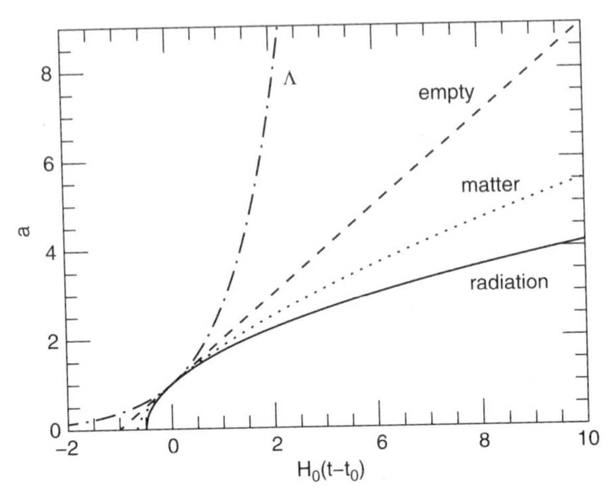

Figure 5.2 Scale factor versus time for an expanding, empty universe (dashed), a flat, matter-dominated universe (dotted), a flat, radiation-dominated universe (solid), and a flat, Λ-dominated universe (dot-dash).

When observing a galaxy with a redshift z, in addition to asking, "When was the light from that galaxy emitted?" you may also ask, "How far away is that galaxy?" In Section 3.5 we saw that in any universe described by a Robertson–Walker metric, the current proper distance from an observer at the origin to a galaxy at coordinate location (r, θ, ϕ) is (see Equation 3.44)

$$d_p(t_0) = a(t_0) \int_0^r dr = r. \tag{5.31}$$

Moreover, if light is emitted by the galaxy at time t_e and detected by the observer at time t_0, the null geodesic followed by the light satisfies Equation 3.54:

$$c \int_{t_e}^{t_0} \frac{dt}{a(t)} = \int_0^r dr = r. \tag{5.32}$$

Thus, the current proper distance from you (the observer) to the galaxy (the light source) is

$$d_p(t_0) = c \int_{t_e}^{t_0} \frac{dt}{a(t)}. \tag{5.33}$$

Equation 5.33 holds true in any universe whose geometry is described by a Robertson–Walker metric. In the specific case of an empty expanding universe, $a(t) = t/t_0$, and thus

$$d_p(t_0) = ct_0 \int_{t_e}^{t_0} \frac{dt}{t} = ct_0 \ln\left(\frac{t_0}{t_e}\right). \tag{5.34}$$

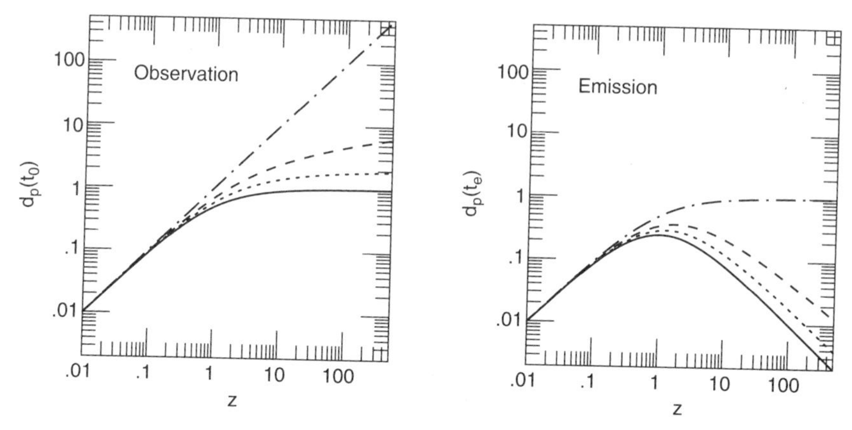

Figure 5.3 The proper distance to an object with observed redshift z, measured in units of the Hubble distance, c/H_0. Left panel: the proper distance at the time the light is observed. Right panel: proper distance at the time the light was emitted. Line types are the same as those of Figure 5.2.

Expressed in terms of the redshift z of the observed galaxy,

$$d_p(t_0) = \frac{c}{H_0} \ln(1+z). \tag{5.35}$$

This relation is plotted as the dashed line in the left panel of Figure 5.3. In the limit $z \ll 1$, there is a linear relation between d_p and z. In the limit $z \gg 1$, however, $d_p \propto \ln z$ in an empty expanding universe.

In an empty expanding universe, we can see objects that are currently at an arbitrarily large distance. At first glance, it may seem counterintuitive that we can see a light source at a proper distance much greater than c/H_0 when the age of the universe is only $1/H_0$. However, remember that $d_p(t_0)$ is the proper distance to the light source at the time of observation; at the time of *emission*, the proper distance $d_p(t_e)$ was smaller by a factor $a(t_e)/a(t_0) = 1/(1+z)$. In an empty expanding universe, the proper distance at the time of emission was

$$d_p(t_e) = \frac{c}{H_0} \frac{\ln(1+z)}{1+z}, \tag{5.36}$$

shown as the dashed line in the right panel of Figure 5.3. In an empty expanding universe, $d_p(t_e)$ has a maximum for objects with a redshift $z = e - 1 \approx 1.72$, where $d_p(t_e) = (1/e)\, c/H_0 \approx 0.37\, c/H_0$. Objects with much higher redshifts are seen as they were very early in the history of the universe, when their proper distance from the observer was very small.

5.3 Single-component Universes

Setting the energy density ε equal to zero is one way of simplifying the Friedmann equation. Another way is to set $\kappa = 0$ and to demand that the universe contain only a single component, with a single value of w. In such a spatially flat, single-component universe, the Friedmann equation takes the form

$$\dot{a}^2 = \frac{8\pi G\varepsilon_0}{3c^2} a^{-(1+3w)}. \tag{5.37}$$

To solve this equation, we first make the educated guess that the scale factor has the power-law form $a \propto t^q$. The left-hand side of Equation 5.37 is then $\propto t^{2q-2}$, and the right-hand side is $\propto t^{-(1+3w)q}$, yielding the solution

$$q = \frac{2}{3+3w}, \tag{5.38}$$

with the restriction $w \neq -1$. With the proper normalization, the scale factor in a spatially flat, single-component universe is

$$a(t) = \left(\frac{t}{t_0}\right)^{2/(3+3w)}. \tag{5.39}$$

The age of the universe, t_0, is linked to the present energy density by the relation

$$t_0 = \frac{1}{1+w}\left(\frac{c^2}{6\pi G\varepsilon_0}\right)^{1/2}. \tag{5.40}$$

The Hubble constant in such a universe is

$$H_0 \equiv \left(\frac{\dot{a}}{a}\right)_{t=t_0} = \frac{2}{3(1+w)} t_0^{-1}. \tag{5.41}$$

The age of the universe, in terms of the Hubble time, is then

$$t_0 = \frac{2}{3(1+w)} H_0^{-1}. \tag{5.42}$$

In a spatially flat universe, if $w > -1/3$, the universe is *younger* than the Hubble time. If $w < -1/3$, the universe is *older* than the Hubble time.

As a function of scale factor, the energy density of a component with equation-of-state parameter w is

$$\varepsilon(a) = \varepsilon_0 a^{-3(1+w)}, \tag{5.43}$$

so in a spatially flat universe with only a single component, the energy density as a function of time is (combining Equations 5.39 and 5.43)

$$\varepsilon(t) = \varepsilon_0 \left(\frac{t}{t_0}\right)^{-2}, \tag{5.44}$$

regardless of the value of w. Making the substitution

$$\varepsilon_0 = \varepsilon_{c,0} = \frac{3c^2}{8\pi G} H_0^2 = \frac{c^2}{6\pi(1+w)^2} t_0^{-2}, \tag{5.45}$$

Equation 5.44 can be written in the form

$$\varepsilon(t) = \frac{1}{6\pi(1+w)^2}\frac{c^2}{G}t^{-2} = \frac{1}{6\pi(1+w)^2}\frac{E_P}{\ell_P^3}\left(\frac{t}{t_P}\right)^{-2}. \tag{5.46}$$

Suppose yourself to be in a spatially flat, single-component universe. If you see a galaxy, or other distant light source, with a redshift z, you can use the relation

$$1+z = \frac{a(t_0)}{a(t_e)} = \left(\frac{t_0}{t_e}\right)^{2/(3+3w)} \tag{5.47}$$

to compute the time t_e at which the light from the distant galaxy was emitted:

$$t_e = \frac{t_0}{(1+z)^{3(1+w)/2}} = \frac{2}{3(1+w)H_0}\frac{1}{(1+z)^{3(1+w)/2}}. \tag{5.48}$$

The current proper distance to the galaxy is

$$d_p(t_0) = c\int_{t_e}^{t_0}\frac{dt}{a(t)} = ct_0\frac{3(1+w)}{1+3w}\left[1-(t_e/t_0)^{(1+3w)/(3+3w)}\right], \tag{5.49}$$

when $w \neq -1/3$. In terms of H_0 and z rather than t_0 and t_e, the current proper distance is

$$d_p(t_0) = \frac{c}{H_0}\frac{2}{1+3w}\left[1-(1+z)^{-(1+3w)/2}\right]. \tag{5.50}$$

The most distant object you can see (in theory) is one for which the light emitted at $t=0$ is just now reaching us at $t=t_0$. The proper distance (at the time of observation) to such an object is called the *horizon distance*.[5] Here on Earth, the horizon is a circle centered on you, beyond which you cannot see because of the Earth's curvature. In the universe, the horizon is a spherical surface centered on you, beyond which you cannot see because light from more distant objects has not had time to reach you. In a universe described by a Robertson–Walker metric, the current horizon distance is

$$d_{\rm hor}(t_0) = c\int_0^{t_0}\frac{dt}{a(t)}. \tag{5.51}$$

In a spatially flat universe, the horizon distance has a finite value if $w > -1/3$. In such a case, computing the value of $d_p(t_0)$ in the limit $t_e \to 0$ (or, equivalently, $z \to \infty$) yields

$$d_{\rm hor}(t_0) = ct_0\frac{3(1+w)}{1+3w} = \frac{c}{H_0}\frac{2}{1+3w}. \tag{5.52}$$

In a flat universe dominated by matter ($w=0$) or by radiation ($w=1/3$), an observer can see only a finite portion of the infinite volume of the universe.

[5] More technically, this is what's called the *particle horizon distance*; we'll continue to call it the horizon distance, for short.

The portion of the universe lying within the horizon for a particular observer is referred to as the *visible universe* for that observer. The visible universe consists of all points in space that have had sufficient time to send information, in the form of photons or other relativistic particles, to the observer. In other words, the visible universe consists of all points that are *causally connected* to the observer.

In a flat universe with $w \leq -1/3$, the horizon distance is infinite, and all of space is causally connected to any observer. In such a universe with $w \leq -1/3$, you could see every point in space – assuming the universe was transparent, of course. However, for extremely distant points, you would see extremely redshifted versions of what they looked like extremely early in the history of the universe.

5.3.1 Matter only

Let's now look at specific examples of spatially flat universes, starting with a universe containing only nonrelativistic matter ($w = 0$).[6] The age of such a universe is

$$t_0 = \frac{2}{3H_0}, \tag{5.53}$$

and the horizon distance is

$$d_{\rm hor}(t_0) = 3ct_0 = 2c/H_0. \tag{5.54}$$

The scale factor, as a function of time, is

$$a_m(t) = \left(\frac{t}{t_0}\right)^{2/3}, \tag{5.55}$$

illustrated as the dotted line in Figure 5.2. If you see a galaxy with redshift z in a flat, matter-only universe, the proper distance to that galaxy, at the time of observation, is

$$d_p(t_0) = c\int_{t_e}^{t_0} \frac{dt}{(t/t_0)^{2/3}} = 3ct_0\left[1 - \left(\frac{t_e}{t_0}\right)^{1/3}\right] = \frac{2c}{H_0}\left[1 - \frac{1}{\sqrt{1+z}}\right], \tag{5.56}$$

illustrated as the dotted line in the left panel of Figure 5.3. The proper distance at the time the light was emitted was smaller by a factor $1/(1+z)$:

$$d_p(t_e) = \frac{2c}{H_0(1+z)}\left[1 - \frac{1}{\sqrt{1+z}}\right], \tag{5.57}$$

illustrated as the dotted line in the right panel of Figure 5.3. In a flat, matter-only universe, $d_p(t_e)$ has a maximum for galaxies with a redshift $z = 5/4$, where $d_p(t_e) = (8/27)c/H_0 \approx 0.30c/H_0$.

[6] Such a universe is sometimes called an *Einstein–de Sitter universe*, after Albert Einstein and the cosmologist Willem de Sitter, who jointly wrote a paper on flat, matter-dominated universes in 1932.

5.3.2 Radiation only

The case of a spatially flat universe containing only *radiation* is of particular interest, since early in the history of our own universe, the radiation ($w = 1/3$) term dominated the right-hand side of the Friedmann equation (see Equation 5.25). Thus, at early times – long before the time of radiation–matter equality – the universe was well described by a spatially flat, radiation-only model. In an expanding, flat universe containing only radiation, the age of the universe is

$$t_0 = \frac{1}{2H_0}, \tag{5.58}$$

and the horizon distance at t_0 is

$$d_{\rm hor}(t_0) = 2ct_0 = \frac{c}{H_0}. \tag{5.59}$$

In the special case of a flat, radiation-only universe, the horizon distance is exactly equal to the Hubble distance, which is not generally the case. The scale factor of a flat, radiation-only universe is

$$a(t) = \left(\frac{t}{t_0}\right)^{1/2}, \tag{5.60}$$

illustrated as the solid line in Figure 5.2. If at a time t_0 you observe a distant light source with redshift z in a flat, radiation-only universe, the proper distance to the light source will be

$$d_p(t_0) = c\int_{t_e}^{t_0} \frac{dt}{(t/t_0)^{1/2}} = 2ct_0\left[1 - \left(\frac{t_e}{t_0}\right)^{1/2}\right] = \frac{c}{H_0}\frac{z}{1+z}, \tag{5.61}$$

illustrated as the solid line in the left panel of Figure 5.3. The proper distance at the time the light was emitted was

$$d_p(t_e) = \frac{c}{H_0}\frac{z}{(1+z)^2}, \tag{5.62}$$

illustrated as the solid line in the right panel of Figure 5.3. In a flat, radiation-dominated universe, $d_p(t_e)$ has a maximum for light sources with a redshift $z = 1$, where $d_p(t_e) = 0.25c/H_0$.

From Equation (5.46), the energy density in a flat, radiation-only universe is

$$\varepsilon_r(t) = \frac{3}{32\pi}\frac{E_P}{\ell_P^3}\left(\frac{t}{t_P}\right)^{-2} \approx 0.030\frac{E_P}{\ell_P^3}\left(\frac{t}{t_P}\right)^{-2}. \tag{5.63}$$

Using the blackbody relation between energy density and temperature, given in Equations 2.28 and 2.29, we may assign a temperature to a universe dominated by blackbody radiation:

$$T(t) = \left(\frac{45}{32\pi^2}\right)^{1/4} T_P \left(\frac{t}{t_P}\right)^{-1/2} \approx 0.61T_P\left(\frac{t}{t_P}\right)^{-1/2}. \tag{5.64}$$

Here T_P is the Planck temperature, $T_P = 1.42 \times 10^{32}$ K. The mean energy per photon in a radiation-dominated universe is then

$$E_{\text{mean}}(t) \approx 2.7kT(t) \approx 1.7E_P \left(\frac{t}{t_P}\right)^{-1/2}, \tag{5.65}$$

and the number density of photons is (combining Equations 5.63 and 5.65)

$$n(t) = \frac{\varepsilon_r(t)}{E_{\text{mean}}(t)} \approx \frac{0.018}{\ell_P^3}\left(\frac{t}{t_P}\right)^{-3/2}. \tag{5.66}$$

In a flat, radiation-only universe, as $t \to 0$, $\varepsilon_r \to \infty$ (Equation 5.63). Thus, at the instant $t = 0$, the energy density of our own universe (well approximated as a flat, radiation-only model in its early stages) was infinite, according to this analysis; this infinite energy density was provided by an infinite number density of particles (Equation 5.66), each of infinite energy (Equation 5.65). Should we take these infinities seriously? Not really, since the assumptions of general relativity, on which the Friedmann equation is based, break down at $t \approx t_P$.

Why can't general relativity be used at times earlier than the Planck time? General relativity is a classical theory; that is, it does not take into account the effects of quantum mechanics. In cosmological contexts, general relativity assumes that the energy content of the universe is smooth down to arbitrarily small scales, instead of being parceled into individual quanta. As long as a radiation-dominated universe has many quanta, or photons, within a horizon distance, then the approximation of a smooth, continuous energy density is justifiable, and we may safely use the results of general relativity. However, if there are only a few photons within the visible universe, then quantum mechanical effects *must* be taken into account, and the classical results of general relativity no longer apply. In a flat, radiation-only universe, the horizon distance grows linearly with time:

$$d_{\text{hor}}(t) = 2ct = 2\ell_P \left(\frac{t}{t_P}\right), \tag{5.67}$$

so the volume of the visible universe at time t is

$$V_{\text{hor}}(t) = \frac{4\pi}{3} d_{\text{hor}}^3 \approx 34\ell_P^3 \left(\frac{t}{t_P}\right)^3. \tag{5.68}$$

Combining Equations 5.68 and 5.66, we find that the number of photons inside the horizon at time t is

$$N(t) = V_{\text{hor}}(t)n(t) \approx 0.6\left(\frac{t}{t_P}\right)^{3/2}. \tag{5.69}$$

The quantization of the universe can no longer be ignored when $N(t) \approx 1$, equivalent to a time $t \approx 1.4t_P$.

To accurately describe the universe at its very earliest stages, prior to the Planck time, a theory of quantum gravity is needed. Unfortunately, a complete theory of quantum gravity does not yet exist. Consequently, in this book, we will not deal with times earlier than the Planck time, $t \sim t_P \sim 10^{-43}$ s, when the number density of photons was $n \sim \ell_P^{-3} \sim 10^{104}\,\text{m}^{-3}$, and the mean photon energy was $E_{\text{mean}} \sim E_P \sim 10^{28}$ eV.

5.3.3 Lambda only

Consider a spatially flat universe in which the energy density is contributed by a cosmological constant Λ.[7] For a flat, lambda-dominated universe, the Friedmann equation takes the form

$$\dot{a}^2 = \frac{8\pi G\varepsilon_\Lambda}{3c^2}a^2, \tag{5.70}$$

where ε_Λ is constant with time. This equation can be rewritten in the form

$$\dot{a} = H_0 a, \tag{5.71}$$

where

$$H_0 = \left(\frac{8\pi G\varepsilon_\Lambda}{3c^2}\right)^{1/2}. \tag{5.72}$$

The solution to Equation 5.71 in an expanding universe is

$$a(t) = e^{H_0(t-t_0)}. \tag{5.73}$$

This scale factor is shown as the dot-dashed line in Figure 5.2. A spatially flat universe with nothing but a cosmological constant is exponentially expanding; we've seen an exponentially expanding universe before, in Section 2.3, under the label "Steady State universe." In a Steady State universe, the density ε of the universe remains constant because of the continuous creation of real particles. If the cosmological constant Λ is provided by the vacuum energy, then the density ε of a lambda-dominated universe remains constant because of the continuous creation and annihilation of virtual particle–antiparticle pairs.

A flat universe containing nothing but a cosmological constant is infinitely old, and has an infinite horizon distance d_{hor}. If, in a flat, lambda-only universe, you see a light source with a redshift z, the proper distance to the light source, at the time you observe it, is

$$d_p(t_0) = c\int_{t_e}^{t_0} e^{H_0(t_0-t)}dt = \frac{c}{H_0}[e^{H_0(t_0-t_e)} - 1] = \frac{c}{H_0}z, \tag{5.74}$$

shown as the dot-dashed line in the left panel of Figure 5.3. The proper distance at the time the light was emitted was

[7] Such a universe is sometimes called a *de Sitter universe*, after Willem de Sitter, who pioneered its study in the year 1917.

$$d_p(t_e) = \frac{c}{H_0} \frac{z}{1+z}, \tag{5.75}$$

shown as the dot-dashed line in the right panel of Figure 5.3.

An exponentially growing universe, such as the flat lambda-dominated model, is the only universe for which $d_p(t_0)$ is linearly proportional to z for all values of z. In other universes, the relation $d_p(t_0) \propto z$ holds true only in the limit $z \ll 1$. In a flat lambda-dominated universe, a light source with $z \gg 1$ is at a distance $d_p(t_0) \gg c/H_0$ at the time of observation; however, the observed photons were emitted by the light source when it was at a distance $d_p(t_e) \approx c/H_0$. Once the light source is more than a Hubble distance from the observer, its recession velocity is greater than the speed of light, and photons from the light source can no longer reach the observer.

5.4 Multiple-component Universes

The simple models that we've examined so far – empty universes, or flat universes with a single component – continue to expand forever if they are expanding at $t = t_0$. Is it possible to have universes that stop expanding, then start to collapse? Is it possible to have universes in which the scale factor is not a simple power-law or exponential function of time? The short answer to these questions is "yes." To study universes with more complicated behavior, however, it is necessary to put aside our simple toy universes, with a single term on the right-hand side of the Friedmann equation, and look at complicated toy universes, with multiple terms on the right-hand side of the Friedmann equation.

The Friedmann equation, in general, can be written in the form

$$H(t)^2 = \frac{8\pi G}{3c^2}\varepsilon(t) - \frac{\kappa c^2}{R_0^2 a(t)^2}, \tag{5.76}$$

where $H \equiv \dot{a}/a$, and $\varepsilon(t)$ is the energy density contributed by all the components of the universe, including the cosmological constant. Equation 4.36 tells us the relation between κ, R_0, H_0, and Ω_0,

$$\frac{\kappa}{R_0^2} = \frac{H_0^2}{c^2}(\Omega_0 - 1), \tag{5.77}$$

so we can rewrite the Friedmann equation without explicitly including the curvature:

$$H(t)^2 = \frac{8\pi G}{3c^2}\varepsilon(t) - \frac{H_0^2}{a(t)^2}(\Omega_0 - 1). \tag{5.78}$$

Dividing by H_0^2, this becomes

$$\frac{H(t)^2}{H_0^2} = \frac{\varepsilon(t)}{\varepsilon_{c,0}} + \frac{1-\Omega_0}{a(t)^2}, \tag{5.79}$$

where the critical density today is

$$\varepsilon_{c,0} \equiv \frac{3c^2H_0^2}{8\pi G}. \tag{5.80}$$

We know that our universe contains matter, for which the energy density ε_m has the dependence $\varepsilon_m = \varepsilon_{m,0}/a^3$, and radiation, for which the energy density has the dependence $\varepsilon_r = \varepsilon_{r,0}/a^4$. Current evidence indicates the presence of a cosmological constant, with energy density $\varepsilon_\Lambda = \varepsilon_{\Lambda,0} =$ constant. We will therefore consider a universe with contributions from matter ($w = 0$), radiation ($w = 1/3$), and a cosmological constant ($w = -1$).[8]

In our universe, we expect the Friedmann equation to take the form

$$\frac{H^2}{H_0^2} = \frac{\Omega_{r,0}}{a^4} + \frac{\Omega_{m,0}}{a^3} + \Omega_{\Lambda,0} + \frac{1-\Omega_0}{a^2}, \tag{5.81}$$

where $\Omega_{r,0} = \varepsilon_{r,0}/\varepsilon_{c,0}$, $\Omega_{m,0} = \varepsilon_{m,0}/\varepsilon_{c,0}$, $\Omega_{\Lambda,0} = \varepsilon_{\Lambda,0}/\varepsilon_{c,0}$, and $\Omega_0 = \Omega_{r,0} + \Omega_{m,0} + \Omega_{\Lambda,0}$. The Benchmark Model has $\Omega_0 = 1$, and hence is spatially flat. However, although a perfectly flat universe is consistent with the data, it is not *demanded* by the data. Thus, prudence dictates that we should keep in mind the possibility that the curvature term, $(1 - \Omega_0)/a^2$ in Equation 5.81, might be nonzero.

Since $H = \dot{a}/a$, multiplying Equation 5.81 by a^2, then taking the square root, yields

$$H_0^{-1}\dot{a} = \left[\frac{\Omega_{r,0}}{a^2} + \frac{\Omega_{m,0}}{a} + \Omega_{\Lambda,0}a^2 + (1-\Omega_0)\right]^{1/2}. \tag{5.82}$$

The cosmic time t as a function of scale factor a can then be found by performing the integral

$$\int_0^a \frac{da}{[\Omega_{r,0}/a^2 + \Omega_{m,0}/a + \Omega_{\Lambda,0}a^2 + (1-\Omega_0)]^{1/2}} = H_0 t. \tag{5.83}$$

This is not a user-friendly integral: in the general case, it doesn't have a simple analytic solution. However, for given values of $\Omega_{r,0}$, $\Omega_{m,0}$, and $\Omega_{\Lambda,0}$, it can be integrated numerically.

In many circumstances, the integral in Equation 5.83 has a simple analytic approximation to its solution. For instance, in the limit that $a \ll a_{rm} \approx 2.9 \times 10^{-4}$, the Benchmark Model can be approximated as a flat, radiation-only universe. In the limit that $a \gg a_{m\Lambda} \approx 0.77$, it can be approximated as a lambda-only universe. However, during some epochs of the universe's expansion, two of the components are of comparable density, and provide terms of roughly equal size in the Friedmann equation. During these epochs, a single-component model is a

[8] We can't rule out the possibility that the dark energy has $w \neq -1$, or the possibility that the universe contains even more exotic contributions to its energy density ($w =?$). These possible developments are left as an exercise for the reader.

poor description of the universe, and a two-component model must be utilized. For instance, at scale factors $a \sim a_{rm} \approx 2.9 \times 10^{-4}$, the Benchmark Model is approximated by a flat universe containing only radiation and matter. Such a universe is examined in Section 5.4.4. For scale factors $a \sim a_{m\Lambda} \approx 0.77$, the Benchmark Model is approximated by a flat universe containing only matter and a cosmological constant. Such a universe is examined in Section 5.4.2.

First, however, we will examine a universe that is of great historical interest to cosmology; a universe containing both matter and curvature (either negative or positive). During the mid-twentieth century, when the cosmological constant was out of fashion, cosmologists concentrated much of their interest on the study of curved, matter-dominated universes. In addition to being of historical interest, these curved, matter-dominated universes provide useful physical insight into the interplay between curvature, expansion, and density.

5.4.1 Matter + Curvature

Consider a universe containing nothing but pressureless matter, with $w = 0$. If such a universe is spatially flat, then it expands with time, as demonstrated in Section 5.3.1, with a scale factor

$$a(t) = \left(\frac{t}{t_0}\right)^{2/3}. \tag{5.84}$$

Such a flat, matter-only universe expands outward forever. Such a fate is sometimes known as the "Big Chill," since the temperature of the universe decreases monotonically with time as the universe expands. At this point, it is nearly obligatory for a cosmology text to quote T. S. Eliot: "This is the way the world ends / Not with a bang but a whimper."[9]

In a *curved* universe containing nothing but matter, the ultimate fate of the cosmos is intimately linked to the density parameter Ω_0. The Friedmann equation in a curved, matter-dominated universe (Equation 5.81) can be written in the form

$$\frac{H(t)^2}{H_0^2} = \frac{\Omega_0}{a^3} + \frac{1 - \Omega_0}{a^2}, \tag{5.85}$$

since $\Omega_{m,0} = \Omega_0$ in such a universe. Suppose you are in a universe that is currently expanding ($H_0 > 0$) and contains nothing but nonrelativistic matter. If you ask the question, "Will the universe ever cease to expand?" then Equation 5.85 enables you to answer that question. For the universe to cease expanding, there must be some moment at which $H(t) = 0$. Since the first term on the right-hand side of

[9] Interestingly, this quote is from Eliot's poem *The Hollow Men*, written, for the most part, in 1924, the year when Friedmann published his second paper on the expansion of the universe. However, this coincidence seems to be just that – a coincidence. Eliot did not keep up to date on the technical literature of cosmology.

Equation 5.85 is always positive, $H(t) = 0$ requires the second term on the right-hand side to be negative. This means that a matter-dominated universe will cease to expand if $\Omega_0 > 1$, and hence $\kappa = +1$. At the time of maximum expansion, $H(t) = 0$ and thus

$$0 = \frac{\Omega_0}{a_{\max}^3} + \frac{1 - \Omega_0}{a_{\max}^2}. \tag{5.86}$$

The scale factor at the time of maximum expansion will therefore be

$$a_{\max} = \frac{\Omega_0}{\Omega_0 - 1}, \tag{5.87}$$

where Ω_0, remember, is the density parameter as measured at a scale factor $a = 1$.

Note that in Equation 5.85, the Hubble parameter enters only as H^2. Thus, the contraction phase, after the universe reaches maximum expansion, is just the time reversal of the expansion phase. (More precisely, the contraction is a perfect time reversal of the expansion only when the universe is perfectly homogeneous and the expansion is perfectly adiabatic, or entropy-conserving. In a real, lumpy universe, entropy is not conserved on small scales. Stars, for instance, generate entropy as they emit photons. During the contraction phase of an $\Omega_0 > 1$ universe, small-scale entropy-producing processes will NOT be reversed. Stars will not absorb the photons they previously emitted; people will not live backward from grave to cradle.) Eventually, the $\Omega_0 > 1$ universe will collapse down to $a = 0$, in an event sometimes called the "Big Crunch," after a finite time $t = t_{\text{crunch}}$. A matter-dominated universe with $\Omega_0 > 1$ not only has finite spatial extent, but also has a finite duration in time; just as it began in a hot, dense state, so it will end in a hot, dense state.

A matter-dominated universe with $\Omega_0 > 1$ will expand to a maximum scale factor $a_{\max}$, then collapse in a Big Crunch. What is the ultimate fate of a matter-dominated universe with $\Omega_0 < 1$ and $\kappa = -1$? In the Friedmann equation for such a universe (Equation 5.85), *both* terms on the right-hand side are positive. Thus if such a universe is expanding at a time $t = t_0$, it will continue to expand forever. At early times, when the scale factor is small ($a \ll \Omega_0/[1 - \Omega_0]$), the matter term of the Friedmann equation will dominate, and the scale factor will grow at the rate $a \propto t^{2/3}$. Ultimately, however, the density of matter will be diluted far below the critical density, and the universe will expand like the negatively curved empty universe, with $a \propto t$.

If a universe contains nothing but matter, its curvature, its density, and its ultimate fate are closely linked, as shown in Table 5.1. At this point, the obligatory quote is from Robert Frost: "Some say the world will end in fire / Some say in ice."[10] In a matter-dominated universe, if the density is greater than the critical

[10] This is from Frost's poem *Fire and Ice*, first published in Harper's Magazine in December 1920. Unlike T. S. Eliot, Frost was keenly interested in astronomy, and frequently wrote poems on astronomical themes.

Table 5.1 Curved, matter-dominated universes.

Density	*Curvature*	*Ultimate fate*
$\Omega_0 < 1$	$\kappa = -1$	Big Chill ($a \propto t$)
$\Omega_0 = 1$	$\kappa = 0$	Big Chill ($a \propto t^{2/3}$)
$\Omega_0 > 1$	$\kappa = +1$	Big Crunch

density, the universe will end in a fiery Big Crunch; if the density is less than or equal to the critical density, the universe will end in an icy Big Chill.

In a curved universe containing only matter, the scale factor $a(t)$ can be computed explicitly. The Friedmann equation can be written in the form

$$\frac{\dot{a}^2}{H_0^2} = \frac{\Omega_0}{a} + (1 - \Omega_0), \tag{5.88}$$

so the age t of the universe at a given scale factor a is given by the integral

$$H_0 t = \int_0^a \frac{da}{[\Omega_0/a + (1 - \Omega_0)]^{1/2}}. \tag{5.89}$$

When $\Omega_0 \neq 1$, the solution to this integral is most compactly written in a parametric form. The solution when $\Omega_0 > 1$ is

$$a(\theta) = \frac{1}{2}\frac{\Omega_0}{\Omega_0 - 1}(1 - \cos\theta) \tag{5.90}$$

and

$$t(\theta) = \frac{1}{2H_0}\frac{\Omega_0}{(\Omega_0 - 1)^{3/2}}(\theta - \sin\theta), \tag{5.91}$$

where the parameter θ runs from 0 to 2π. Given this parametric form, the time that elapses between the Big Bang at $\theta = 0$ and the Big Crunch at $\theta = 2\pi$ can be computed as

$$t_{\text{crunch}} = \frac{\pi}{H_0}\frac{\Omega_0}{(\Omega_0 - 1)^{3/2}}. \tag{5.92}$$

A plot of a versus t in the case $\Omega_0 = 1.1$ is shown as the dotted line in Figure 5.4. The $a \propto t^{2/3}$ behavior of an $\Omega_0 = 1$ universe is shown as the solid line.

The solution of Equation 5.89 for the case $\Omega_0 < 1$ can be written in parametric form as

$$a(\eta) = \frac{1}{2}\frac{\Omega_0}{1 - \Omega_0}(\cosh\eta - 1) \tag{5.93}$$

and

$$t(\eta) = \frac{1}{2H_0}\frac{\Omega_0}{(1 - \Omega_0)^{3/2}}(\sinh\eta - \eta), \tag{5.94}$$

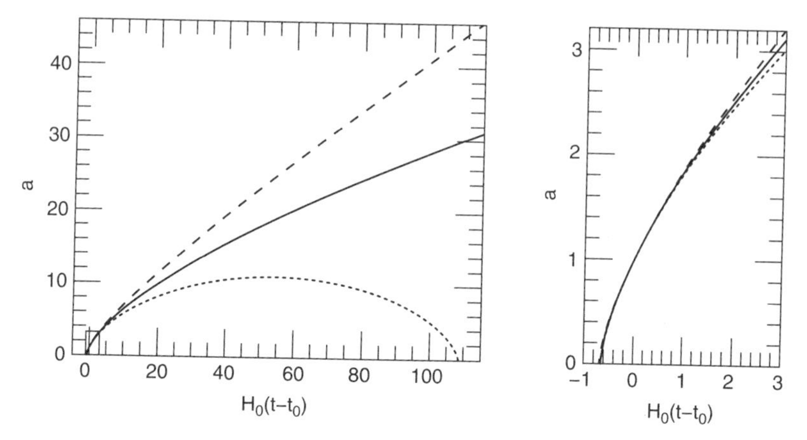

Figure 5.4 Scale factor versus time for universes containing only matter. Solid line: $a(t)$ for a universe with $\Omega_0 = 1$ (flat). Dashed line: $a(t)$ for a universe with $\Omega_0 = 0.9$ (negatively curved). Dotted line: $a(t)$ for a universe with $\Omega_0 = 1.1$ (positively curved). The right panel is a blow-up of the small rectangle near the lower left corner of the left panel.

where the parameter η runs from 0 to infinity. A plot of a versus t in the case $\Omega_0 = 0.9$ is shown as the dashed line in Figure 5.4. Although the ultimate fate of an $\Omega_0 = 0.9$ universe is very different from that of an $\Omega_0 = 1.1$ universe, as shown graphically in the left panel of Figure 5.4, it is very difficult, at $t = t_0$, to tell a universe with Ω_0 slightly less than one from that with Ω_0 slightly greater than one. As shown in the right panel of Figure 5.4, the scale factors of the $\Omega_0 = 1.1$ universe and the $\Omega_0 = 0.9$ universe start to diverge significantly only after a Hubble time or more.

Scientists sometimes joke that they are searching for a theory of the universe that is compact enough to fit on the front of a T-shirt. If the energy content of the universe were contributed almost entirely by nonrelativistic matter, then an appropriate T-shirt slogan would be:

DENSITY
IS
DESTINY!

If the density of matter is less than the critical value, then the destiny of the universe is an ever-expanding Big Chill; if the density is greater than the critical value, then the destiny is a recollapsing Big Crunch. Like all terse summaries of complex concepts, the slogan "Density is Destiny!" requires a qualifying footnote. In this case, the required footnote is "*if $\Lambda = 0$." If the universe has a cosmological constant (or more generally, any component with $w < -1/3$), then the equation Density = Destiny = Curvature no longer applies.

5.4.2 Matter + Lambda

Consider a universe that is spatially flat, but contains both matter and a cosmological constant. (Such a universe is of particular interest to us, since it is a close approximation to our own universe at the present day.) If, at a given time $t = t_0$, the density parameter in matter is $\Omega_{m,0}$ and the density parameter in a cosmological constant Λ is $\Omega_{\Lambda,0}$, the requirement that space be flat tells us that

$$\Omega_{\Lambda,0} = 1 - \Omega_{m,0}, \tag{5.95}$$

and the Friedmann equation for the flat "matter plus lambda" universe reduces to

$$\frac{H^2}{H_0^2} = \frac{\Omega_{m,0}}{a^3} + (1 - \Omega_{m,0}). \tag{5.96}$$

The first term on the right-hand side of Equation 5.96 represents the contribution of matter, and is always positive. The second term represents the contribution of a cosmological constant; it is positive if $\Omega_{m,0} < 1$, implying $\Omega_{\Lambda,0} > 0$, and negative if $\Omega_{m,0} > 1$, implying $\Omega_{\Lambda,0} < 0$. Thus, a flat universe with $\Omega_{\Lambda,0} > 0$ will continue to expand forever if it is expanding at $t = t_0$; this is another example of a Big Chill universe. In a universe with $\Omega_{\Lambda,0} < 0$, however, the negative cosmological constant provides an *attractive* force, not the repulsive force of a positive cosmological constant. A flat universe with $\Omega_{\Lambda,0} < 0$ will cease to expand at a maximum scale factor

$$a_{\max} = \left(\frac{\Omega_{m,0}}{\Omega_{m,0} - 1}\right)^{1/3}, \tag{5.97}$$

and will collapse back down to $a = 0$ at a cosmic time

$$t_{\text{crunch}} = \frac{2\pi}{3H_0} \frac{1}{\sqrt{\Omega_{m,0} - 1}}. \tag{5.98}$$

For a given value of H_0, the larger the value of $\Omega_{m,0}$, the shorter the lifetime of the universe. For a flat, $\Omega_{\Lambda,0} < 0$ universe, the Friedmann equation can be integrated to yield the analytic solution

$$H_0 t = \frac{2}{3\sqrt{\Omega_{m,0} - 1}} \sin^{-1}\left[\left(\frac{a}{a_{\max}}\right)^{3/2}\right]. \tag{5.99}$$

A plot of a versus t in the case $\Omega_{m,0} = 1.1$, $\Omega_{\Lambda,0} = -0.1$ is shown as the dotted line in Figure 5.5. The $a \propto t^{2/3}$ behavior of an $\Omega_{m,0} = 1$, $\Omega_{\Lambda,0} = 0$ universe is shown, for comparison, as the solid line. A flat universe with $\Omega_{\Lambda,0} < 0$ ends in a Big Crunch, reminiscent of that for a positively curved, matter-only universe. However, with a negative cosmological constant providing an attractive force, the lifetime of a flat universe with $\Omega_{\Lambda,0} < 0$ is exceptionally short. For instance, we have seen that a positively curved universe with $\Omega_{m,0} = 1.1$ undergoes a Big

Crunch after a lifetime $t_{\text{crunch}} \approx 110H_0^{-1}$ (Figure 5.4). However, a flat universe with $\Omega_{m,0} = 1.1$ and $\Omega_{\Lambda,0} = -0.1$ has a lifetime of only $t_{\text{crunch}} \approx 7H_0^{-1}$.

Although a negative cosmological constant is permitted by the laws of physics, it appears that we live in a universe with a positive cosmological constant. In a flat universe with $\Omega_{m,0} < 1$ and $\Omega_{\Lambda,0} > 0$, the density contributions of matter and the cosmological constant are equal at the scale factor (Equation 5.21):

$$a_{m\Lambda} = \left(\frac{\Omega_{m,0}}{\Omega_{\Lambda,0}}\right)^{1/3} = \left(\frac{\Omega_{m,0}}{1-\Omega_{m,0}}\right)^{1/3}. \tag{5.100}$$

For a flat, $\Omega_{\Lambda,0} > 0$ universe, the Friedmann equation can be integrated to yield the analytic solution

$$H_0 t = \frac{2}{3\sqrt{1-\Omega_{m,0}}} \ln\left[\left(\frac{a}{a_{m\Lambda}}\right)^{3/2} + \sqrt{1+\left(\frac{a}{a_{m\Lambda}}\right)^{3}}\right]. \tag{5.101}$$

A plot of a versus t in the case $\Omega_{m,0} = 0.9$, $\Omega_{\Lambda,0} = 0.1$ is shown as the dashed line in Figure 5.5. At early times, when $a \ll a_{m\Lambda}$, Equation 5.101 reduces to the relation

$$a(t) \approx \left(\frac{3}{2}\sqrt{\Omega_{m,0}}\, H_0 t\right)^{2/3}, \tag{5.102}$$

giving the $a \propto t^{2/3}$ dependence required for a flat, matter-dominated universe. At late times, when $a \gg a_{m\Lambda}$, Equation 5.101 reduces to

$$a(t) \approx a_{m\Lambda} \exp(\sqrt{1-\Omega_{m,0}}\, H_0 t), \tag{5.103}$$

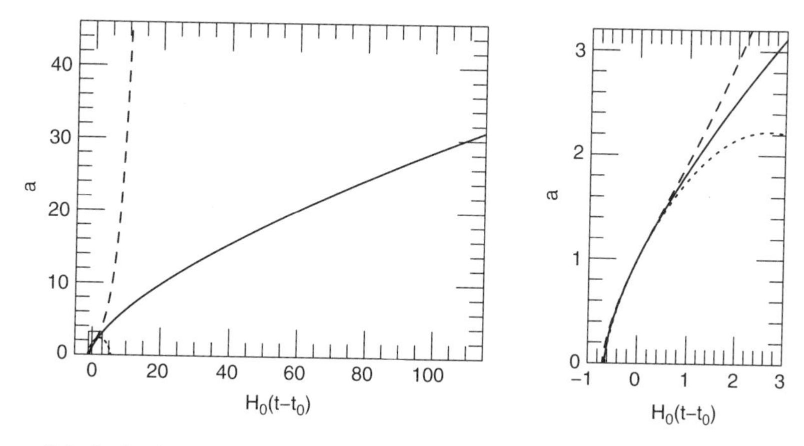

Figure 5.5 Scale factor versus time for flat universes containing both matter and a cosmological constant. Solid line: $a(t)$ for a universe with $\Omega_{m,0} = 1$, $\Omega_{\Lambda,0} = 0$. Dashed line: $a(t)$ for a universe with $\Omega_{m,0} = 0.9$, $\Omega_{\Lambda,0} = 0.1$. Dotted line: $a(t)$ for a universe with $\Omega_{m,0} = 1.1$, $\Omega_{\Lambda,0} = -0.1$. The right panel is a blow-up of the small rectangle near the lower left corner of the left panel.

giving the $a \propto e^{Kt}$ dependence required for a flat, lambda-dominated universe. Suppose you are in a flat universe containing nothing but matter and a cosmological constant; if you measure H_0 and $\Omega_{m,0}$, then Equation 5.101 tells you that the age of the universe is

$$t_0 = \frac{2H_0^{-1}}{3\sqrt{1-\Omega_{m,0}}} \ln\left[\frac{\sqrt{1-\Omega_{m,0}}+1}{\sqrt{\Omega_{m,0}}}\right]. \tag{5.104}$$

If we approximate our own universe as having $\Omega_{m,0} = 0.31$ and $\Omega_{\Lambda,0} = 0.69$, we find that its current age is

$$t_0 = 0.955 H_0^{-1} = 13.74 \pm 0.40\,\text{Gyr}, \tag{5.105}$$

assuming $H_0 = 68 \pm 2\,\text{km}\,\text{s}^{-1}\,\text{Mpc}^{-1}$. (We'll see in Section 5.5 that ignoring the radiation content of the universe has an insignificant effect on our estimate of t_0.) The age at which matter and the cosmological constant had equal energy density was

$$t_{m\Lambda} = \frac{2H_0^{-1}}{3\sqrt{1-\Omega_{m,0}}} \ln[1+\sqrt{2}] = 0.707 H_0^{-1} = 10.17 \pm 0.30\,\text{Gyr}. \tag{5.106}$$

Thus, if our universe is well described by the Benchmark Model, with $\Omega_{m,0} = 0.31$ and $\Omega_{\Lambda,0} \approx 0.69$, then the cosmological constant has been the dominant component of the universe for the last 3.6 billion years or so.

5.4.3 Matter + Curvature + Lambda

By choosing different values of $\Omega_{m,0}$ and $\Omega_{\Lambda,0}$, without constraining the universe to be flat, we can create model universes with scale factors $a(t)$ that exhibit very interesting behavior. Start by writing down the Friedmann equation for a curved universe with both matter and a cosmological constant:

$$\frac{H^2}{H_0^2} = \frac{\Omega_{m,0}}{a^3} + \frac{1-\Omega_{m,0}-\Omega_{\Lambda,0}}{a^2} + \Omega_{\Lambda,0}. \tag{5.107}$$

If $\Omega_{m,0} > 0$ and $\Omega_{\Lambda,0} > 0$, then both the first and last term on the right-hand side of Equation 5.107 are positive. However, if $\Omega_{m,0} + \Omega_{\Lambda,0} > 1$, so that the universe is positively curved, then the central term on the right-hand side is negative. As a result, for some choices of $\Omega_{m,0}$ and $\Omega_{\Lambda,0}$, the value of H^2 will be positive for small values of a (where matter dominates) and for large values of a (where Λ dominates), but will be negative for intermediate values of a (where the curvature term dominates). Since negative values of H^2 are unphysical, this means that these universes have a forbidden range of scale factors. Suppose such a universe starts out with $a \gg 1$ and $H < 0$; that is, it is contracting from a low-density, Λ-dominated state. As the universe contracts, however, the negative curvature term in Equation 5.107 becomes dominant, causing the contraction to stop at

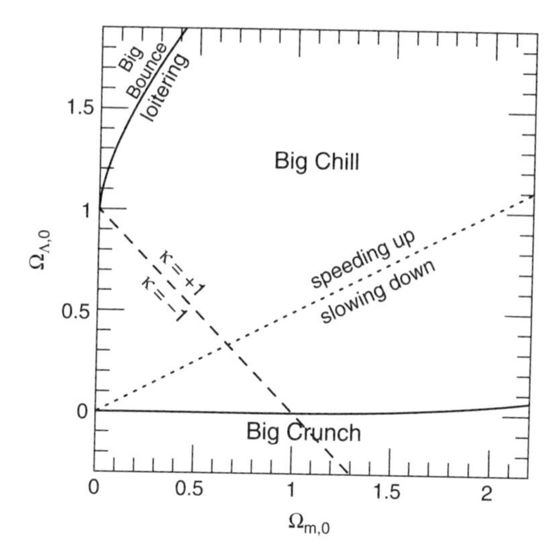

Figure 5.6 Properties of universes containing matter and a cosmological constant. The dashed line indicates flat universes ($\kappa = 0$). The dotted line indicates universes that are not accelerating today ($q_0 = 0$ at $a = 1$). Also shown are the regions where the universe has a "Big Chill" expansion ($a \to \infty$ as $t \to \infty$), a "Big Crunch" recollapse ($a \to 0$ as $t \to t_{\rm crunch}$), a loitering phase ($a \approx$ constant for an extended period), or a "Big Bounce" ($a = a_{\rm min} > 0$ at $t = t_{\rm bounce}$).

a minimum scale factor $a = a_{\rm min}$, and then expand outward again in a "Big Bounce." Thus, it is possible to have a universe that expands outward at late times, but never had an initial Big Bang, with $a = 0$ at $t = 0$.

Another possibility, if the values of $\Omega_{m,0}$ and $\Omega_{\Lambda,0}$ are chosen just right, is a "loitering" universe. Such a universe starts in a matter-dominated state, expanding outward with $a \propto t^{2/3}$. Then, however, it enters a stage (called the loitering stage) in which a is very nearly constant for a long period of time. During this time it is almost – but not quite – Einstein's static universe. After the loitering stage, the cosmological constant takes over, and the universe starts to expand exponentially.[11]

Figure 5.6 shows the general behavior of the scale factor $a(t)$ as a function of $\Omega_{m,0}$ and $\Omega_{\Lambda,0}$. In the region labeled "Big Crunch," the universe starts with $a = 0$ at $t = 0$, reaches a maximum scale factor $a_{\rm max}$, then recollapses to $a = 0$ at a finite time $t = t_{\rm crunch}$. Note that Big Crunch universes can be positively curved, negatively curved, or flat. In the region labeled "Big Chill," the universe starts with $a = 0$ at $t = 0$, then expands outward forever, with $a \to \infty$ as $t \to \infty$. Like Big Crunch universes, Big Chill universes can have any sign for their curvature.

[11] A loitering universe is sometimes referred to as a *Lemaître universe*, since Georges Lemaître discussed, in his 1927 paper on the expanding universe, the possibility of a loitering stage extending into the indefinitely distant past.

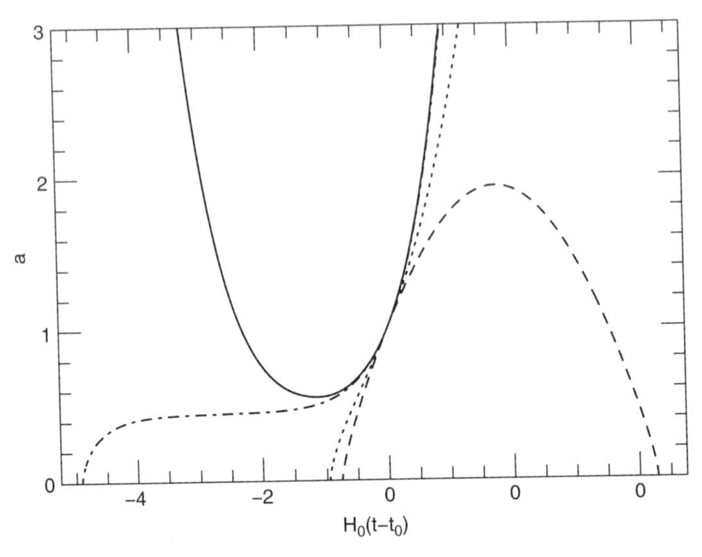

Figure 5.7 Scale factor versus time in four different universes, each with $\Omega_{m,0} = 0.31$. Dotted line: a flat "Big Chill" universe ($\Omega_{\Lambda,0} = 0.69$, $\kappa = 0$). Dashed line: a "Big Crunch" universe ($\Omega_{\Lambda,0} = -0.31$, $\kappa = -1$). Dot-dash line: a loitering universe ($\Omega_{\Lambda,0} = 1.7289$, $\kappa = +1$). Solid line: a "Big Bounce" universe ($\Omega_{\Lambda,0} = 1.8$, $\kappa = +1$).

In the region labeled "Big Bounce," the universe starts in a contracting state, reaches a minimum scale factor $a = a_{\rm min} > 0$ at some time $t_{\rm bounce}$, then expands outward forever, with $a \to \infty$ as $t \to \infty$. Universes that fall just below the dividing line between Big Bounce universes and Big Chill universes are loitering universes. The closer such a universe lies to the Big Bounce–Big Chill dividing line in Figure 5.6, the longer its loitering stage lasts.

To illustrate the possible types of expansion and contraction, Figure 5.7 shows $a(t)$ for a set of four model universes. Each of these universes has the same current density parameter for matter: $\Omega_{m,0} = 0.31$, measured at $t = t_0$ and $a = 1$. These universes cannot be distinguished from each other by measuring their current matter density and Hubble constant. Nevertheless, thanks to their different values for the cosmological constant, they have very different pasts and very different futures. The dotted line in Figure 5.7 shows the scale factor $a(t)$ for a universe with $\Omega_{\Lambda,0} = 0.69$; this universe is spatially flat, and is destined to end in an exponentially expanding Big Chill. The dashed line shows $a(t)$ for a universe with $\Omega_{\Lambda,0} = -0.31$; this universe has an energy density of zero, and is negatively curved. After expanding to a maximum scale factor $a_{\rm max} \approx 1.93$, it will recollapse in a Big Crunch. The dot-dash line shows the scale factor for a universe with $\Omega_{\Lambda,0} = 1.7289$; this is a positively curved loitering universe, which spends a long time with a scale factor $a \approx a_{\rm loiter} \approx 0.45$. Finally, the solid line shows a universe with $\Omega_{\Lambda,0} = 1.8$. This universe lies above the Big Chill–Big Bounce dividing line in Figure 5.6; it is a positively curved universe that "bounced" at a scale factor $a = a_{\rm bounce} \approx 0.552$. If we lived in this Big Bounce universe, the largest redshift we could see would be $z_{\rm max} = 1/a_{\rm bounce} - 1 \approx 0.81$. Extremely distant light sources would actually be blueshifted.

5.4.4 Radiation + Matter

In our universe, radiation–matter equality took place at a scale factor $a_{rm} \equiv \Omega_{r,0}/\Omega_{m,0} \approx 2.9 \times 10^{-4}$. At scale factors $a \ll a_{rm}$, the universe is well described by a flat, radiation-only model, as described in Section 5.3.2. At scale factors $a \sim a_{rm}$, the universe is better described by a flat model containing both radiation and matter. The Friedmann equation around the time of radiation–matter equality can be written in the approximate form

$$\frac{H^2}{H_0^2} = \frac{\Omega_{r,0}}{a^4} + \frac{\Omega_{m,0}}{a^3}. \quad (5.108)$$

This can be rearranged in the form

$$H_0 dt = \frac{a da}{\Omega_{r,0}^{1/2}} \left[1 + \frac{a}{a_{rm}} \right]^{-1/2}. \quad (5.109)$$

Integration yields a fairly simple relation for t as a function of a during the epoch when only radiation and matter are significant:

$$H_0 t = \frac{4a_{rm}^2}{3\sqrt{\Omega_{r,0}}} \left[1 - \left(1 - \frac{a}{2a_{rm}} \right) \left(1 + \frac{a}{a_{rm}} \right)^{1/2} \right]. \quad (5.110)$$

In the limit $a \ll a_{rm}$, this gives the appropriate result for the radiation-dominated phase of evolution,

$$a \approx \left(2\sqrt{\Omega_{r,0}}\, H_0 t \right)^{1/2} \qquad [a \ll a_{rm}]. \quad (5.111)$$

In the limit $a \gg a_{rm}$ (but before curvature or Λ contributes significantly to the Friedmann equation), the approximate result for $a(t)$ becomes

$$a \approx \left(\frac{3}{2}\sqrt{\Omega_{m,0}}\, H_0 t \right)^{2/3} \qquad [a \gg a_{rm}]. \quad (5.112)$$

The time of radiation–matter equality, t_{rm}, can be found by setting $a = a_{rm}$ in Equation 5.110:

$$t_{rm} = \frac{4}{3} \left(1 - \frac{1}{\sqrt{2}} \right) \frac{a_{rm}^2}{\sqrt{\Omega_{r,0}}} H_0^{-1} \approx 0.391 \frac{\Omega_{r,0}^{3/2}}{\Omega_{m,0}^2} H_0^{-1}. \quad (5.113)$$

For the Benchmark Model, with $\Omega_{r,0} = 9.0 \times 10^{-5}$, $\Omega_{m,0} = 0.31$, and $H_0^{-1} = 14.4\,\mathrm{Gyr}$, the time of radiation–matter equality was

$$t_{rm} = 3.47 \times 10^{-6} H_0^{-1} = 50\,000\,\mathrm{yr}. \tag{5.114}$$

The epoch when the universe was radiation-dominated was only about 50 millennia long. This is sufficiently brief that it justifies our ignoring the effects of radiation when computing the age of the universe. The age $t_0 = 0.955 H_0^{-1} = 13.7\,\mathrm{Gyr}$ that we computed in Section 5.4.2 (ignoring radiation) would only be altered by a few parts per million if we included the effects of radiation. This minor correction is dwarfed by the uncertainty in the value of H_0^{-1}.

5.5 Benchmark Model

The Benchmark Model, which we have adopted as a good fit to the currently available observational data, is spatially flat, and contains radiation, matter, and a cosmological constant. Some of its properties are listed, for ready reference, in Table 5.2. The Hubble constant of the Benchmark Model is assumed to be $H_0 = 68\,\mathrm{km\,s^{-1}\,Mpc^{-1}}$. The radiation in the Benchmark Model consists of photons and neutrinos. The photons are assumed to be provided solely by a cosmic microwave background with current temperature $T_0 = 2.7255\,\mathrm{K}$ and density parameter $\Omega_{\gamma,0} = 5.35 \times 10^{-5}$. The energy density of the cosmic neutrino background is theoretically calculated to be 68.1% of that of the cosmic microwave background, as long as neutrinos are relativistic. If a neutrino has a nonzero mass m_ν, Equation 5.17 tells us that it defects from the "radiation" column to

Table 5.2 Properties of the Benchmark Model.

List of ingredients		
Photons:	$\Omega_{\gamma,0} = 5.35 \times 10^{-5}$	
Neutrinos:	$\Omega_{\nu,0} = 3.65 \times 10^{-5}$	
Total radiation:	$\Omega_{r,0} = 9.0 \times 10^{-5}$	
Baryonic matter:	$\Omega_{\mathrm{bary},0} = 0.048$	
Nonbaryonic dark matter:	$\Omega_{\mathrm{dm},0} = 0.262$	
Total matter:	$\Omega_{m,0} = 0.31$	
Cosmological constant:	$\Omega_{\Lambda,0} \approx 0.69$	
Important epochs		
Radiation–matter equality:	$a_{rm} = 2.9 \times 10^{-4}$	$t_{rm} = 0.050\,\mathrm{Myr}$
Matter–lambda equality:	$a_{m\Lambda} = 0.77$	$t_{m\Lambda} = 10.2\,\mathrm{Gyr}$
Now:	$a_0 = 1$	$t_0 = 13.7\,\mathrm{Gyr}$

the "matter" column when the scale factor is $a \sim 5 \times 10^{-4}\,\mathrm{eV}/(m_\nu c^2)$. The matter content of the Benchmark Model consists partly of baryonic matter (that is, matter composed of protons and neutrons, with associated electrons) and partly of nonbaryonic dark matter. The baryonic material that we are familiar with from our everyday existence has a density parameter $\Omega_{\mathrm{bary},0} \approx 0.048$ today. The density parameter of the nonbaryonic dark matter is over five times greater: $\Omega_{\mathrm{dm},0} \approx 0.262$. The bulk of the energy density in the Benchmark Model, however, is not provided by radiation or matter, but by a cosmological constant, with $\Omega_{\Lambda,0} = 1 - \Omega_{m,0} - \Omega_{r,0} \approx 0.69$.

With $\Omega_{r,0}$, $\Omega_{m,0}$, and $\Omega_{\Lambda,0}$ known, the scale factor $a(t)$ can be computed numerically using the Friedmann equation, in the form of Equation 5.81. Figure 5.8 shows the scale factor, thus computed, for the Benchmark Model. Note that the transition from the $a \propto t^{1/2}$ radiation-dominated phase to the $a \propto t^{2/3}$ matter-dominated phase is not an abrupt one; neither is the later transition from the matter-dominated phase to the exponentially growing lambda-dominated phase. One curious feature of the Benchmark Model illustrated vividly in Figure 5.8 is that we are living very close to the time of matter–lambda equality (at least, as plotted on a logarithmic scale).

Once $a(t)$ is known, other properties of the Benchmark Model can be computed readily. For instance, the left panel of Figure 5.9 shows the current proper distance to a galaxy with redshift z. The heavy solid line is the result for the

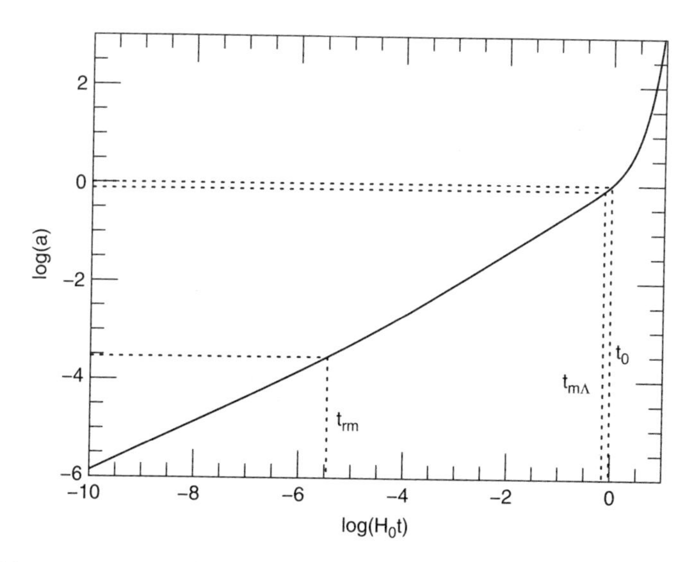

Figure 5.8 The scale factor a as a function of time t (measured in units of the Hubble time), computed for the Benchmark Model. The dotted lines indicate the time of radiation–matter equality, $a_{rm} = 2.9 \times 10^{-4}$, the time of matter–lambda equality, $a_{m\Lambda} = 0.77$, and the present moment, $a_0 = 1$.

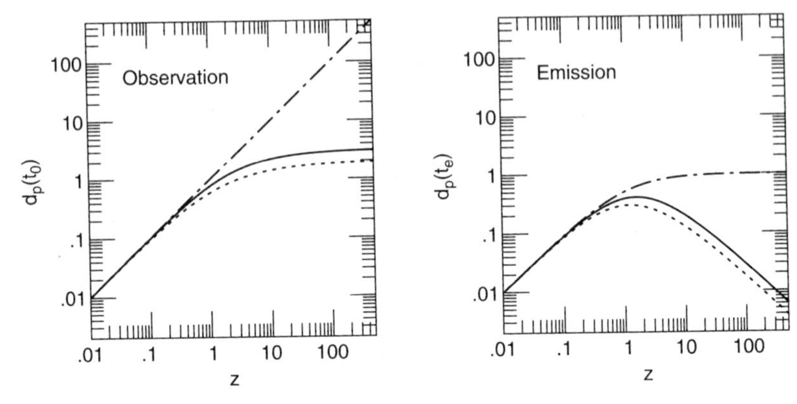

Figure 5.9 The proper distance to a light source with redshift z, in units of the Hubble distance, c/H_0. The left panel shows the distance at the time of observation; the right panel shows the distance at the time of emission. The bold solid line indicates the Benchmark Model. For comparison, the dot-dash line indicates a flat, lambda-only universe, and the dotted line a flat, matter-only universe.

Benchmark Model; for purposes of comparison, the result for a flat lambda-only universe is shown as a dot-dash line and the result for a flat matter-only universe is shown as the dotted line. In the limit $z \to \infty$, the proper distance $d_p(t_0)$ approaches a limiting value $d_p \to 3.20c/H_0$, in the case of the Benchmark Model. Thus, the Benchmark Model has a finite horizon distance,

$$d_{\rm hor}(t_0) = 3.20c/H_0 = 3.35ct_0 = 14\,000\,{\rm Mpc}. \tag{5.115}$$

If the Benchmark Model is a good description of our own universe, then we can't see objects more than 14 gigaparsecs away because light from them has not yet had time to reach us. The right panel of Figure 5.9 shows $d_p(t_e)$, the distance to a galaxy with observed redshift z at the time the observed photons were emitted. For the Benchmark Model, $d_p(t_e)$ has a maximum for galaxies with redshift $z = 1.6$, where $d_p(t_e) = 0.405c/H_0$.

When astronomers observe a distant galaxy, they ask the related, but not identical, questions, "How far away is that galaxy?" and "How long has the light from that galaxy been traveling?" In the Benchmark Model, or any other model, we can answer the question "How far away is that galaxy?" by computing the proper distance $d_p(t_0)$. We can answer the question "How long has the light from that galaxy been traveling?" by computing the *lookback time*. If light emitted at time t_e is observed at time t_0, the lookback time is simply $t_0 - t_e$. In the limit of very small redshifts, $t_0 - t_e \approx z/H_0$. However, as shown in Figure 5.10, at larger redshifts the relation between lookback time and redshift becomes nonlinear. The exact dependence of lookback time on redshift depends on the cosmological model used. For example, consider a galaxy with redshift $z = 2$.

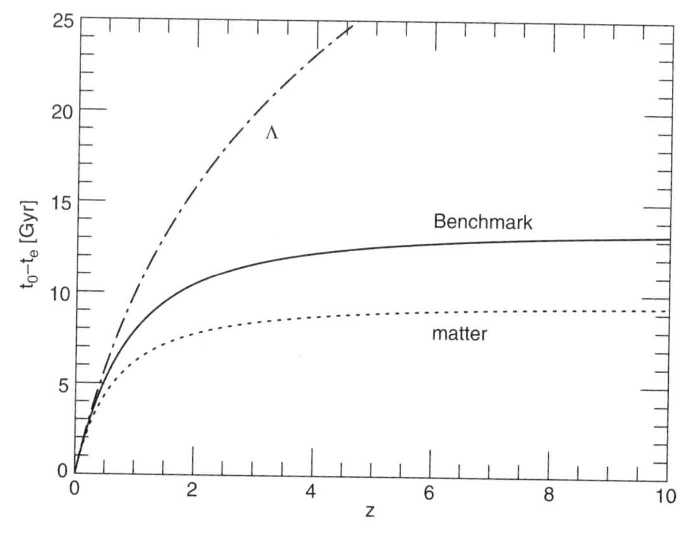

Figure 5.10 The lookback time, $t_0 - t_e$, for galaxies with observed redshift z. The Hubble time is assumed to be $H_0^{-1} = 14.4\,\text{Gyr}$. The bold solid line shows the result for the Benchmark Model. For comparison, the dot-dash line indicates a flat, lambda-only universe, and the dotted line a flat, matter-only universe.

In the Benchmark Model, the lookback time to that galaxy is 10.5 Gyr; we are seeing a redshifted image of that galaxy as it was 10.5 billion years ago. In a flat, lambda-only universe, however, the lookback time to a $z = 2$ galaxy is 15.8 Gyr, assuming $H_0^{-1} = 14.4\,\text{Gyr}$. In a flat, matter-dominated universe, the lookback time to a $z = 2$ galaxy is a mere 7.7 Gyr, with the same assumed Hubble constant.

The most distant galaxies that have been observed (at the time of writing) are at a redshift $z \approx 10$. Consider such a high-redshift galaxy. Using the Benchmark Model, we find that the current proper distance to a galaxy with $z = 10$ is $d_p(t_0) = 2.18c/H_0 = 9500\,\text{Mpc}$, about two-thirds of the current horizon distance. The proper distance at the time the light was emitted was $d_p(t_e) = d_p(t_0)/(1 + z) = 0.20c/H_0 = 870\,\text{Mpc}$. The light we observe now was emitted when the age of the universe was $t_e = 0.033H_0^{-1} = 0.47\,\text{Gyr}$; this is less than 4% of the universe's current age, $t_0 = 0.955H_0^{-1} = 13.74\,\text{Gyr}$. The lookback time to a $z = 10$ galaxy in the Benchmark Model is thus $t_0 - t_e = 0.922H_0^{-1} = 13.27\,\text{Gyr}$. Astronomers are fond of saying, "A telescope is a time machine."[12] As you look further and further out into the universe, to objects with larger and larger values of $d_p(t_0)$, you are looking back to objects with smaller and smaller values of t_e. When you observe a galaxy with a redshift $z = 10$, according to the Benchmark Model, you are glimpsing the universe as it was as a youngster, less than half a billion years old.

[12] Or, as William Herschel phrased it over two centuries ago, "A telescope with a power of penetrating into space...has also, as it may be called, a power of penetrating into time past."

Exercises

5.1 A light source in a flat, single-component universe has a redshift z when observed at a time t_0. Show that the observed redshift z changes at a rate

$$\frac{dz}{dt_0} = H_0(1+z) - H_0(1+z)^{3(1+w)/2}. \tag{5.116}$$

For what values of w does the observed redshift increase with time?

5.2 Suppose you are in a flat, matter-only universe that has a Hubble constant $H_0 = 68\,\mathrm{km\,s^{-1}\,Mpc^{-1}}$. You observe a galaxy with $z = 1$. How long will you have to keep observing the galaxy to see its redshift change by one part in 10^6? [Hint: use the result from the previous problem.]

5.3 In a positively curved universe containing only matter ($\Omega_0 > 1$, $\kappa = +1$), show that the present age of the universe is given by the formula

$$H_0 t_0 = \frac{\Omega_0}{2(\Omega_0 - 1)^{3/2}} \cos^{-1}\left(\frac{2-\Omega_0}{\Omega_0}\right) - \frac{1}{\Omega_0 - 1}. \tag{5.117}$$

Assuming $H_0 = 68\,\mathrm{km\,s^{-1}\,Mpc^{-1}}$, plot t_0 as a function of Ω_0 in the range $1 \leq \Omega_0 \leq 3$.

5.4 In a negatively curved universe containing only matter ($\Omega_0 < 1$, $\kappa = -1$), show that the present age of the universe is given by the formula

$$H_0 t_0 = \frac{1}{1-\Omega_0} - \frac{\Omega_0}{2(1-\Omega_0)^{3/2}} \cosh^{-1}\left(\frac{2-\Omega_0}{\Omega_0}\right). \tag{5.118}$$

Assuming $H_0 = 68\,\mathrm{km\,s^{-1}\,Mpc^{-1}}$, plot t_0 as a function of Ω_0 in the range $0 \leq \Omega_0 \leq 1$.

5.5 One speculation in cosmology is that the dark energy may take the form of "phantom energy" with an equation-of-state parameter $w < -1$. Suppose that the universe is spatially flat and contains matter with a density parameter $\Omega_{m,0}$, and phantom energy with a density parameter $\Omega_{p,0} = 1 - \Omega_{m,0}$ and equation-of-state parameter $w_p < -1$. At what scale factor a_{mp} are the energy density of phantom energy and matter equal? Write down the Friedmann equation for this universe in the limit that $a \gg a_{mp}$. Integrate the Friedmann equation to show that the scale factor a goes to infinity at a finite cosmic time t_{rip}, given by the relation

$$H_0(t_{\mathrm{rip}} - t_0) \approx \frac{2}{3|1+w_p|}(1-\Omega_{m,0})^{-1/2}. \tag{5.119}$$

This fate for the universe is called the "Big Rip." Current observations of our own universe are consistent with $H_0 = 68\,\mathrm{km\,s^{-1}\,Mpc^{-1}}$, $\Omega_{m,0} = 0.3$, and $w_p = -1.1$. If these numbers are correct, how long do we have remaining until the "Big Rip"?

5.6 Suppose you wanted to "pull an Einstein," and create a static universe ($\dot{a} = 0$, $\ddot{a} = 0$) in which the gravitational attraction of matter is exactly balanced by the gravitational repulsion of dark energy with equation-of-state parameter $-1/3 < w_q < -1$ and energy density ε_q. What is the necessary matter density (ε_m) required to produce a static universe, expressed in terms of ε_q and w_q? Will the curvature of this static universe be negative or positive? What will be its radius of curvature, expressed in terms of ε_q and w_q?

5.7 Consider a positively curved universe containing only matter (the "Big Crunch" model discussed in Section 5.4.1). At some time $t_0 > t_{\rm crunch}/2$, during the contraction phase of this universe, an astronomer named Elbbuh Niwde discovers that nearby galaxies have blueshifts ($-1 \leq z < 0$) proportional to their distance. He then measures H_0 and Ω_0, finding $H_0 < 0$ and $\Omega_0 > 1$. Given H_0 and Ω_0, how long a time will elapse between Dr. Niwde's observations at $t = t_0$ and the final Big Crunch at $t = t_{\rm crunch}$? What is the highest amplitude blueshift that Dr. Niwde is able to observe? What is the lookback time to an object with this blueshift?

5.8 Consider an expanding, positively curved universe containing only a cosmological constant ($\Omega_0 = \Omega_{\Lambda,0} > 1$). Show that such a universe underwent a "Big Bounce" at a scale factor

$$a_{\rm bounce} = \left(\frac{\Omega_0 - 1}{\Omega_0}\right)^{1/2}, \tag{5.120}$$

and that the scale factor as a function of time is

$$a(t) = a_{\rm bounce} \cosh[\sqrt{\Omega_0} H_0 (t - t_{\rm bounce})], \tag{5.121}$$

where $t_{\rm bounce}$ is the time at which the Big Bounce occurred. What is the time $t_0 - t_{\rm bounce}$ that has elapsed since the Big Bounce, expressed as a function of H_0 and Ω_0?

5.9 A universe is spatially flat, and contains both matter and a cosmological constant. For what value of $\Omega_{m,0}$ is t_0 exactly equal to H_0^{-1}?

5.10 In the Benchmark Model, what is the total mass of all the matter within our horizon? What is the total energy of all the photons within our horizon? How many baryons are within the horizon?

6

Measuring Cosmological Parameters

Cosmologists would like to know the scale factor $a(t)$ for the universe. For a model universe whose contents are known with precision, the scale factor can be computed from the Friedmann equation. Finding $a(t)$ for the real universe, however, is much more difficult. The scale factor is not directly observable; it can only be deduced indirectly from the imperfect and incomplete observations that we make of the universe around us.

In the previous chapter, I pointed out that if we knew the energy density ε for each component of the universe, we could use the Friedmann equation to find the scale factor $a(t)$. The argument works in the other direction, as well; if we could determine $a(t)$ from observations, we could use that knowledge to find ε for each component. Let's see, then, what constraints we can put on the scale factor by making observations of distant astronomical objects.

6.1 "A Search for Two Numbers"

Since determining the exact functional form of $a(t)$ is difficult, it is useful, instead, to do a Taylor series expansion for $a(t)$ around the present moment. The complete Taylor series is

$$a(t) = a(t_0) + \left.\frac{da}{dt}\right|_{t=t_0} (t - t_0) + \frac{1}{2} \left.\frac{d^2a}{dt^2}\right|_{t=t_0} (t - t_0)^2 + \cdots \tag{6.1}$$

To exactly reproduce an arbitrary function $a(t)$ for all values of t, an infinite number of terms is required in the expansion. However, the usefulness of a Taylor series expansion resides in the fact that if a doesn't fluctuate wildly with t, using only the first few terms of the expansion gives a good approximation in the immediate vicinity of t_0. The scale factor $a(t)$ is a good candidate for a Taylor expansion. The different model universes examined in the previous two chapters all had smoothly varying scale factors, and there's no evidence that the real universe has a wildly oscillating scale factor.

Keeping the first three terms of the Taylor expansion, the scale factor in the recent past and the near future can be approximated as

$$a(t) \approx a(t_0) + \left.\frac{da}{dt}\right|_{t=t_0}(t-t_0) + \frac{1}{2}\left.\frac{d^2a}{dt^2}\right|_{t=t_0}(t-t_0)^2. \tag{6.2}$$

Dividing by the current scale factor, $a(t_0)$,

$$\frac{a(t)}{a(t_0)} \approx 1 + \left.\frac{\dot{a}}{a}\right|_{t=t_0}(t-t_0) + \frac{1}{2}\left.\frac{\ddot{a}}{a}\right|_{t=t_0}(t-t_0)^2. \tag{6.3}$$

Using the normalization $a(t_0) = 1$, this expansion for the scale factor is customarily written in the form

$$a(t) \approx 1 + H_0(t-t_0) - \frac{1}{2}q_0 H_0^2 (t-t_0)^2. \tag{6.4}$$

In Equation 6.4, the parameter H_0 is our old acquaintance the Hubble constant,

$$H_0 \equiv \left.\frac{\dot{a}}{a}\right|_{t=t_0}, \tag{6.5}$$

and the parameter q_0 is a dimensionless number called the *deceleration parameter*, defined as

$$q_0 \equiv -\left(\frac{\ddot{a}a}{\dot{a}^2}\right)_{t=t_0} = -\left(\frac{\ddot{a}}{aH^2}\right)_{t=t_0}. \tag{6.6}$$

Note the choice of sign in defining q_0. A positive value of q_0 corresponds to $\ddot{a} < 0$, meaning that the universe's expansion is decelerating (that is, the relative velocity of any two points is decreasing). A negative value of q_0 corresponds to $\ddot{a} > 0$, meaning that the relative velocity of any two points is increasing with time. The choice of sign for q_0, and the fact that it's named the *deceleration* parameter, is because it was first defined during the mid-1950s, when the limited information available favored a matter-dominated universe with $\ddot{a} < 0$. If the universe contains a sufficiently large cosmological constant, however, the deceleration parameter q_0 can have either sign.

The Taylor expansion of Equation 6.4 is physics-free. It is simply a mathematical description of how the universe expands at times $t \sim t_0$, and says nothing at all about what forces act to accelerate the expansion (to take a Newtonian viewpoint of the physics involved). In a famous 1970 review article, the observational cosmologist Allan Sandage described all of cosmology as "a search for two numbers." Those two numbers were H_0 and q_0. Although the scope of cosmology has widened considerably since Sandage wrote his article, it is still possible to describe the recent expansion of the universe in terms of H_0 and q_0.

Although H_0 and q_0 are themselves free of the theoretical assumptions underlying the Friedmann and acceleration equations, we can use the acceleration equation to predict what q_0 will be in a given model universe. If our model universe contains N components, each with a different value of the equation-of-state parameter w_i, the acceleration equation can be written

$$\frac{\ddot{a}}{a} = -\frac{4\pi G}{3c^2} \sum_{i=1}^{N} \varepsilon_i (1 + 3w_i). \tag{6.7}$$

Divide each side of the acceleration equation by the square of the Hubble parameter $H(t)$ and change sign:

$$-\frac{\ddot{a}}{aH^2} = \frac{1}{2}\left[\frac{8\pi G}{3c^2H^2}\right] \sum_{i=1}^{N} \varepsilon_i (1 + 3w_i). \tag{6.8}$$

The quantity in square brackets in Equation 6.8 is the inverse of the critical energy density ε_c. Thus, we can rewrite the acceleration equation in the form

$$-\frac{\ddot{a}}{aH^2} = \frac{1}{2} \sum_{i=1}^{N} \Omega_i (1 + 3w_i). \tag{6.9}$$

Evaluating Equation 6.9 at the present moment, $t = t_0$, tells us the relation between the deceleration parameter q_0 and the density parameters of the different components of the universe:

$$q_0 = \frac{1}{2} \sum_{i=1}^{N} \Omega_{i,0} (1 + 3w_i). \tag{6.10}$$

For a universe containing radiation, matter, and a cosmological constant,

$$q_0 = \Omega_{r,0} + \frac{1}{2}\Omega_{m,0} - \Omega_{\Lambda,0}. \tag{6.11}$$

Such a universe will currently be accelerating outward ($q_0 < 0$) if $\Omega_{\Lambda,0} > \Omega_{r,0} + \Omega_{m,0}/2$. The Benchmark Model, for instance, has $q_0 \approx -0.53$.

In principle, determining H_0 should be easy. For small redshifts, the relation between a galaxy's distance d and its redshift z is linear Equation (2.8):

$$cz = H_0 d. \tag{6.12}$$

Thus, if you measure the distance d and redshift z for a large sample of galaxies, and fit a straight line to a plot of cz versus d, the slope of the plot gives you the value of H_0.[1] In practice, the distance to a galaxy is not only difficult to measure,

1 The peculiar velocities of galaxies cause a significant amount of scatter in the plot, but by using a large number of galaxies, you can beat down the statistical errors. If you use galaxies at $d < 100$ Mpc, you must also make allowances for the local inhomogeneity and anisotropy.

but also somewhat difficult to define. In Section 3.5, the proper distance $d_p(t)$ between two points was defined as the length of the spatial geodesic between the points when the scale factor is fixed at the value $a(t)$. The proper distance is perhaps the most straightforward definition of the spatial distance between two points in an expanding universe. Moreover, there is a helpful relation between scale factor and proper distance. If we observe, at time t_0, light that was emitted by a distant galaxy at time t_e, the current proper distance to that galaxy is (Equation 5.33):

$$d_p(t_0) = c \int_{t_e}^{t_0} \frac{dt}{a(t)}. \tag{6.13}$$

For the model universes examined in Chapter 5, we knew the exact functional form of $a(t)$, and hence could exactly compute $d_p(t_0)$ for a galaxy of any redshift. If we have only partial knowledge of the scale factor, in the form of the Taylor expansion of Equation 6.4, we may use the expansion

$$\frac{1}{a(t)} \approx 1 - H_0(t - t_0) + \left(1 + \frac{q_0}{2}\right) H_0^2 (t - t_0)^2 \tag{6.14}$$

in Equation 6.13. Including the two lowest-order terms in the lookback time, $t_0 - t_e$, we find that the proper distance to the galaxy is

$$d_p(t_0) \approx c(t_0 - t_e) + \frac{cH_0}{2}(t_0 - t_e)^2. \tag{6.15}$$

The first term in the above equation, $c(t_0 - t_e)$, is what the proper distance would be in a static universe – the lookback time times the speed of light. The second term is a correction due to the expansion of the universe during the time the light was traveling.

Equation 6.15 would be extremely useful if the photons from distant galaxies carried a stamp telling us the lookback time, $t_0 - t_e$. They don't; instead, they carry a stamp telling us the scale factor $a(t_e)$ at the time the light was emitted. The observed redshift z of a galaxy, remember, is

$$z = \frac{1}{a(t_e)} - 1. \tag{6.16}$$

Using Equation 6.14, we may write an approximate relation between redshift and lookback time:

$$z \approx H_0(t_0 - t_e) + \left(1 + \frac{q_0}{2}\right) H_0^2 (t_0 - t_e)^2. \tag{6.17}$$

Inverting Equation 6.17 to give the lookback time as a function of redshift, we find

$$t_0 - t_e \approx H_0^{-1}\left[z - \left(1 + \frac{q_0}{2}\right) z^2\right]. \tag{6.18}$$

Substituting Equation 6.18 into Equation 6.15 gives us an approximate relation for the current proper distance to a galaxy with redshift z:

$$d_p(t_0) \approx \frac{c}{H_0}\left[z - \left(1 + \frac{q_0}{2}\right)z^2\right] + \frac{cH_0}{2}\frac{z^2}{H_0^2} = \frac{c}{H_0}z\left[1 - \frac{1+q_0}{2}z\right]. \tag{6.19}$$

The linear Hubble relation $d_p \propto z$ thus holds true only in the limit $z \ll 2/(1+q_0)$. If $q_0 > -1$, then the proper distance to a galaxy of moderate redshift ($z \sim 0.1$, say) is less than would be predicted from the linear Hubble relation.

6.2 Luminosity Distance

Unfortunately, the current proper distance to a galaxy, $d_p(t_0)$, is not a measurable property. If you tried to measure the distance to a galaxy with a tape measure, for instance, the distance would be continuously increasing as you extended the tape. To measure the proper distance at time t_0, you would need a tape measure that could be extended with infinite speed; alternatively, you would need to stop the expansion of the universe at its current scale factor while you measured the distance at your leisure. Neither of these alternatives is physically possible.

Since cosmology is ultimately based on observations, if we want to find the distance to a galaxy, we need some way of computing a distance from that galaxy's observed properties. In devising ways of computing the distance to galaxies, astronomers have found it useful to adopt and adapt the techniques used to measure shorter distances. Let's examine, then, the techniques used to measure relatively short distances. Within the solar system, astronomers measure the distance to planets by reflecting radar signals from them. If δt is the time taken for a photon to complete the round-trip, then the distance to the reflecting body is $d = c\,\delta t/2$. (Since the relative speeds of objects within the solar system are much smaller than c, the corrections due to relative motion during the time δt are minuscule.) The accuracy with which distances have been determined with this technique is impressive; the length of the astronomical unit, for instance, is $1\,\text{AU} = 149\,597\,870.7\,\text{km}$. The radar technique is useful only within the solar system. Beyond $\sim 10\,\text{AU}$, the reflected radio waves are too faint to detect.

A favorite method for determining distances to other stars within our galaxy is the method of trigonometric parallax. When a star is observed from two points separated by a distance b, the star's apparent position will shift by an angle θ. If the baseline of observation is perpendicular to the line of sight to the star, the *parallax distance* will be

$$d_\pi = 1\,\text{pc}\left(\frac{b}{1\,\text{AU}}\right)\left(\frac{\theta}{1\,\text{arcsec}}\right)^{-1}. \tag{6.20}$$

Measuring the distances to stars using the Earth's orbit ($b = 2\,\text{AU}$) as a baseline is a standard technique. Since the size of the Earth's orbit is known with great accuracy from radar measurements, the accuracy with which the parallax distance can be determined is limited by the accuracy with which the parallax angle θ can be measured. The *Gaia* satellite, launched by the European Space Agency in 2013, was designed to measure the parallax of stars with an error as small as ~ 10 microarcseconds. However, to measure θ for a galaxy 100 Mpc away, an error of < 0.01 microarcseconds would be required, using the Earth's orbit as a baseline. The trigonometric parallaxes of galaxies at cosmological distances are too small to be measured with current technology.

Let's focus on the properties that we *can* measure for objects at cosmological distances. We can measure the flux of light, f, from the object, in units of watts per square meter. The complete flux, integrated over all wavelengths of light, is called the *bolometric* flux. The adjective "bolometric" is a reference to the scientific instrument known as a bolometer, an extremely sensitive thermometer capable of detecting electromagnetic radiation over a wide range of wavelengths. The bolometer was invented around the year 1880 by the astronomer Samuel Langley, who used it to measure solar radiation. As expressed more poetically in an anonymous limerick:

Oh, Langley devised the bolometer:
It's really a kind of thermometer
Which measures the heat
From a polar bear's feet
At a distance of half a kilometer.[2]

More prosaically, given the technical difficulties of measuring the true bolometric flux, the flux over a limited range of wavelengths is measured. If the light from the object has emission or absorption lines, we can measure the redshift, z. If the object is an extended source rather than a point of light, we can measure its angular diameter, $\delta\theta$.

One way of using measured properties to assign a distance is the *standard candle* method. A standard candle is an object whose luminosity L is known. For instance, if some class of astronomical object had luminosities that were the same throughout all of spacetime, they would act as excellent standard candles – if their unique luminosity L were known. If you know, by some means or other, the luminosity of an object, then you can use its measured flux f to compute a function known as the *luminosity distance*, defined as

$$d_L \equiv \left(\frac{L}{4\pi f}\right)^{1/2}. \tag{6.21}$$

[2] The earliest version of this poem that I can find (in the May 1950 issue of *Electronics*) refers to the polar bear's seat, rather than its feet. I leave it to you to choose your favorite bit of the bear's anatomy.

The function d_L is called a "distance" because its dimensionality is that of a distance, and because it is what the proper distance to the standard candle would be *if* the universe were static and Euclidean. In a static Euclidean universe, propagation of light follows the inverse square law: $f = L/(4\pi d^2)$.

Suppose, though, that you are in a universe described by a Robertson–Walker metric Equation (3.41):

$$ds^2 = -c^2dt^2 + a(t)^2[dr^2 + S_\kappa(r)^2 d\Omega^2], \tag{6.22}$$

with

$$S_\kappa(r) = \begin{cases} R_0 \sin(r/R_0) & (\kappa = +1) \\ r & (\kappa = 0) \\ R_0 \sinh(r/R_0) & (\kappa = -1). \end{cases} \tag{6.23}$$

You are at the origin. At the present moment, $t = t_0$, you see light that was emitted by a standard candle at comoving coordinate location (r, θ, ϕ) at a time t_e (Figure 6.1). The photons emitted at time t_e are, at the present moment, spread over a sphere of proper radius $d_p(t_0) = r$ and proper surface area $A_p(t_0)$. If space is flat ($\kappa = 0$), then the proper area of the sphere is given by the Euclidean relation $A_p(t_0) = 4\pi d_p(t_0)^2 = 4\pi r^2$. More generally, however,

$$A_p(t_0) = 4\pi S_\kappa(r)^2. \tag{6.24}$$

When space is positively curved, $A_p(t_0) < 4\pi r^2$, and the photons are spread over a smaller area than they would be in flat space. When space is negatively curved, $A_p(t_0) > 4\pi r^2$, and photons are spread over a larger area than they would be in flat space.

In addition to these geometric effects, which apply even in a static universe, the expansion of the universe causes the observed flux of light from a standard candle of redshift z to be decreased by a factor of $(1+z)^{-2}$. First, the expansion of the universe causes the energy of each photon from the standard candle to decrease. If a photon starts with an energy $E_e = hc/\lambda_e$ when the scale factor is

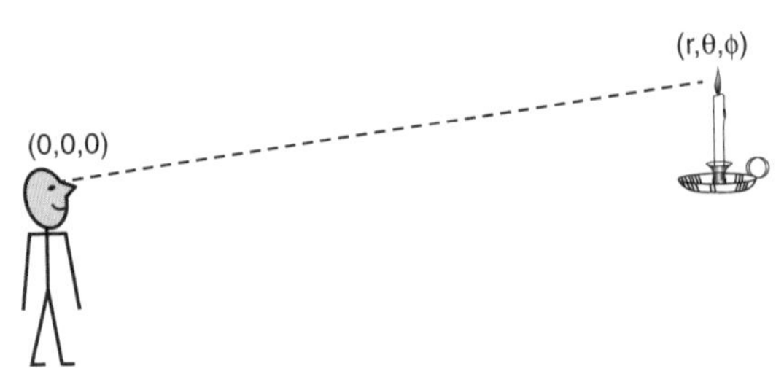

Figure 6.1 An observer at the origin observes a standard candle, of known luminosity L, at comoving coordinate location (r, θ, ϕ).

$a(t_e)$, by the time we observe it, when the scale factor is $a(t_0) = 1$, the wavelength will have grown to

$$\lambda_0 = \frac{1}{a(t_e)}\lambda_e = (1+z)\lambda_e, \tag{6.25}$$

and the energy will have fallen to

$$E_0 = \frac{E_e}{1+z}. \tag{6.26}$$

Second, thanks to the expansion of the universe, the time between photon detections will be greater. If two photons are emitted in the same direction separated by a time interval δt_e, the proper distance between them will initially be $c(\delta t_e)$; by the time we detect the photons at time t_0, the proper distance between them will be stretched to $c(\delta t_e)(1+z)$, and we will detect them separated by a time interval $\delta t_0 = \delta t_e(1+z)$.

The net result is that in an expanding, spatially curved universe, the relation between the observed flux f and the luminosity L of a distant light source is

$$f = \frac{L}{4\pi S_\kappa(r)^2(1+z)^2}, \tag{6.27}$$

and the luminosity distance is

$$d_L = S_\kappa(r)(1+z). \tag{6.28}$$

The available evidence indicates that our universe is nearly flat, with a radius of curvature R_0 much larger than the current horizon distance $d_{\rm hor}(t_0)$. Objects with finite redshift are at proper distances smaller than the horizon distance, and hence much smaller than the radius of curvature. Thus, it is safe to make the approximation $r \ll R_0$, implying $S_\kappa(r) \approx r$. With our assumption that space is very close to being flat, the relation between the luminosity distance and the current proper distance becomes very simple:

$$d_L = r(1+z) = d_p(t_0)(1+z) \qquad [\kappa = 0]. \tag{6.29}$$

Thus, even if space is perfectly flat, if you estimate the distance to a standard candle by using a naïve inverse square law, you will overestimate the actual proper distance by a factor $(1+z)$, where z is the standard candle's redshift.

Figure 6.2 shows the luminosity distance d_L as a function of redshift for the Benchmark Model and for two other flat universes, one dominated by matter and one dominated by a cosmological constant. When $z \ll 1$, the current proper distance may be approximated as

$$d_P(t_0) \approx \frac{c}{H_0}z\left(1 - \frac{1+q_0}{2}z\right). \tag{6.30}$$

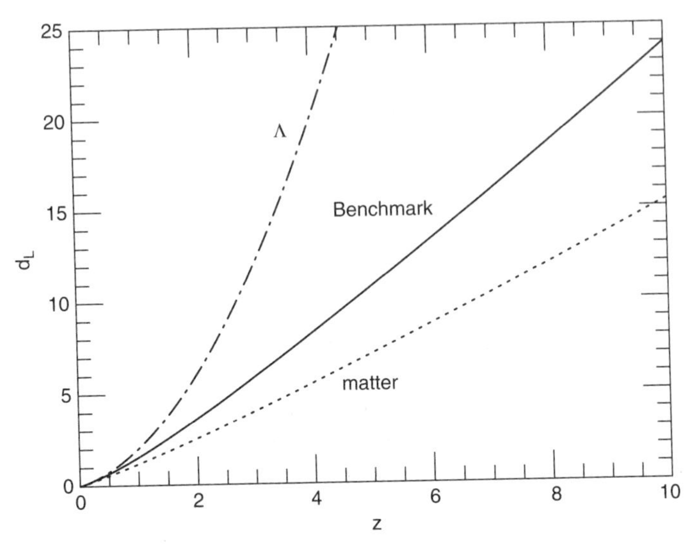

Figure 6.2 Luminosity distance of a standard candle with observed redshift z, in units of the Hubble distance, c/H_0. The bold solid line gives the result for the Benchmark Model. For comparison, the dot-dash line indicates a flat, lambda-only universe, and the dotted line a flat, matter-only universe.

In a nearly flat universe, the luminosity distance may thus be approximated as

$$d_L \approx \frac{c}{H_0}z\left(1 - \frac{1+q_0}{2}z\right)(1+z) \approx \frac{c}{H_0}z\left(1 + \frac{1-q_0}{2}z\right). \tag{6.31}$$

6.3 Angular-diameter Distance

The luminosity distance d_L is not the only distance measure that can be computed using the observable properties of cosmological objects. Suppose that instead of a standard candle, you observed a *standard yardstick*. A standard yardstick is an object whose proper length ℓ is known. In many cases, it is convenient to choose as your yardstick an object that is tightly bound together, by gravity or duct tape or some other influence, and hence is not expanding along with the universe as a whole.

Suppose a yardstick of constant proper length ℓ is aligned perpendicular to your line of sight, as shown in Figure 6.3. You measure an angular distance $\delta\theta$ between the ends of the yardstick, and a redshift z for the light that the yardstick emits. If $\delta\theta \ll 1$, and if you know the length ℓ of the yardstick, you can compute a distance to the yardstick using the small-angle formula

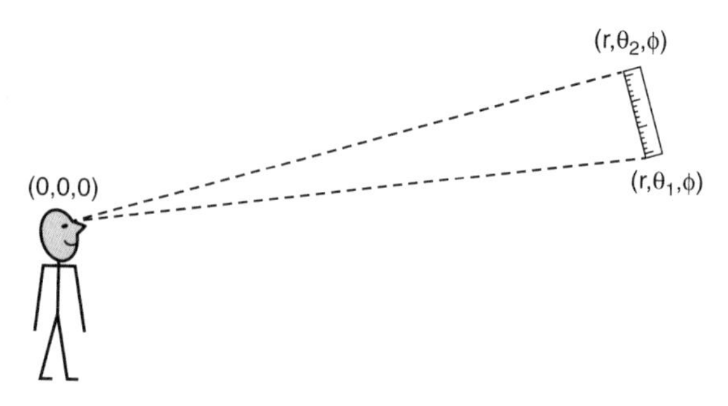

Figure 6.3 An observer at the origin observes a standard yardstick, of known proper length ℓ, at comoving coordinate distance r.

$$d_A \equiv \frac{\ell}{\delta\theta}. \tag{6.32}$$

This function of ℓ and $\delta\theta$ is called the *angular-diameter distance*. The angular-diameter distance is equal to the proper distance to the yardstick if the universe is static and Euclidean.

In general, though, if the universe is expanding or curved, the angular-diameter distance will not be equal to the current proper distance. Suppose you are in a universe described by the Robertson–Walker metric given in Equation 6.22. Choose your comoving coordinate system so that you are at the origin. The yardstick is at a comoving coordinate distance r. At a time t_e, the yardstick emitted the light that you observe at time t_0. The comoving coordinates of the two ends of the yardstick, at the time the light was emitted, were (r, θ_1, ϕ) and (r, θ_2, ϕ). As the light from the yardstick moves toward the origin, it travels along geodesics with $\theta =$ constant and $\phi =$ constant. Thus, the angular size you measure for the yardstick will be $\delta\theta = \theta_2 - \theta_1$. The distance ds between the two ends of the yardstick, measured at the time t_e when the light was emitted, can be found from the Robertson–Walker metric:

$$ds = a(t_e)S_\kappa(r)\delta\theta. \tag{6.33}$$

However, for a standard yardstick whose length ℓ is known, we can set $ds = \ell$, and thus find that

$$\ell = a(t_e)S_\kappa(r)\delta\theta = \frac{S_\kappa(r)\delta\theta}{1+z}. \tag{6.34}$$

Thus, the angular-diameter distance d_A to a standard yardstick is

$$d_A \equiv \frac{\ell}{\delta\theta} = \frac{S_\kappa(r)}{1+z}. \tag{6.35}$$

Comparison with Equation 6.28 shows that the relation between the angular-diameter distance and the luminosity distance is

$$d_A = \frac{d_L}{(1+z)^2}. \tag{6.36}$$

Thus, if you observe a redshifted object that is both a standard candle and a standard yardstick, the angular-diameter distance that you compute for the object will be smaller than the luminosity distance. Moreover, if the universe is spatially flat,

$$d_A(1+z) = d_p(t_0) = \frac{d_L}{1+z} \qquad [\kappa = 0]. \tag{6.37}$$

In a flat universe, therefore, if you compute the angular-diameter distance d_A of a standard yardstick, it isn't equal to the current proper distance $d_p(t_0)$; rather, it is equal to the proper distance at the time the light from the object was emitted: $d_A = d_p(t_0)/(1+z) = d_p(t_e)$.

Figure 6.4 shows the angular-diameter distance d_A for the Benchmark Model, and for two other spatially flat universes, one dominated by matter and one dominated by a cosmological constant. [Since d_A is, for these flat universes, equal to $d_p(t_e)$, Figure 6.4 is simply a replotting of the right panel in Figure 5.9.] When $z \ll 1$, the approximate value of d_A is given by the expansion

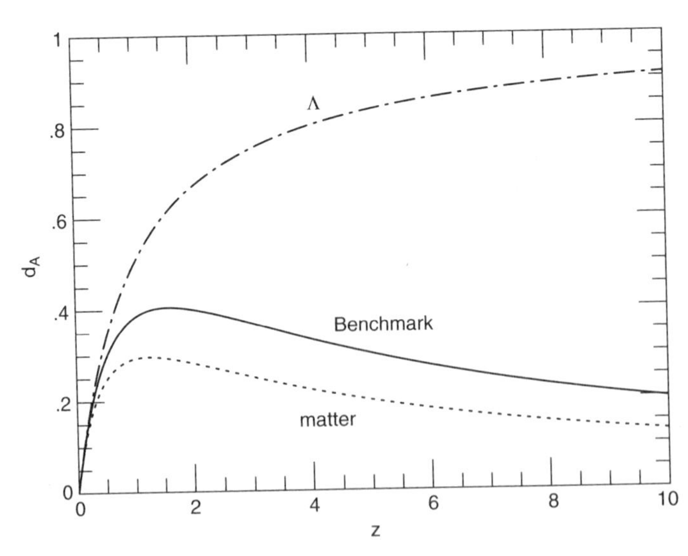

Figure 6.4 Angular-diameter distance of a standard yardstick with observed redshift z, in units of the Hubble distance, c/H_0. The bold solid line gives the result for the Benchmark Model. For comparison, the dot-dash line indicates a flat, lambda-only universe, and the dotted line a flat, matter-only universe.

$$d_A \approx \frac{c}{H_0} z \left(1 - \frac{3+q_0}{2} z\right). \tag{6.38}$$

Thus, comparing Equations 6.30, 6.31, and 6.38, we find that in the limit $z \to 0$, $d_A \approx d_L \approx d_p(t_0) \approx (c/H_0)z$. However, the state of affairs is very different in the limit $z \to \infty$. In models with a finite horizon size, $d_p(t_0) \to d_{\rm hor}(t_0)$ as $z \to \infty$. The luminosity distance to highly redshifted objects, in this case, diverges as $z \to \infty$, with

$$d_L(z \to \infty) \approx z d_{\rm hor}(t_0). \tag{6.39}$$

However, the angular-diameter distance to highly redshifted objects approaches zero as $z \to \infty$, with

$$d_A(z \to \infty) \approx \frac{d_{\rm hor}(t_0)}{z}. \tag{6.40}$$

In model universes other than the lambda-only model, the angular-diameter distance d_A has a maximum for standard yardsticks at some critical redshift z_c. For instance, the Benchmark Model has a critical redshift $z_c = 1.6$, where $d_A(\rm max) = 0.405c/H_0 = 1770\,\rm Mpc$. If the universe were full of glow-in-the-dark yardsticks, all of the same size ℓ, their angular size $\delta\theta$ would decrease with redshift out to $z = z_c$, but then would increase at larger redshifts. The sky would be full of big, faint, redshifted yardsticks.

In principle, standard yardsticks, like standard candles, can be used to measure cosmological parameters such as H_0, $\Omega_{\Lambda,0}$, and $\Omega_{m,0}$. In practice, the use of standard yardsticks to determine cosmological parameters was long plagued with observational difficulties. For instance, a standard yardstick must have an angular size large enough to be resolved by your telescope. A yardstick of physical size ℓ will have its angular size $\delta\theta$ minimized when it is at the critical redshift z_c. For the Benchmark Model,

$$\delta\theta(\rm min) = \frac{\ell}{d_A(\rm max)} = \frac{\ell}{1770\,\rm Mpc} \approx 0.1\,\rm arcsec \left(\frac{\ell}{1\,\rm kpc}\right). \tag{6.41}$$

Both galaxies and clusters of galaxies are large enough to be useful standard candles. Unfortunately for cosmologists, galaxies and clusters of galaxies do not have sharply defined edges, so assigning a particular angular size $\delta\theta$, and a corresponding physical size ℓ, to these objects is a somewhat tricky task. Moreover, galaxies and clusters of galaxies are not isolated, rigid yardsticks of fixed length. Galaxies tend to become larger with time as they undergo mergers with their neighbors. Clusters, too, tend to become larger with time, as galaxies fall into them. Correcting for these evolutionary trends is a difficult task. Given the historical difficulties involved in using standard yardsticks to determine cosmological parameters, let's first look at how standard *candles* can be used to determine H_0.

6.4 Standard Candles and H_0

Using standard candles to determine the Hubble constant has a long and honorable history; it's the method used by Hubble himself. The recipe for finding the Hubble constant is a simple one:

- Identify a population of standard candles with luminosity L.
- Measure the redshift z and flux f for each standard candle.
- Compute $d_L = (L/4\pi f)^{1/2}$ for each standard candle.
- Plot cz versus d_L.
- Measure the slope of the cz versus d_L relation when $z \ll 1$; this gives H_0.

As with the apocryphal recipe for rabbit stew that begins "First catch your rabbit," the hardest step is the first one. A good standard candle is hard to find. For cosmological purposes, a standard candle should be bright enough to be detected at large redshifts. It should also have a luminosity that is well determined.[3]

One time-honored variety of standard candle is the class of *Cepheid variable stars*. Cepheids, as they are known, are highly luminous supergiant stars, with mean luminosities in the range $\bar{L} = 400 \rightarrow 40\,000\,\mathrm{L}_\odot$. Cepheids are pulsationally unstable. As they pulsate radially, their luminosity varies in response, partially due to the changes in their surface area, and partially due to the changes in the surface temperature as the star pulsates. The pulsational periods, as reflected in the observed brightness variations of the star, lie in the range $P = 1.5 \rightarrow 60$ days.

On the face of it, Cepheids don't seem sufficiently standardized to be standard candles; their mean luminosities range over two orders of magnitude. How can you tell whether you are looking at an intrinsically faint Cepheid ($L \approx 400\,\mathrm{L}_\odot$) or at an intrinsically bright Cepheid ($L \approx 40\,000\,\mathrm{L}_\odot$) ten times farther away? The key to calibrating Cepheids was discovered by Henrietta Leavitt, at Harvard College Observatory. In the years prior to World War I, Leavitt was studying variable stars in the Large and Small Magellanic Clouds, a pair of relatively small satellite galaxies orbiting our own galaxy. For each Cepheid in the Small Magellanic Cloud (SMC), she measured the period P by finding the time between maxima in the observed brightness, and found the mean flux $\bar{f}$, averaged over one complete period. She noted that there was a clear relation between P and $\bar{f}$, with stars having the longest period of variability also having the largest flux. Since the depth of the SMC, front to back, is small compared to its distance from us, she was justified in assuming that the difference in mean flux for the Cepheids was due to differences in their mean luminosity, not differences in their luminosity distance. Leavitt had discovered a period–luminosity relation for Cepheid variable stars.

[3] A useful cautionary tale in this regard is the saga of Edwin Hubble. In the 1929 paper that first demonstrated that $d_L \propto z$ when $z \ll 1$, Hubble underestimated the luminosity distances to galaxies by a factor of ~ 7 because he underestimated the luminosity of his standard candles by a factor of ~ 49.

If the same period–luminosity relation holds true for all Cepheids, in all galaxies, then Cepheids can act as a standard candle.

Suppose, for instance, you find a Cepheid star in the Large Magellanic Cloud (LMC) and another in M31. They both have a pulsational period of 10 days, so you assume, from the period–luminosity relation, that they have the same mean luminosity $\bar{L}$. By careful measurement, you determine that

$$\frac{\bar{f}_{\rm LMC}}{\bar{f}_{\rm M31}} = 230. \tag{6.42}$$

Thus, you conclude that the luminosity distance to M31 is greater than that to the LMC by a factor

$$\frac{d_L({\rm M31})}{d_L({\rm LMC})} = \left(\frac{\bar{f}_{\rm LMC}}{\bar{f}_{\rm M31}}\right)^{1/2} = \sqrt{230} = 15.2. \tag{6.43}$$

(In practice, given the intrinsic scatter in the period–luminosity relation, and the inevitable error in measuring fluxes, astronomers don't rely on a single Cepheid in each galaxy. Rather, they measure $\bar{f}$ and P for as many Cepheids as possible in each galaxy, then find the ratio of luminosity distances that makes the period–luminosity relations for the two galaxies coincide.)

Note that if you only know the relative fluxes of the two Cepheids, and not their luminosity $\bar{L}$, you will only know the *relative* distances of M31 and the LMC. To fix an absolute distance to M31, to the LMC, and to other galaxies containing Cepheids, you need to know the luminosity $\bar{L}$ for a Cepheid of a given period P. If, for instance, you could measure the parallax distance d_π to a Cepheid within our own galaxy, you could then compute its luminosity $\bar{L} = 4\pi d_\pi^2 \bar{f}$, and use it to normalize the period–luminosity relation for Cepheids.[4] Unfortunately, Cepheids are rare stars; only the very nearest Cepheids in our galaxy have had their distances measured accurately. The nearest Cepheid is Polaris, as it turns out, at $d_\pi = 130 \pm 10\,{\rm pc}$. The next nearest is δ Cephei (the prototype after which all Cepheids are named), at $d_\pi = 270 \pm 10\,{\rm pc}$. Historically, given the lack of Cepheid parallaxes, astronomers have relied on alternative methods of normalizing the period–luminosity relation for Cepheids. The most usual method involved finding the distance to the Large Magellanic Cloud by secondary methods, then using this distance to compute the mean luminosity of the LMC Cepheids. The current consensus is that the Large Magellanic Cloud has a luminosity distance $d_L = 50 \pm 2\,{\rm kpc}$, implying a distance to M31 of $d_L = 760 \pm 30\,{\rm kpc}$.

The fluxes and periods of Cepheids can be accurately measured out to luminosity distances $d_L \sim 30\,{\rm Mpc}$. Observation of Cepheid stars in the Virgo cluster of galaxies, for instance, has yielded a distance $d_L({\rm Virgo}) = 300\,d_L({\rm LMC}) =$

[4] Within our galaxy, which is not expanding, the parallax distance, the luminosity distance, and the proper distance are identical.

15 Mpc. One of the motivating reasons for building the *Hubble Space Telescope* in the first place was to use Cepheids to determine H_0. The net result of the Hubble Key Project to measure H_0 is displayed in Figure 2.5, showing that the Cepheid data are best fitted with a Hubble constant of $H_0 = 75 \pm 8\,\mathrm{km\,s^{-1}\,Mpc^{-1}}$.

There is a hidden difficulty involved in using Cepheid stars to determine H_0. Cepheids can take you out only to a distance $d_L \sim 30\,\mathrm{Mpc}$; on this scale, the universe cannot be assumed to be homogeneous and isotropic. In fact, the Local Group is gravitationally attracted toward the Virgo cluster, causing it to have a peculiar motion in that direction. It is estimated, from dynamical models, that the recession velocity cz that we measure for the Virgo cluster is $250\,\mathrm{km\,s^{-1}}$ less than it would be if the universe were perfectly homogeneous. The plot of cz versus d_L given in Figure 2.5 uses recession velocities that are corrected for this "Virgocentric flow," as it is called.

6.5 Standard Candles and Acceleration

To determine the value of H_0 without having to worry about Virgocentric flow and other peculiar velocities, we need to determine the luminosity distance to standard candles with $d_L > 100\,\mathrm{Mpc}$, or $z > 0.02$. To determine the acceleration of the universe, we need to view standard candles for which the relation between d_L and z deviates significantly from the linear relation that holds true at lower redshifts. In terms of H_0 and q_0, the luminosity distance at small redshift is, from Equation 6.31,

$$d_L \approx \frac{c}{H_0} z \left[1 + \frac{1 - q_0}{2} z \right]. \tag{6.44}$$

At a redshift $z = 0.2$, for instance, the luminosity distance d_L in the Benchmark Model (with $q_0 = -0.53$) is 5 percent larger than d_L in an empty universe (with $q_0 = 0$).

For a standard candle to be seen at $d_L > 1000\,\mathrm{Mpc}$, it must be very luminous. In recent years, the standard candle of choice among cosmologists has been *type Ia supernovae*. A supernova may be loosely defined as an exploding star. Early in the history of supernova studies, when little was known about their underlying physics, supernovae were divided into two classes, on the basis of their spectra. Type I supernovae contain no hydrogen absorption lines in their spectra; type II supernovae contain strong hydrogen absorption lines. Gradually, it was realized that all type II supernovae are the same species of beast; they are massive stars ($M > 8\,\mathrm{M_\odot}$) whose cores collapse to form a black hole or neutron star when their nuclear fuel is exhausted. During the rapid collapse of the core, the outer layers of the star are thrown off into space. Type I supernovae are actually two separate species, called type Ia and type Ib. Type Ib supernovae, it is thought, are massive stars whose cores collapse after the hydrogen-rich outer layers of the star have been blown away in strong stellar winds. Thus, type Ib and type II supernovae are driven by very similar mechanisms – their differences are superficial, in the most literal sense. Type Ia supernovae, however, are something completely different. They begin as white dwarfs; that is, stellar remnants that are supported against gravity by the quantum mechanical effect known as electron degeneracy pressure. The maximum mass at which a white dwarf can be supported against its self-gravity is called the Chandrasekhar mass; the value of the Chandrasekhar mass is $M \approx 1.4\,\mathrm{M}_\odot$. A white dwarf can go over this limit by merging with another white dwarf, or by accreting gas from a stellar companion. If the Chandrasekhar limit is approached or exceeded, the white dwarf starts to collapse until its increased density triggers a runaway nuclear fusion reaction. The entire white dwarf becomes a fusion bomb, blowing itself to smithereens; unlike type II supernovae, type Ia supernovae do not leave a condensed stellar remnant behind.

Within our galaxy, type Ia supernovae occur roughly once per century, on average. Although type Ia supernovae are not frequent occurrences locally, they are extraordinarily luminous, and hence can be seen to large distances. The luminosity of an average type Ia supernova, at peak brightness, is $L = 4 \times 10^9\,\mathrm{L}_\odot$; that's 100 000 times more luminous than even the brightest Cepheid. For a few days, a type Ia supernova in a moderately bright galaxy can outshine all the other stars in the galaxy combined. Since moderately bright galaxies can be seen at $z \sim 1$, this means that type Ia supernovae can also be seen at $z \sim 1$.

So far, type Ia supernovae sound like ideal standard candles; very luminous and all produced by the same mechanism. There's one complication, however. Observation of supernovae in galaxies whose distances have been well determined by Cepheids reveals that type Ia supernovae do not have identical luminosities. Instead of all having $L = 4 \times 10^9\,\mathrm{L}_\odot$, their peak luminosities lie in the fairly broad range $L \approx (3 \to 5) \times 10^9\,\mathrm{L}_\odot$. However, it has also been noted that the peak luminosity of a type Ia supernova is tightly correlated with the shape of its light curve. Type Ia supernovae with luminosities that shoot up rapidly and decline rapidly are less luminous than average at their peak; supernovae with luminosities that rise and fall in a more leisurely manner are more luminous than average. Thus, just as the period of a Cepheid tells you its luminosity, the rise and fall time of a type Ia supernova tells you its peak luminosity.

At the end of the 20th century, two research teams, the "Supernova Cosmology Project" and the "High-z Supernova Search Team," conducted searches for supernovae in distant galaxies, using the observed fluxes of the supernovae to constrain the acceleration of the expansion of the universe. To present the supernova results, I will have to introduce the "magnitude" system used by astronomers to express fluxes and luminosities. The magnitude system, like much else in astronomy, has its roots in ancient Greece. The Greek astronomer Hipparchus, in the second century BC, divided the stars into six classes, according to their apparent brightness. The brightest stars were of "first magnitude," the faintest stars visible to the naked eye were of "sixth magnitude," and intermediate stars were ranked as second, third, fourth, and fifth magnitude. Long after the time of Hipparchus, it was realized that the response of the human eye is roughly logarithmic, and that stars of the first magnitude have fluxes (at visible wavelengths) about 100 times greater than stars of the sixth magnitude. On the basis of this realization, the magnitude system was placed on a more rigorous mathematical basis.

Nowadays, the bolometric *apparent magnitude* of a light source is defined in terms of the source's bolometric flux as

$$m \equiv -2.5 \log_{10}(f/f_x), \tag{6.45}$$

where the reference flux f_x is set at the value $f_x = 2.53 \times 10^{-8}$ watt m^{-2}. Thanks to the negative sign in the definition, a small value of m corresponds to a large flux f. For instance, the flux of sunlight at the Earth's location is $f = 1361$ watts m^{-2}; the Sun thus has a bolometric apparent magnitude of $m = -26.8$. The choice of reference flux f_x constitutes a tip of the hat to Hipparchus, since for stars visible to the naked eye it typically yields $0 < m < 6$.

The bolometric *absolute magnitude* of a light source is defined as the apparent magnitude that it would have if it were at a luminosity distance of $d_L = 10$ pc. Thus, a light source with luminosity L has a bolometric absolute magnitude

$$M \equiv -2.5 \log_{10}(L/L_x), \tag{6.46}$$

where the reference luminosity is $L_x = 78.7\,\mathrm{L}_\odot$, since that is the luminosity of an object that produces a flux $f_x = 2.53 \times 10^{-8}$ watt m^{-2} when viewed from a distance of 10 parsecs. The bolometric absolute magnitude of the Sun is thus $M = 4.74$. Although the system of apparent and absolute magnitudes seems strange to the uninitiated, the apparent magnitude is really nothing more than a logarithmic measure of the flux, and the absolute magnitude is a logarithmic measure of the luminosity.

Given the definitions of apparent and absolute magnitude, the relation between an object's apparent magnitude and its absolute magnitude can be written in the form

$$M = m - 5 \log_{10}\left(\frac{d_L}{10\,\mathrm{pc}}\right), \tag{6.47}$$

where d_L is the luminosity distance to the light source. If the luminosity distance is given in units of megaparsecs, this relation becomes

$$M = m - 5 \log_{10}\left(\frac{d_L}{1\,\mathrm{Mpc}}\right) - 25. \tag{6.48}$$

Since astronomers frequently quote fluxes and luminosities in terms of apparent and absolute magnitudes, they find it convenient to quote luminosity distances in terms of the *distance modulus* to a light source. The distance modulus is defined as $m - M$, and is related to the luminosity distance by the relation

$$m - M = 5\log_{10}\left(\frac{d_L}{1\,\mathrm{Mpc}}\right) + 25. \tag{6.49}$$

The distance modulus of the Large Magellanic Cloud, for instance, at $d_L = 0.050\,\mathrm{Mpc}$, is $m - M = 18.5$. The distance modulus of the Virgo cluster, at $d_L = 15\,\mathrm{Mpc}$, is $m - M = 30.9$. When $z \ll 1$, the luminosity distance to a light source is

$$d_L \approx \frac{c}{H_0} z \left(1 + \frac{1 - q_0}{2} z\right). \tag{6.50}$$

Substituting this relation into Equation 6.49, we have an equation that gives the relation between distance modulus and redshift at low redshift:

$$m - M \approx 43.23 - 5\log_{10}\left(\frac{H_0}{68\,\mathrm{km\,s^{-1}\,Mpc^{-1}}}\right) + 5\log_{10} z + 1.086(1 - q_0)z. \tag{6.51}$$

For a population of standard candles with known luminosity L (and hence of known bolometric absolute magnitude M), we measure the flux f (or equivalently the bolometric apparent magnitude m) and the redshift z. In the limit $z \to 0$, a plot of $m - M$ versus $\log z$ gives a straight line whose amplitude at a fixed z tells us the value of H_0. At slightly larger values of z, the deviation of the plot from a straight line tells us whether the expansion of the universe is speeding up or slowing down. At a given value of z, standard candles have a lower flux in an accelerating universe (with $q_0 < 0$) than in a decelerating universe (with $q_0 > 0$).

Figure 6.5 shows the plot of distance modulus versus redshift for a compilation of actual supernova observations from a variety of sources. The solid line running through the data is the result expected for the Benchmark Model. At a redshift $z \approx 1$, supernovae in the Benchmark Model are about 0.6 magnitudes fainter than they would be in a flat, matter-only universe; it was the observed faintness of Type Ia supernovae at $z > 0.3$ that led to the conclusion that the universe is accelerating. However, at $z \approx 1$, supernovae in the Benchmark Model are about 0.6 magnitudes brighter than they would be in a flat, lambda-only universe. Thus, the observations of Type Ia supernovae that tell us that the universe is accelerating also place useful upper limits on the magnitude of the acceleration.

Figure 6.6 shows the results of fitting the supernova data with different model universes; these models contain both matter and a cosmological constant, but are not required to be spatially flat. The bold ellipse represents the 95% confidence interval; that is, given the available set of supernova data, there is a 95% chance that the plotted ellipse contains the true values of $\Omega_{m,0}$ and $\Omega_{\Lambda,0}$. Notice from the plot that decelerating universes with $q_0 > 0$ (below the dotted line) are strongly excluded by the supernova data, as are Big Crunch universes and Big Bounce

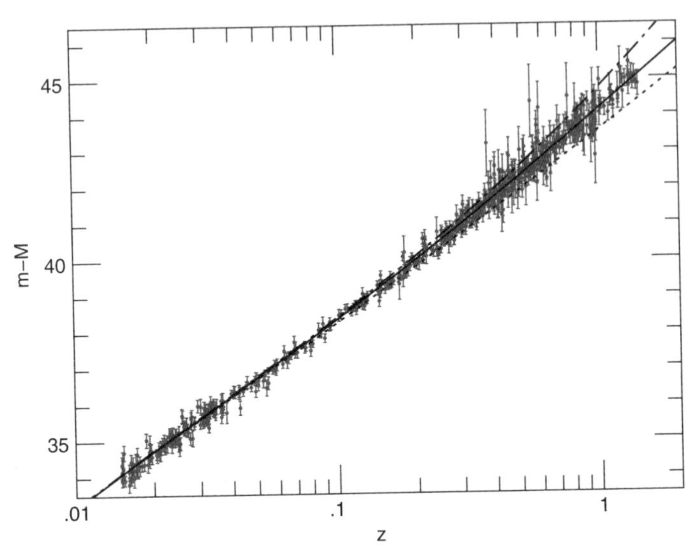

Figure 6.5 Distance modulus versus redshift for a set of 580 type Ia supernovae. The bold solid line gives the expected relation for the Benchmark Model. For comparison, the dot-dash line indicates a flat, lambda-only universe, and the dotted line a flat, matter-only universe. [data from Suzuki *et al.* 2012, *ApJ*, **716**, 85]

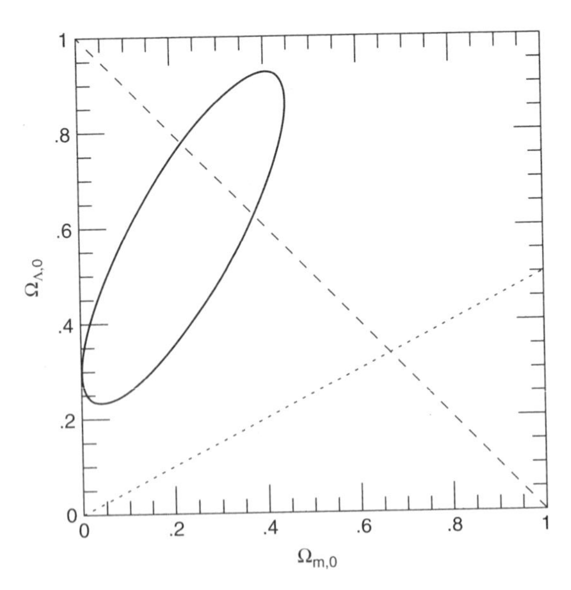

Figure 6.6 The values of $\Omega_{m,0}$ and $\Omega_{\Lambda,0}$ that best fit the supernova data. The bold elliptical contour represents the 95% confidence interval. For reference, the dashed line represents flat universes, and the dotted line represents coasting ($q_0 = 0$) universes: compare to Figure 5.6. [Anže Slosar & José Alberto Vázquez, Brookhaven National Laboratory]

universes. However, the supernova data, taken by themselves, are consistent with positively curved or negatively curved universes, as well as with a flat universe. We will see in Chapter 8 how observations of the cosmic microwave background combine with the supernova results to suggest that we live in a universe that is both accelerating and spatially flat, with $\Omega_{m,0} \sim 0.3$ and $\Omega_{\Lambda,0} \sim 0.7$.

Exercises

6.1 Suppose that a polar bear's foot has a luminosity of $L = 10$ watts. What is the bolometric absolute magnitude of the bear's foot? What is the bolometric apparent magnitude of the foot at a luminosity distance of $d_L = 0.5$ km? If a bolometer can detect the bear's foot at a maximum luminosity distance of $d_L = 0.5$ km, what is the maximum luminosity distance at which it could detect the Sun? What is the maximum luminosity distance at which it could detect a supernova with $L = 4 \times 10^9\,\mathrm{L}_\odot$?

6.2 Suppose that a polar bear's foot has a diameter of $\ell = 0.16$ m. What is the angular size $\delta\theta$ of the foot at an angular-diameter distance of $d_A = 0.5$ km? In the Benchmark Model, what is the minimum possible angular size of the polar bear's foot?

6.3 Suppose that you are in a spatially flat universe containing a single component with a unique equation-of-state parameter w. What are the current proper distance $d_P(t_0)$, the luminosity distance d_L, and the angular-diameter distance d_A as a function of z and w? At what redshift will d_A have a maximum value? What will this maximum value be, in units of the Hubble distance?

6.4 Verify that Equation 6.51 is correct in the limit of small z. (You will probably want to use the relation $\log_{10}(1+x) \approx 0.4343 \ln(1+x) \approx 0.4343x$ in the limit $|x| \ll 1$.)

6.5 The surface brightness Σ of an astronomical object is its observed flux divided by its observed angular area; thus, $\Sigma \propto f/(\delta\theta)^2$. For a class of objects that are both standard candles and standard yardsticks, what is Σ as a function of redshift? Would observing the surface brightness of this class of objects be a useful way of determining the value of the deceleration parameter q_0? Why or why not?

6.6 You observe a quasar at a redshift $z = 5.0$, and determine that the observed flux of light from the quasar varies on a timescale $\delta t_0 = 3$ days. If the observed variation in flux is due to a variation in the intrinsic luminosity of the quasar, what was the variation timescale δt_e at the time the light was emitted? For the light from the quasar to vary on a timescale δt_e, the bulk of the light must come from a region of physical size $R \leq R_{\max} = c(\delta t_e)$. What is $R_{\max}$ for the observed quasar? What is the angular size of $R_{\max}$ in the Benchmark Model?

6.7 Derive the relation $A_p(t_0) = 4\pi S_\kappa(r)^2$, as given in Equation 6.24, starting from the Robertson–Walker metric of Equation 6.22.

6.8 A spatially flat universe contains a single component with equation-of-state parameter w. In this universe, standard candles of luminosity L are distributed homogeneously in space. The number density of the standard candles is n_0 at $t = t_0$, and the standard candles are neither created nor destroyed. Show that the observed flux from a single standard candle at redshift z is

$$f(z) = \frac{L(1+3w)^2}{16\pi(c/H_0)^2} \frac{1}{(1+z)^2} \left[1 - (1+z)^{-(1+3w)/2}\right]^{-2}, \tag{6.52}$$

when $w \neq -1/3$. What is the corresponding relation when $w = -1/3$? Show that the observed intensity (that is, the power per unit area per steradian of sky) from standard candles with redshifts in the range $z \to z+dz$ is

$$dJ(z) = \frac{n_0 L(c/H_0)}{4\pi}(1+z)^{-(7+3w)/2} dz. \tag{6.53}$$

What will be the total intensity J of all standard candles integrated over all redshifts? Explain why the night sky is of finite brightness even in universes with $w \leq -1/3$, which have an infinite horizon distance.

6.9 In the Benchmark Model, at what scale factor a did $\ddot{a} = 0$? [This represents the moment when expansion switched from slowing down ($\ddot{a} < 0$) to speeding up ($\ddot{a} > 0$).] Is this scale factor larger or smaller than the scale factor $a_{m\Lambda}$ at which the energy density of matter equaled the energy density of the cosmological constant?

7

Dark Matter

Cosmologists, over the years, have dedicated much time and effort to determining the matter density of the universe. There are many reasons for this obsession. First, the density parameter in matter, $\Omega_{m,0}$, is important in determining the spatial curvature and expansion rate of the universe. Even if the cosmological constant is nonzero, the matter content of the universe is not negligible today, and was the dominant component in the fairly recent past. Another reason for wanting to know the matter density of the universe is to find out what the universe is made of. What fraction of the density is made of stars, and other familiar types of baryonic matter? What fraction of the density is made of dark matter? What constitutes the dark matter – cold stellar remnants, black holes, exotic elementary particles, or some other substance too dim for us to see? These questions, and others, have driven astronomers to take a census of the universe, to find out what types of matter it contains, and in what quantities.

We have already seen in the previous chapter one method of putting limits on $\Omega_{m,0}$. The apparent magnitude (or flux) of type Ia supernovae as a function of redshift is consistent with a flat universe having $\Omega_{m,0} \approx 0.3$ and $\Omega_{\Lambda,0} \approx 0.7$. However, neither $\Omega_{m,0}$ nor $\Omega_{\Lambda,0}$ is individually well-constrained by the supernova observations. The supernova data are consistent with $\Omega_{m,0} = 0$ if $\Omega_{\Lambda,0} \approx 0.3$; they are also consistent with $\Omega_{m,0} = 0.45$ if $\Omega_{\Lambda,0} \approx 0.9$. In order to determine $\Omega_{m,0}$ more accurately, we will have to adopt alternate methods of estimating the matter content of the universe.

7.1 Visible Matter

Some types of matter, such as stars, help astronomers to detect them by broadcasting photons in all directions. Stars emit light primarily in the infrared, visible, and ultraviolet range of the electromagnetic spectrum. Suppose, for instance, you install a V-band filter on your telescope. Such a filter allows photons in the wavelength range 500 nm $< \lambda <$ 590 nm to pass through. The "V" in V-band stands for "visual"; although your eyes can detect the broader wavelength range 400 nm $< \lambda <$ 700 nm, a V-band filter lets through the green and yellow wavelengths of light to which your retina is most sensitive. About 12 percent of the Sun's luminosity can pass through a V-band filter; thus, the Sun's luminosity in the V band is $L_{\odot,V} \approx 0.12 L_\odot \approx 4.6 \times 10^{25}$ watts.[1]

Surveys of galaxies reveal that in the local universe (out to $d \sim 0.1c/H_0$), the luminosity density in the V band is

$$\Psi_V = 1.1 \times 10^8 L_{\odot,V}\,\mathrm{Mpc}^{-3}. \tag{7.1}$$

To convert a luminosity density into a mass density $\rho_\star$ of stars, we need to know the *mass-to-light ratio* of the stars. If all stars were identical to the Sun, we could simply say that there is one solar mass of stars for each solar luminosity of output power, or $\langle M/L_V \rangle = 1\,\mathrm{M}_\odot/\,\mathrm{L}_{\odot,V}$; this corresponds to about 43 metric tons for every watt of yellow-green light. However, stars are not uniform in their properties.

Consider, for instance, the stars that astronomers refer to as "main sequence" stars; these are stars that are powered, like the Sun, by hydrogen fusion in their cores. The surface temperature and luminosity of a main sequence star are determined by its mass, with the most massive stars being the hottest and brightest. Astronomers find it useful to encode the surface temperature of a star as a letter, called the *spectral type* of the star. For historical reasons, these spectral types are not in alphabetical order: from hottest to coolest, they are O, B, A, F, G, K, and M. (Although the sequence of spectral types looks like an explosion in an alphabet soup factory, it does provide us with a useful shorthand: hot, luminous, massive main sequence stars can be called "O stars" for short, while cool, dim, low-mass main sequence stars are "M stars.") An O star with mass $M = 60\,\mathrm{M}_\odot$ has a V-band luminosity $L_V \approx 20\,000\,\mathrm{L}_{\odot,V}$, and thus a mass-to-light ratio $M/L_V \approx 0.003\,\mathrm{M}_\odot/\,\mathrm{L}_{\odot,V}$. By contrast, an M star with mass $M = 0.1\,\mathrm{M}_\odot$ has $L_V \approx 5 \times 10^{-5}\,\mathrm{L}_{\odot,V}$, and thus a mass-to-light ratio $M/L_V \approx 2000\,\mathrm{M}_\odot/\,\mathrm{L}_{\odot,V}$.

The mass-to-light ratio of the stars in a galaxy will therefore depend on the mix of stars that it contains. The physical processes that form stars are found empirically to favor low-mass stars over high-mass stars. In a star-forming region, the *initial mass function* $\chi(M)$ is defined so that $\chi(M)dM$ is the number of stars created with masses in the range $M \to M + dM$. At masses $M > 1\,\mathrm{M}_\odot$, the initial mass function is well fitted by a power law,

$$\chi(M) \propto M^{-\beta} \qquad [M > 1\,\mathrm{M}_\odot]. \tag{7.2}$$

[1] Although old-fashioned incandescent light bulbs are castigated for their inefficiency at producing visible light, the Sun isn't hyper-efficient at producing visible light either. (Or rather, to get the causality right, our eyes haven't evolved to be hyper-efficient at detecting sunlight.) About 10% of the Sun's luminosity is in the ultraviolet range and 50% is in the infrared, leaving only 40% in the wavelength range $\lambda = 400 \to 700$ nm.

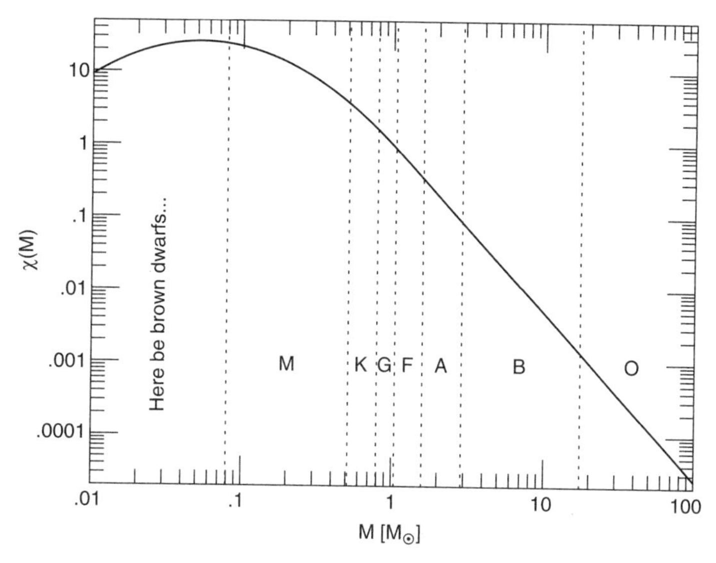

Figure 7.1 An initial mass function for stars and brown dwarfs. Mass ranges corresponding to the standard stellar spectral types O through M are indicated, as well as the low-mass realm of brown dwarfs. The values $\beta = 2.3$, $M_c = 0.2\,\mathrm{M}_\odot$ and $\sigma = 0.5$ are assumed in the Chabrier mass function of Equations 7.2 and 7.3.

The power-law index β varies from location to location, but a value $\beta = 2.3$ is typical. At lower masses, the shape of the initial mass function is less well determined, but a log-normal distribution is found to give a reasonable fit:

$$\chi(M) \propto \frac{1}{M} \exp\left(-\frac{(\log M - \log M_c)^2}{2\sigma^2}\right) \qquad [M < 1\,\mathrm{M}_\odot]. \tag{7.3}$$

The characteristic mass M_c and the width σ of the distribution vary from location to location. However, typical values, when masses are measured in units of the solar mass, are $M_c \approx 0.2$ and $\sigma \approx 0.5$.

The initial mass function found by combining Equations 7.2 and 7.3 is plotted in Figure 7.1.[2] Gaseous spheres less massive than $M = 0.08\,\mathrm{M}_\odot$ are actually *brown dwarfs* rather than stars. The difference between a brown dwarf and a star is that a brown dwarf is too low in mass for hydrogen fusion to be ignited at its center. Since brown dwarfs are not powered by nuclear fusion, they tend to be even cooler and dimmer than M stars. The initial mass function for stars and brown dwarfs is highest in the mass range $0.02\,\mathrm{M}_\odot \rightarrow 0.2\,\mathrm{M}_\odot$; O stars, with $M > 18\,\mathrm{M}_\odot$, are far out on the power-law tail of the initial mass function. At the time of formation, there will be about 250 low-mass M stars for every O star.

[2] An initial mass function that takes the form of a log-normal distribution with a power-law tail to high masses is called a Chabrier function, after the astronomer Gilles Chabrier.

Although the total mass of 250 M stars is comparable to that of a single O star, their total V-band luminosity is negligible compared to the O star's luminosity. In galaxies actively forming stars today, the mass-to-light ratio of the stellar population is found to be as small as $M/L_V \approx 0.3\,\mathrm{M}_\odot/\,\mathrm{L}_{\odot,V}$.

Although O stars are extremely luminous, they are also short-lived. An O star with a mass $M = 60\,\mathrm{M}_\odot$ will run out of fuel for fusion in a time $t \approx 3\,\mathrm{Myr}$; it will then explode as a type II supernova. Thus, a galaxy that is quiescent (that is, one that has long since stopped forming new stars) will lack O stars. The mass-to-light ratio of quiescent galaxies can rise to as large as $M/L_V \approx 8\,\mathrm{M}_\odot/\,\mathrm{L}_{\odot,V}$. In the local universe, there is a mix of star-forming and quiescent galaxies, so we can't go too badly wrong if we take an averaged mass-to-light ratio of $\langle M/L_V\rangle \approx 4\,\mathrm{M}_\odot/\,\mathrm{L}_{\odot,V}$. With this value, we find that the mass density of stars in the universe today is

$$\rho_{\star,0} = \langle M/L_V\rangle \Psi_V \approx 4\times 10^8\,\mathrm{M}_\odot\,\mathrm{Mpc}^{-3}. \tag{7.4}$$

Since the current critical density of the universe, expressed as a mass density, is $\rho_{c,0} = 1.28\times 10^{11}\,\mathrm{M}_\odot\,\mathrm{Mpc}^{-3}$, the current density parameter of stars is

$$\Omega_{\star,0} = \frac{\rho_{\star,0}}{\rho_{c,0}} \approx \frac{4\times 10^8\,\mathrm{M}_\odot\,\mathrm{Mpc}^{-3}}{1.28\times 10^{11}\,\mathrm{M}_\odot\,\mathrm{Mpc}^{-3}} \approx 0.003. \tag{7.5}$$

By this accounting, stars make up just 0.3% of the density needed to flatten the universe. The density parameter in stars is boosted slightly if you broaden the category of stars to include stellar remnants such as white dwarfs, neutron stars, and black holes, as well as substellar objects such as brown dwarfs. However, even when you add ex-stars and not-quite-stars to the total, you still find a density parameter $\Omega_{\star,0} < 0.005$.

Galaxies also contain baryonic matter that is not in the form of stars, stellar remnants, or brown dwarfs. In our galaxy and in M31, for instance, the mass of interstellar gas is about 20 percent of the mass in stars. In irregular galaxies such as the Magellanic Clouds, the ratio of gas to stars is even higher. In addition, there is a significant amount of gas between galaxies. Consider a rich cluster of galaxies such as the Coma cluster, located about 100 Mpc from our galaxy, in the direction of the constellation Coma Berenices. At visible wavelengths, as shown in Figure 7.2, most of the light comes from the stars within the cluster's galaxies. The two brightest galaxies in the Coma cluster, NGC 4889 (on the left in Figure 7.2) and NGC 4874 (on the right), each have a luminosity $L_V \approx 2.5\times 10^{11} L_{\odot,V}$.[3] The Coma cluster contains thousands of galaxies, most of them far less luminous than NGC 4889 and NGC 4874; their summed luminosity in the V band comes to $L_{\mathrm{Coma},V} \approx 5\times 10^{12}\,\mathrm{L}_{\odot,V}$. If the mass-to-light ratio of the stars in the Coma

[3] The bright star (with diffraction spikes) just above NGC 4874 in Figure 7.2 is HD 112887, a main sequence F star at a distance $d \approx 77$ pc, less than a millionth the distance to the Coma cluster. Almost every other light source in Figure 7.2 is a galaxy within the Coma cluster.

Figure 7.2 The Coma cluster as seen in visible light. The region shown is 36 arcminutes by 24 arcminutes, equivalent to 1.1 Mpc by 0.7 Mpc at the distance of the Coma cluster. [Sloan Digital Sky Survey]

Figure 7.3 The Coma cluster as seen in X-ray light. The location, orientation, and scale are the same as in the visible light image of Figure 7.2. [NASA SkyView: data from *ROSAT* orbiting X-ray observatory]

cluster is $\langle M/L_V \rangle \approx 4\,\mathrm{M}_\odot/\,\mathrm{L}_{\odot,V}$, then the total mass of stars in the Coma cluster is $M_{\mathrm{Coma},\star} \approx 2 \times 10^{13}\,\mathrm{M}_\odot$. Although 20 trillion solar masses represents a lot of stars, the stellar mass in the Coma cluster is small compared to the mass of the hot, intracluster gas between the galaxies in the cluster. X-ray images, such as the one shown in Figure 7.3, reveal that hot, low-density gas, with a typical temperature of $T \approx 10^8\,\mathrm{K}$, fills the space between clusters, emitting X-rays with a typical energy of $E \sim kT_{\mathrm{gas}} \sim 9\,\mathrm{keV}$. The total amount of X-ray emitting gas in the Coma cluster is estimated to be $M_{\text{Coma,gas}} \approx 2 \times 10^{14}\,\text{M}_\odot$, roughly ten times the mass in stars.

Not all the baryonic matter in the universe is easy to detect. About 85% of the baryons in the universe are in the extremely tenuous gas of intergalactic space, outside galaxies and clusters of galaxies. Much of this intergalactic gas is too low in density to be readily detected with current technology. The best limits on the baryon density of the universe actually come from observations of the cosmic microwave background and from the predictions of primordial nucleosynthesis in the early universe. The cosmic microwave background has temperature fluctuations whose properties depend on the baryon-to-photon ratio when the universe was a quarter of a million years old. In addition, the efficiency with which nucleosynthesis takes place in the early universe, converting hydrogen into deuterium, helium, lithium, and other elements, depends on the baryon-to-photon ratio when the universe was a few minutes old. Both these sources of information about the early universe indicate that the density parameter of baryonic matter today must be

$$\Omega_{\text{bary},0} = 0.048 \pm 0.003, \tag{7.6}$$

ten to twenty times the density parameter for stars. When you stare up at the night sky and marvel at the glory of the stars, you are actually marveling at a minority of the baryonic matter in the universe.

7.2 Dark Matter in Galaxies

The situation, in fact, is even more extreme than stated in the previous section. Not only is most of the baryonic matter undetectable by our eyes, but most of the matter is not even baryonic. The majority of the matter in the universe is *nonbaryonic dark matter*, which doesn't absorb, emit, or scatter light of any wavelength. One way of detecting dark matter is to look for its gravitational influence on visible matter. A classic method of detecting dark matter involves looking at the orbital speeds of stars in spiral galaxies such as our own galaxy and M31. Spiral galaxies contain flattened disks of stars; within the disk, stars are on nearly circular orbits around the center of the galaxy. The Sun, for instance, is on such an orbit – it is $R = 8.2\,\text{kpc}$ from the galactic center, and has an orbital speed of $v = 235\,\text{km}\,\text{s}^{-1}$.

Suppose that a star is on a circular orbit around the center of its galaxy. If the radius of the orbit is R and the orbital speed is v, then the star experiences an acceleration

$$a = \frac{v^2}{R}, \tag{7.7}$$

directed toward the center of the galaxy. If the acceleration is provided by the gravitational attraction of the galaxy, then

$$a = \frac{GM(R)}{R^2}, \tag{7.8}$$

where $M(R)$ is the mass contained within a sphere of radius R centered on the galactic center.[4] The relation between v and M is found by setting Equation 7.7 equal to Equation 7.8:

$$\frac{v^2}{R} = \frac{GM(R)}{R^2}, \tag{7.9}$$

or

$$v = \sqrt{\frac{GM(R)}{R}}. \tag{7.10}$$

The surface brightness I of the disk of a spiral galaxy typically falls off exponentially with distance from the center:

$$I(R) = I(0) \exp\left(-\frac{R}{R_s}\right), \tag{7.11}$$

with the scale length R_s typically being a few kiloparsecs. For our galaxy, the scale length measured in the V band is $R_s \approx 4\,\text{kpc}$; for M31, a somewhat larger disk galaxy, $R_s \approx 6\,\text{kpc}$. Once you are a few scale lengths from the center of the spiral galaxy, the mass of stars inside R becomes essentially constant. Thus, if stars contributed all, or most, of the mass in a galaxy, the velocity would fall as $v \propto 1/\sqrt{R}$ at large radii. This relation between orbital speed and orbital radius, $v \propto 1/\sqrt{R}$, is referred to as "Keplerian rotation," since it's what Kepler found for orbits in the solar system, where 99.8 percent of the mass is contained within the Sun.

The first astronomer to detect the rotation of M31 was Vesto Slipher, in 1914, two years after he measured the blueshift resulting from its motion toward our own galaxy. However, given the difficulty of measuring the spectra at low surface brightness, the orbital speed v at $R > 3R_s = 18\,\text{kpc}$ was not accurately measured until more than half a century later. In 1970, Vera Rubin and Kent Ford looked at emission lines from regions of hot ionized gas in M31, and were able to find the orbital speed $v(R)$ out to a radius $R = 24\,\text{kpc} = 4R_s$. Their results gave no sign of a Keplerian decrease in the orbital speed. At $R > 4R_s$, a small amount of atomic hydrogen is still in the disk of M31, which can be detected by means of its emission line at $\lambda = 21\,\text{cm}$. From observations of the Doppler shift of this emission line, the orbital speed is found to be nearly constant at $v(R) \approx 230\,\text{km}\,\text{s}^{-1}$ out to $R = 35\,\text{kpc} \approx 6R_s$. Since the orbital speed of the stars and

[4] Equation 7.8 assumes that the mass distribution of the galaxy is spherically symmetric. This is not, strictly speaking, true (the stars in the disk obviously have a flattened distribution), but the flattening of the galaxy provides only a small correction to the equation for the gravitational acceleration.

gas at large radii ($R > 3R_s$) is greater than it would be if stars and gas were the only matter present, we deduce the presence of a *dark halo* within which the visible stellar disk is embedded. The mass of the dark halo provides the necessary gravitational "anchor" to keep the high-speed stars and gas from being flung out into intergalactic space.

M31 is not a freak; most, if not all, spiral galaxies have comparable dark halos. For instance, our own galaxy has an orbital speed that actually seems to be roughly constant at $R > 15\,\text{kpc}$, instead of decreasing in a Keplerian fashion. If we approximate the orbital speed v as being constant with radius, the mass of a spiral galaxy, including both the luminous disk and the dark halo, can be found from Equation 7.10:

$$M(R) = \frac{v^2 R}{G} = 1.05 \times 10^{11}\,\text{M}_\odot \left(\frac{v}{235\,\text{km}\,\text{s}^{-1}}\right)^2 \left(\frac{R}{8.2\,\text{kpc}}\right). \tag{7.12}$$

The values of v and R in the above equation are scaled to the Sun's location in our galaxy. Since our galaxy's luminosity in the V band is estimated to be $L_{\text{gal},V} = 2.0 \times 10^{10}\,\text{L}_{\odot,V}$, this means that the mass-to-light ratio of our galaxy, taken as a whole, is

$$\langle M/L_V \rangle_{\text{gal}} \approx 64\,\text{M}_\odot/\,\text{L}_{\odot,V} \left(\frac{R_{\text{halo}}}{100\,\text{kpc}}\right), \tag{7.13}$$

using $v = 235\,\text{km}\,\text{s}^{-1}$ in Equation 7.12. The quantity R_{halo} is the radius of the dark halo surrounding the luminous disk of our galaxy. The exact value of R_{halo} is poorly known. A rough estimate of the halo size can be made by looking at the velocities of the globular clusters and satellite galaxies (such as the Magellanic Clouds) that orbit our galaxy. For these hangers-on to remain gravitationally bound to our galaxy, the halo must extend as far as $R_{\text{halo}} \approx 75\,\text{kpc}$, implying a total mass for our galaxy of $M_{\text{gal}} \approx 9.6 \times 10^{11}\,\text{M}_\odot$, and a total mass-to-light ratio $\langle M/L_V \rangle_{\text{gal}} \approx 48\,\text{M}_\odot/\,\text{L}_{\odot,V}$. This mass-to-light ratio is an order of magnitude greater than that of the stars in our galaxy, implying a dark halo much more massive than the stellar disk. Some astronomers have speculated that the dark halo is actually four times larger in radius, with $R_{\text{halo}} \approx 300\,\text{kpc}$; this would mean that our halo stretches nearly halfway to M31. With $R_{\text{halo}} \approx 300\,\text{kpc}$, the mass of our galaxy would be $M_{\text{gal}} \approx 3.8 \times 10^{12}\,\text{M}_\odot$, and the total mass-to-light ratio would be $\langle M/L_V \rangle_{\text{gal}} \approx 190\,\text{M}_\odot/\,\text{L}_{\odot,V}$.

7.3 Dark Matter in Clusters

The first astronomer to make a compelling case for the existence of large quantities of dark matter was Fritz Zwicky, in the 1930s. In studying the Coma cluster of galaxies (shown in Figure 7.2), he noted that the dispersion in the radial velocity of the cluster's galaxies was very large – around $1000\,\mathrm{km\,s^{-1}}$. The stars and gas visible within the galaxies simply did not provide enough gravitational attraction to hold the cluster together. In order to keep the galaxies in the Coma cluster from flying off into the surrounding voids, Zwicky concluded, the cluster must contain a large amount of "dunkle Materie," or (translated into English) "dark matter."[5]

To follow Zwicky's reasoning at a more mathematical level, let us suppose that a cluster of galaxies consists of N galaxies, each of which can be approximated as a point mass, with a mass m_i ($i = 1, 2, \ldots, N$), a position $\vec{x}_i$, and a velocity $\dot{\vec{x}}_i$. Clusters of galaxies are gravitationally bound objects, not expanding with the Hubble flow. The motion of individual galaxies within the cluster is well described by Newtonian physics; the acceleration of the ith galaxy is thus given by the formula

$$\ddot{\vec{x}}_i = G \sum_{j \neq i} m_j \frac{\vec{x}_j - \vec{x}_i}{|\vec{x}_j - \vec{x}_i|^3}. \tag{7.14}$$

Note that Equation 7.14 assumes that the cluster is an isolated system, with the gravitational acceleration due to matter outside the cluster being negligibly small.

The gravitational *potential energy* of the system of N galaxies is

$$W = -\frac{G}{2} \sum_{\substack{i,j \\ j \neq i}} \frac{m_i m_j}{|\vec{x}_j - \vec{x}_i|}. \tag{7.15}$$

This is the energy that would be required to pull the N galaxies away from each other so that they would all be at infinite distance from each other. (The factor of $1/2$ in front of the double summation ensures that each pair of galaxies is only counted once in computing the potential energy.) The potential energy of the cluster can also be written in the form

$$W = -\alpha \frac{GM^2}{r_h}, \tag{7.16}$$

where $M = \sum m_i$ is the total mass of all the galaxies in the cluster, α is a numerical factor of order unity that depends on the density profile of the cluster, and r_h is the *half-mass* radius of the cluster – that is, the radius of a sphere centered on the cluster's center of mass and containing a mass $M/2$. For observed clusters of galaxies, it is found that $\alpha \approx 0.45$ gives a good fit to the potential energy.

The *kinetic energy* associated with the relative motion of the galaxies in the cluster is

$$K = \frac{1}{2} \sum_i m_i |\dot{\vec{x}}_i|^2. \tag{7.17}$$

[5] Although Zwicky's work popularized the phrase "dark matter," he was not the first to use it in an astronomical context. For instance, in 1908, Henri Poincaré discussed the possible existence within our galaxy of "matière obscure" (rendered as "dark matter" in the standard English translation of Poincaré's works).

The kinetic energy K can also be written in the form

$$K = \frac{1}{2}M\langle v^2\rangle, \tag{7.18}$$

where

$$\langle v^2\rangle \equiv \frac{1}{M}\sum_i m_i|\dot{\vec{x}}_i|^2 \tag{7.19}$$

is the mean square velocity (weighted by galaxy mass) of all the galaxies in the cluster.

It is also useful to define the *moment of inertia* of the cluster as

$$I \equiv \sum_i m_i|\vec{x}_i|^2. \tag{7.20}$$

The moment of inertia I can be linked to the kinetic energy and the potential energy if we start by taking the second time derivative of I:

$$\ddot{I} = 2\sum_i m_i(\vec{x}_i \cdot \ddot{\vec{x}}_i + \dot{\vec{x}}_i \cdot \dot{\vec{x}}_i). \tag{7.21}$$

Using Equation 7.17, we can rewrite this as

$$\ddot{I} = 2\sum_i m_i(\vec{x}_i \cdot \ddot{\vec{x}}_i) + 4K. \tag{7.22}$$

To introduce the potential energy W into the above relation, we can use Equation 7.14 to write

$$\sum_i m_i(\vec{x}_i \cdot \ddot{\vec{x}}_i) = G\sum_{\substack{i,j\\ j\neq i}} m_i m_j \frac{\vec{x}_i \cdot (\vec{x}_j - \vec{x}_i)}{|\vec{x}_j - \vec{x}_i|^3}. \tag{7.23}$$

However, we could equally well switch around the i and j subscripts to find the equally valid equation

$$\sum_j m_j(\vec{x}_j \cdot \ddot{\vec{x}}_j) = G\sum_{\substack{j,i\\ i\neq j}} m_j m_i \frac{\vec{x}_j \cdot (\vec{x}_i - \vec{x}_j)}{|\vec{x}_i - \vec{x}_j|^3}. \tag{7.24}$$

Since

$$\sum_i m_i(\vec{x}_i \cdot \ddot{\vec{x}}_i) = \sum_j m_j(\vec{x}_j \cdot \ddot{\vec{x}}_j) \tag{7.25}$$

(it doesn't matter whether we call the variable over which we're summing i or j or k or "Fred"), we can combine Equations 7.23 and 7.24 to find

$$\begin{aligned}\sum_i m_i(\vec{x}_i \cdot \ddot{\vec{x}}_i) &= \frac{1}{2}\left[\sum_i m_i(\vec{x}_i \cdot \ddot{\vec{x}}_i) + \sum_j m_j(\vec{x}_j \cdot \ddot{\vec{x}}_j)\right] \\ &= -\frac{G}{2}\sum_{\substack{i,j\\ j\neq i}} \frac{m_i m_j}{|\vec{x}_j - \vec{x}_i|} = W. \end{aligned} \tag{7.26}$$

Thus, the first term on the right-hand side of Equation 7.22 is simply $2W$, and we may now write down the simple relation

$$\ddot{I} = 2W + 4K. \tag{7.27}$$

This relation is known as the *virial theorem*. It was first derived in the nineteenth century in the context of the kinetic theory of gases, but as we have seen, it applies perfectly well to a self-gravitating system of point masses.

The virial theorem is particularly useful when it is applied to a system in steady state, with a constant moment of inertia. (This implies, among other things, that the system is neither expanding nor contracting, and that we are using a coordinate system in which the center of mass of the cluster is at rest.) If $I =$ constant, then the *steady-state virial theorem* is

$$0 = W + 2K, \tag{7.28}$$

or

$$K = -\frac{W}{2}. \tag{7.29}$$

Using Equations 7.16 and 7.18 in Equation 7.29, we find

$$\frac{1}{2}M\langle v^2\rangle = \frac{\alpha}{2}\frac{GM^2}{r_h}. \tag{7.30}$$

This means we can use the virial theorem to estimate the mass of a cluster of galaxies, or any other self-gravitating steady-state system:

$$M = \frac{\langle v^2\rangle r_h}{\alpha G}. \tag{7.31}$$

Note the similarity between Equation 7.12, used to estimate the mass of a rotating spiral galaxy, and Equation 7.31, used to estimate the mass of a cluster of galaxies. In either case, we estimate the mass of a self-gravitating system by multiplying the square of a characteristic velocity by a characteristic radius, then dividing by the gravitational constant G.

Applying the virial theorem to a real cluster of galaxies, such as the Coma cluster, is complicated by the fact that we have only partial information about the cluster, and thus do not know $\langle v^2\rangle$ and r_h exactly. For instance, we can find the line-of-sight velocity of each galaxy from its redshift, but the velocity perpendicular to the line of sight is unknown. From measurements of the redshifts of hundreds of galaxies in the Coma cluster, the mean redshift of the cluster is found to be

$$\langle z\rangle = 0.0232, \tag{7.32}$$

which can be translated into a distance

$$d_{\text{Coma}} = (c/H_0)\langle z\rangle = 102\,\text{Mpc}. \tag{7.33}$$

The velocity dispersion of the cluster along the line of sight is found to be

$$\sigma_r = \langle (v_r - \langle v_r \rangle)^2 \rangle^{1/2} = 880\,\mathrm{km\,s^{-1}}. \tag{7.34}$$

If we assume that the velocity dispersion is isotropic, then the three-dimensional mean square velocity $\langle v^2 \rangle$ will be equal to three times the one-dimensional mean square velocity σ_r^2, yielding

$$\langle v^2 \rangle = 3(880\,\mathrm{km\,s^{-1}})^2 = 2.32 \times 10^{12}\,\mathrm{m^2\,s^{-2}}. \tag{7.35}$$

Estimating the half-mass radius r_h of the Coma cluster is even more peril-ridden than estimating the mean square velocity $\langle v^2 \rangle$. After all, we don't know the distribution of dark matter in the cluster beforehand; in fact, the total amount of dark matter is what we're trying to find out. However, if we assume that the mass-to-light ratio is constant with radius, then the sphere containing half the mass of the cluster will be the same as the sphere containing half the luminosity of the cluster. If we further assume that the cluster is intrinsically spherical, then the observed distribution of galaxies within the Coma cluster indicates a half-mass radius

$$r_h \approx 1.5\,\mathrm{Mpc} \approx 4.6 \times 10^{22}\,\mathrm{m}. \tag{7.36}$$

After all these assumptions and approximations, we may estimate the mass of the Coma cluster to be

$$M_{\mathrm{Coma}} = \frac{\langle v^2 \rangle r_h}{\alpha G} \approx \frac{(2.32 \times 10^{12}\,\mathrm{m^2\,s^{-2}})(4.6 \times 10^{22}\,\mathrm{m})}{(0.45)(6.67 \times 10^{-11}\,\mathrm{m^3\,s^{-2}\,kg^{-1}})}$$
$$\approx 4 \times 10^{45}\,\mathrm{kg} \approx 2 \times 10^{15}\,\mathrm{M_\odot}. \tag{7.37}$$

Thus, about one percent of the mass of the Coma cluster consists of stars ($M_{\mathrm{Coma},\star} \approx 2 \times 10^{13}\,\mathrm{M_\odot}$), and about ten percent consists of hot intracluster gas ($M_{\mathrm{Coma,gas}} \approx 2 \times 10^{14}\,\mathrm{M_\odot}$). Combined with the luminosity of the Coma cluster, $L_{\mathrm{Coma},V} \approx 5 \times 10^{12}\,\mathrm{L_{\odot,\mathit{V}}}$, the total mass of the Coma cluster implies a mass-to-light ratio

$$\langle M/L_V \rangle_{\mathrm{Coma}} \sim 400\,\mathrm{M_\odot}/\mathrm{L_{\odot,\mathit{V}}}, \tag{7.38}$$

greater than the mass-to-light ratio of our galaxy.

The presence of a vast reservoir of dark matter in the Coma cluster is confirmed by the fact that the hot, X-ray emitting intracluster gas, shown in Figure 7.3, is still in place; if there were no dark matter to anchor the gas gravitationally, the hot gas would have expanded beyond the cluster on time scales much shorter than the Hubble time. The temperature and density of the hot gas in the Coma cluster can be used to make yet another estimate of the cluster's mass. If the hot intracluster gas is supported by its own pressure against gravitational infall, it must obey the equation of hydrostatic equilibrium:

$$\frac{dP_{\mathrm{gas}}}{dr} = -\frac{GM(r)\rho_{\mathrm{gas}}(r)}{r^2}, \tag{7.39}$$

where P_{gas} is the pressure of the gas, ρ_{gas} is the density of the gas, and M is the *total* mass inside a sphere of radius r, including gas, stars, dark matter, lost socks, and anything else.

The pressure of the gas is given by the perfect gas law,

$$P_{\text{gas}} = \frac{\rho_{\text{gas}} k T_{\text{gas}}}{\mu}, \tag{7.40}$$

where T_{gas} is the temperature of the gas, and μ is the mean mass per gas particle. The mass of the cluster, as a function of radius, is found by combining Equations 7.39 and 7.40:

$$M(r) = \frac{kT_{\text{gas}}(r)r}{G\mu}\left[-\frac{d\ln\rho_{\text{gas}}}{d\ln r} - \frac{d\ln T_{\text{gas}}}{d\ln r}\right]. \tag{7.41}$$

The above equation assumes that μ is constant with radius, as we'd expect if the chemical composition and ionization state of the gas are uniform throughout the cluster.

The X-rays emitted from the hot intracluster gas are a combination of bremsstrahlung emission (caused by the acceleration of free electrons by protons and helium nuclei) and line emission from highly ionized iron and other heavy elements. Starting from an X-ray spectrum, it is possible to fit models to the emission and thus compute the temperature $T_{\text{gas}}(r)$, density $\rho_{\text{gas}}(r)$, and chemical composition of the gas. Using this technique, the mass of the Coma cluster is estimated to be $M \approx 1.3 \times 10^{15}\,\text{M}_\odot$ within $r \approx 4$ Mpc of the cluster center. Given the uncertainties, this is consistent with the mass estimate of the virial theorem.

Other clusters of galaxies besides the Coma cluster have had their masses estimated, using the virial theorem applied to their galaxies or the equation of hydrostatic equilibrium applied to their gas. Typical mass-to-light ratios for rich clusters are similar to those of the Coma cluster. If the masses of all the clusters of galaxies are added together, it is found that their density parameter is

$$\Omega_{\text{clus},0} \approx 0.2. \tag{7.42}$$

This provides a *lower limit* to the matter density of the universe, since any smoothly distributed matter in the intercluster voids will not be included in this number.

7.4 Gravitational Lensing

So far, I have outlined the classical methods for detecting dark matter via its gravitational effects on luminous matter.[6] We can detect dark matter around spiral

[6] The roots of these methods can be traced back as far as the year 1846, when Leverrier and Adams deduced the existence of the dim planet Neptune by its effect on the orbit of Uranus.

galaxies because it affects the motions of stars and interstellar gas. We can detect dark matter in clusters of galaxies because it affects the motions of galaxies and intracluster gas. However, as Einstein realized, dark matter will affect not only the trajectory of matter, but also the trajectory of photons. Thus, dark matter can bend and focus light, acting as a *gravitational lens*. The effects of dark matter on photons have been used to search for dark matter within the halo of our own galaxy, as well as in distant clusters of galaxies.

To see how gravitational lensing can be used to detect dark matter, start by considering the dark halo surrounding our galaxy. If there were a population of cold white dwarfs, black holes, brown dwarfs, or similar dim compact objects in the halo, they would be very difficult to detect from the light that they emit. Thus, it was suggested that part of the dark matter in the halo could consist of MACHOs, a slightly strained acronym for MAssive Compact Halo Objects. If a photon passes such a compact massive object at an impact parameter b, as shown in Figure 7.4, the local curvature of spacetime will cause the photon to be deflected by an angle

$$\alpha = \frac{4GM}{c^2 b}, \tag{7.43}$$

where M is the mass of the compact object. For instance, light from a distant star that just grazes the Sun's surface should be deflected through an angle

$$\alpha = \frac{4G\,\mathrm{M}_\odot}{c^2\,\mathrm{R}_\odot} = 1.7\,\mathrm{arcsec}. \tag{7.44}$$

In 1919, after Einstein predicted a deflection of this magnitude, an eclipse expedition photographed stars in the vicinity of the Sun. Comparison of the eclipse photographs with photographs of the same star field taken six months earlier revealed that the apparent positions of the stars were deflected by the amount that Einstein had predicted. This result brought fame to Einstein and experimental support to the theory of general relativity.

Since a massive object can deflect light, it can act as a lens. Suppose a MACHO in the halo of our galaxy passes directly between an observer in our galaxy and a star in the Large Magellanic Cloud. Figure 7.5 shows such a situation, with a MACHO halfway between the observer and the star. As the MACHO deflects the light from the distant star, it produces an image of the star that is both distorted and amplified. If the MACHO is *exactly* along the line of sight between the observer and the lensed star, the image produced is a perfect ring, with angular radius

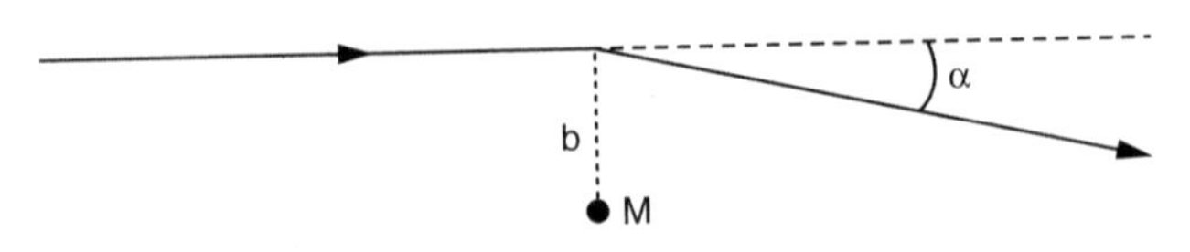

Figure 7.4 Deflection of light by a massive compact object.

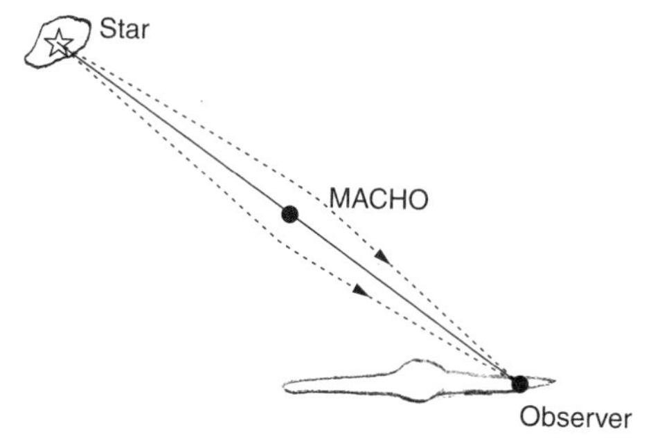

Figure 7.5 Light from a star in the Large Magellanic Cloud is deflected by a MACHO on its way to an observer in the disk of our galaxy (seen edge-on in this figure).

$$\theta_E = \left(\frac{4GM}{c^2 d}\frac{1-x}{x}\right)^{1/2}, \tag{7.45}$$

where M is the mass of the lensing MACHO, d is the distance from the observer to the lensed star, and xd (where $0 < x < 1$) is the distance from the observer to the lensing MACHO. The angle θ_E is known as the *Einstein radius*. If $x \approx 0.5$ (that is, if the MACHO is roughly halfway between the observer and the lensed star), then

$$\theta_E \approx 4 \times 10^{-4}\,\text{arcsec}\left(\frac{M}{1\,\text{M}_\odot}\right)^{1/2}\left(\frac{d}{50\,\text{kpc}}\right)^{-1/2}. \tag{7.46}$$

If the MACHO does not lie perfectly along the line of sight to the star, then the image of the star will be distorted into two or more arcs instead of a single unbroken ring. Although the Einstein radius for an LMC star being lensed by a MACHO is too small to be resolved, it is possible, in some cases, to detect the amplification of the flux from the star. For the amplification to be significant, the angular distance between the MACHO and the lensed star, as seen from Earth, must be comparable to, or smaller than, the Einstein radius. Given the small size of the Einstein radius, the probability of any particular star in the LMC being lensed at any moment is tiny. It has been calculated that if the dark halo of our galaxy were entirely composed of MACHOs, then the probability of any given star in the LMC being lensed at any given time would still only be $P \sim 5 \times 10^{-7}$.

To detect lensing by MACHOs, various research groups took up the daunting task of monitoring millions of stars in the Large Magellanic Cloud to watch for changes in their flux. Since the MACHOs in our dark halo and the stars in the LMC are in constant relative motion, the typical signature of a "lensing event" is a star that becomes brighter as the angular distance between star and MACHO decreases, then becomes dimmer as the angular distance increases again. The typical time scale for a lensing event is the time it takes a MACHO to travel through an angular distance equal to θ_E as seen from Earth; for a MACHO halfway between here and the LMC, this is

$$\Delta t = \frac{d\theta_E}{2v} \approx 90\,\text{days}\left(\frac{M}{1\,\text{M}_\odot}\right)^{1/2}\left(\frac{v}{200\,\text{km}\,\text{s}^{-1}}\right)^{-1}, \tag{7.47}$$

where v is the relative transverse velocity of the MACHO and the lensed star as seen by the observer on Earth. Generally speaking, more massive MACHOs produce larger Einstein rings and thus will amplify the lensed star for a longer time.

The research groups that searched for MACHOs found a scarcity of short duration lensing events, suggesting that there is no significant population of brown dwarfs or freefloating planets (with $M < 0.08\,\text{M}_\odot$) in the dark halo of our galaxy. The total number of lensing events they found suggests that at most 8 percent of the halo mass could be in the form of MACHOs. The general conclusion is that most of the matter in the dark halo of our galaxy is due to a smooth distribution of nonbaryonic dark matter, instead of being congealed into MACHOs of roughly stellar or planetary mass.

Gravitational lensing occurs at all mass scales. Suppose, for instance, that a cluster of galaxies, with $M \sim 10^{14}\,\text{M}_\odot$, at a distance $\sim 500\,\text{Mpc}$ from our galaxy, lenses a background galaxy at $d \sim 1000\,\text{Mpc}$. The Einstein radius for this configuration will be

$$\theta_E \approx 0.5\,\text{arcmin}\left(\frac{M}{10^{14}\,\text{M}_\odot}\right)^{1/2}\left(\frac{d}{1000\,\text{Mpc}}\right)^{-1/2}. \tag{7.48}$$

The arc-shaped images into which the background galaxy is distorted by the lensing cluster can thus be resolved. For instance, Figure 7.6 shows a *Hubble*

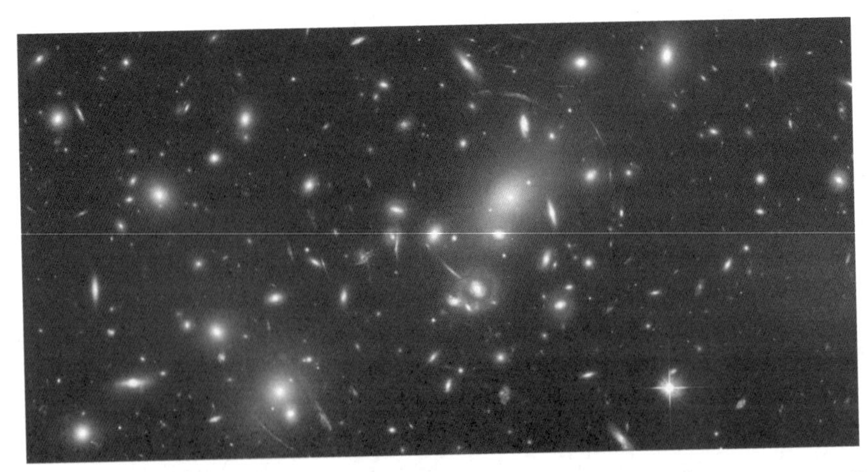

Figure 7.6 The central regions of the rich cluster Abell 2218, displaying gravitationally lensed arcs. The region shown is 3.2 arcminutes by 1.6 arcminutes, equivalent to 0.68 Mpc by 0.34 Mpc at the distance of Abell 2218. [NASA, ESA, and Johan Richard (Caltech)]

Space Telescope image of the cluster Abell 2218, which has a redshift $z = 0.176$, and hence is at a proper distance $d = 740\,\text{Mpc}$. The elongated, slightly curved arcs seen in Figure 7.6 are not oddly shaped galaxies within the cluster; instead, they are background galaxies, at redshifts $z > 0.176$, which are gravitationally lensed by the cluster mass. The mass of clusters can be estimated by the degree to which they lens background galaxies. The masses calculated in this way are in general agreement with the masses found by applying the virial theorem to the motions of galaxies in the cluster or by applying the equation of hydrostatic equilibrium to the hot intracluster gas.

7.5 What's the Matter?

We described how to detect dark matter by its gravitational effects, but have been dodging the essential question: "What is it?" As you might expect, conjecture about the nature of the nonbaryonic dark matter has run rampant (some might even say it has run amok). A component of the universe that is totally invisible is an open invitation to speculation. To give a taste of the variety of speculation, some scientists have proposed that the dark matter might be made of axions, a type of elementary particle with a rest energy of $m_{\text{ax}}c^2 \sim 10^{-5}\,\text{eV}$, equivalent to $m_{\text{ax}} \sim 2 \times 10^{-41}\,\text{kg}$. This is a rather low mass – it would take some 50 billion axions (if they indeed exist) to equal the mass of one electron. On the other hand, some scientists have conjectured that the dark matter might be made of primordial black holes, with masses up to $m_{\text{BH}} \sim 10^5\,\text{M}_\odot$, equivalent to $m_{\text{BH}} \sim 2 \times 10^{35}\,\text{kg}$.[7] This is a rather high mass – it would take some 30 billion Earths to equal the mass of one primordial black hole (if they indeed exist). It is a sign of the vast ignorance concerning nonbaryonic dark matter that two candidates for the role of dark matter differ in mass by 76 orders of magnitude.

One nonbaryonic particle that we know exists, and which has a nonzero mass, is the neutrino. As stated in Section 5.1, there should exist today a cosmic background of neutrinos. Just as the cosmic microwave background is a relic of the time when the universe was opaque to photons, the cosmic neutrino background is a relic of the time when the universe was hot and dense enough to be opaque to neutrinos. The number density of each of the three flavors of neutrinos (ν_e, ν_μ, and ν_τ) has been calculated to be $3/11$ times the number density of CMB photons, yielding a total number density of neutrinos

$$n_\nu = 3\left(\frac{3}{11}\right)n_\gamma = \left(\frac{9}{11}\right)(4.108 \times 10^8\,\text{m}^{-3}) = 3.36 \times 10^8\,\text{m}^{-3}. \tag{7.49}$$

This means that at any instant, about twenty million cosmic neutrinos are zipping through your body, "like photons through a pane of glass." In order to

[7] A *primordial* black hole is one that formed very early in the history of the universe, rather than by the collapse of a massive star later on.

provide *all* the nonbaryonic mass in the universe, the average neutrino mass would have to be

$$m_\nu c^2 = \frac{\Omega_{\mathrm{dm},0}\varepsilon_{c,0}}{n_\nu}. \qquad (7.50)$$

Given a density parameter in nonbaryonic dark matter of $\Omega_{\mathrm{dm},0} \approx 0.262$, this implies that an average neutrino mass of

$$m_\nu c^2 \approx \frac{0.262(4870\,\mathrm{MeV\,m^{-3}})}{3.36\times 10^8\,\mathrm{m^{-3}}} \approx 3.8\,\mathrm{eV} \qquad (7.51)$$

would be necessary to provide all the nonbaryonic dark matter in the universe. Studies of neutrino oscillations and of the large scale structure of the universe (see Equations 2.25 and 2.26) indicate that the average neutrino mass actually lies in the range

$$0.019\,\mathrm{eV} < m_\nu c^2 < 0.1\,\mathrm{eV}. \qquad (7.52)$$

This implies that the current density parameter in massive neutrinos lies in the range

$$0.0013 < \Omega_{\nu,0} < 0.007, \qquad (7.53)$$

and that less than 3 percent of the dark matter takes the form of neutrinos.

Given the insufficient mass density of neutrinos, particle physicists have provided several possible alternative candidates for the role of dark matter. For instance, consider the extension of the Standard Model of particle physics known as supersymmetry. Various supersymmetric models predict the existence of massive nonbaryonic particles such as photinos, gravitinos, axinos, sneutrinos, gluinos, and so forth. Like neutrinos, the hypothetical supersymmetric particles interact with other particles only through gravity and through the weak nuclear force, which makes them intrinsically difficult to detect. Particles that interact via the weak nuclear force, but which are much more massive than the upper limit on the neutrino mass, are known generically as Weakly Interacting Massive Particles, or WIMPs.[8] Since WIMPs, like neutrinos, do interact with atomic nuclei on occasion, experimenters have set up WIMP detectors to discover cosmic WIMPs. So far (to repeat a statement made in the first edition of this book), no convincing detections have been made – but the search goes on.

Exercises

7.1 Suppose it were suggested that black holes of mass $10^{-8}\,\mathrm{M}_\odot$ made up all the dark matter in the halo of our galaxy. How far away would you expect the nearest such black hole to be? How frequently would you expect such a

[8] The acronym "MACHO," encountered in the previous section, was first coined as a humorous riposte to the acronym "WIMP."

black hole to pass within 1 AU of the Sun? (An order-of-magnitude estimate is sufficient.)

Suppose it were suggested that MACHOs of mass $10^{-3}\,M_\odot$ (about the mass of Jupiter) made up all the dark matter in the halo of our galaxy. How far away would you expect the nearest MACHO to be? How frequently would such a MACHO pass within 1 AU of the Sun? (Again, an order-of-magnitude estimate will suffice.)

7.2 The Draco galaxy is a dwarf galaxy within the Local Group. Its luminosity is $L = (1.8 \pm 0.8) \times 10^5\,L_\odot$ and half its total luminosity is contained within a sphere of radius $r_h = 120 \pm 12\,\text{pc}$. The red giant stars in the Draco galaxy are bright enough to have their line-of-sight velocities measured. The measured velocity dispersion of the red giant stars in the Draco galaxy is $\sigma_r = 10.5 \pm 2.2\,\text{km s}^{-1}$. What is the mass of the Draco galaxy? What is its mass-to-light ratio? Describe the possible sources of error in your mass estimate of this galaxy.

7.3 A light ray just grazes the surface of the Earth ($M = 6.0 \times 10^{24}\,\text{kg}$, $R = 6.4 \times 10^6\,\text{m}$). Through what angle α is the light ray bent by gravitational lensing? (Ignore the refractive effects of the Earth's atmosphere.) Repeat your calculation for a white dwarf ($M = 2.0 \times 10^{30}\,\text{kg}$, $R = 1.5 \times 10^7\,\text{m}$) and for a neutron star ($M = 3.0 \times 10^{30}\,\text{kg}$, $R = 1.2 \times 10^4\,\text{m}$).

7.4 If the halo of our galaxy is spherically symmetric, what is the mass density $\rho(r)$ within the halo? If the universe contains a cosmological constant with density parameter $\Omega_{\Lambda,0} = 0.7$, would you expect it to significantly affect the dynamics of our galaxy's halo? Explain why or why not.

7.5 In the previous chapter, we noted that galaxies in rich clusters are poor standard candles because they tend to grow brighter as they merge with other galaxies. Let's estimate the galaxy merger rate in the Coma cluster to see whether it's truly significant. The Coma cluster contains $N \approx 1000$ galaxies within its half-mass radius of $r_h \approx 1.5\,\text{Mpc}$. What is the mean number density of galaxies within the half-mass radius? Suppose that the typical cross-sectional area of a galaxy is $\Sigma \approx 10^{-3}\,\text{Mpc}^2$. How far will a galaxy in the Coma cluster travel, on average, before it collides with another galaxy? The velocity dispersion of the Coma cluster is $\sigma \approx 880\,\text{km s}^{-1}$. What is the average time between collisions for a galaxy in the Coma cluster? Is this time greater than or less than the Hubble time?

7.6 Fusion reactions in the Sun's core produce 2×10^{38} neutrinos per second. If these solar neutrinos radiate isotropically away from the Sun, about how many solar neutrinos are inside your body at any given time? Is this larger or smaller than the number of neutrinos from the cosmic neutrino background that are inside your body at the same moment?

8

The Cosmic Microwave Background

If Heinrich Olbers had lived in intergalactic space and had eyes that operated at millimeter wavelengths (admittedly a very large "if"), he would not have formulated Olbers' paradox. At wavelengths of a few millimeters, thousands of times longer than human eyes can detect, most of the light in the universe comes not from the hot balls of gas we call stars, but from the cosmic microwave background (CMB). Unknown to Olbers, the night sky actually *is* uniformly bright – it's just uniformly bright at a temperature $T_0 = 2.7255\,\mathrm{K}$ rather than at a temperature $T \sim T_\odot \sim 6000\,\mathrm{K}$. The current energy density of the cosmic microwave background,

$$\varepsilon_{\gamma,0} = \alpha T_0^4 = 0.2606\,\mathrm{MeV\,m^{-3}}, \tag{8.1}$$

is only one part in 19 000 of the current critical density. However, since the energy per CMB photon is small ($hf_{\mathrm{mean}} = 6.34 \times 10^{-4}\,\mathrm{eV}$), the number density of CMB photons in the universe is large:

$$n_{\gamma,0} = 4.107 \times 10^8\,\mathrm{m^{-3}}. \tag{8.2}$$

It is particularly enlightening to compare the energy density and number density of photons to those of baryons (that is, protons and neutrons). In the Benchmark Model, the current energy density of baryons is

$$\varepsilon_{\mathrm{bary},0} = \Omega_{\mathrm{bary},0}\varepsilon_{c,0} \approx 234\,\mathrm{MeV\,m^{-3}}. \tag{8.3}$$

The energy density in baryons today is thus 900 times the energy density in CMB photons. Note, though, that the rest energy of a proton or neutron, $E_{\mathrm{bary}} \approx 939\,\mathrm{MeV}$, is more than a trillion times the mean energy of a CMB photon. The number density of baryons, therefore, is much lower than the number density of photons:

$$n_{\mathrm{bary},0} = \frac{\varepsilon_{\mathrm{bary},0}}{E_{\mathrm{bary}}} \approx \frac{234\,\mathrm{MeV\,m^{-3}}}{939\,\mathrm{MeV}} \approx 0.25\,\mathrm{m^{-3}}. \tag{8.4}$$

The ratio of baryons to photons in the universe (a number usually designated by the Greek letter η) is, from Equations 8.2 and 8.4,

$$\eta = \frac{n_{\text{bary},0}}{n_{\gamma,0}} \approx \frac{0.25\,\text{m}^{-3}}{4.107 \times 10^{8}\,\text{m}^{-3}} \approx 6.1 \times 10^{-10}. \tag{8.5}$$

Baryons are badly outnumbered by photons in the universe as a whole, by a ratio of 1.6 billion to one.

8.1 Observing the CMB

Although CMB photons are as common as dirt,[1] Arno Penzias and Robert Wilson were surprised when they serendipitously discovered the cosmic microwave background. At the time of their discovery, Penzias and Wilson were radio astronomers working at Bell Laboratories. The horn-reflector radio antenna they used had previously been utilized to receive microwave signals, of wavelength $\lambda = 7.35\,\text{cm}$, reflected from an orbiting communications satellite. Turning from telecommunications to astronomy, Penzias and Wilson found a slightly stronger signal than they expected when they turned the antenna toward the sky. They did everything they could think of to reduce "noise" in their system. They even shooed away a pair of pigeons that had roosted in the antenna and cleaned up what they later called "the usual white dielectric" generated by pigeons.

The excess signal remained. It was isotropic and constant with time, so it couldn't be associated with an isolated celestial source. Wilson and Penzias were puzzled until they were put in touch with Robert Dicke and his research group at Princeton University. Dicke had deduced that the universe, if it started in a hot dense state, should now be filled with microwave radiation.[2] In fact, Dicke and his group were in the process of building a microwave antenna when Penzias and Wilson told them that they had already detected the predicted microwave radiation. Penzias and Wilson wrote a paper for *The Astrophysical Journal* in which they wrote, "Measurements of the effective zenith noise temperature of the 20-foot horn-reflector antenna ... at 4080 Mc/s have yielded a value about 3.5 K higher than expected. This excess temperature is, within the limits of our observations, isotropic, unpolarized, and free from seasonal variations (July, 1964–April, 1965). A possible explanation for the observed excess noise temperature is the one given by Dicke, Peebles, Roll, and Wilkinson in a companion letter in this issue." The companion paper by Dicke and his collaborators points out that the radiation could be a relic of an early, hot, dense, and opaque state of the universe.

[1] Actually, much commoner than dirt, when you stop to think of it, since dirt is made of baryons.

[2] To give credit where it's due, as early as 1948, Ralph Alpher and Robert Herman had predicted the existence of cosmic background radiation with a temperature of "about 5 K." However, their prediction had fallen into obscurity.

Measuring the spectrum of the CMB, and confirming that it is indeed a blackbody, is not a simple task, even with modern technology. The mean energy per CMB photon (6.34×10^{-4} eV) is tiny compared to the energy required to break up an atomic nucleus (~ 2 MeV) or even the energy required to ionize an atom (~ 10 eV). However, the mean photon energy is comparable to the rotational energy of a small molecule such as H_2O. Thus, CMB photons can zip along for more than 13 billion years through tenuous intergalactic gas, then be absorbed a microsecond away from the Earth's surface by a water molecule in the atmosphere. Microwaves with wavelengths shorter than $\lambda \sim 3$ cm are strongly absorbed by water molecules. Penzias and Wilson observed the CMB at a wavelength $\lambda = 7.35$ cm, corresponding to a photon energy $E = 1.7 \times 10^{-5}$ eV, because that was the wavelength of the signals that Bell Labs had been bouncing off orbiting satellites. Thus, Penzias and Wilson were detecting CMB photons far on the low-energy tail of the blackbody spectrum (Figure 2.7), with an energy just 0.027 times the mean photon energy.

The CMB can be measured at wavelengths shorter than 3 cm by observing from high-altitude balloons or from the South Pole, where the combination of cold temperatures and high altitude[3] keeps the atmospheric humidity low. The best way to measure the spectrum of the CMB, however, is to go completely above the damp atmosphere of the Earth. The CMB spectrum was measured accurately over a wide range of wavelengths by the *Cosmic Background Explorer (COBE)* satellite, launched in 1989, into an orbit 900 km above the Earth's surface. The CMB was then mapped at greater angular resolution by the *Wilkinson Microwave Anisotropy Probe (WMAP)*, launched in 2001, and by the *Planck* satellite, launched in 2009. Both *WMAP* and *Planck* were in orbits librating about the L_2 point of the Sun–Earth system, 1.5 million km from the Earth. Multiple results have come from observations of the cosmic microwave background.

Result number one: At any angular position (θ, ϕ) on the sky, the spectrum of the cosmic microwave background is very close to that of an ideal blackbody, as illustrated in Figure 8.1. How close is very close? *COBE* could have detected fluctuations in the spectrum as small as one part in 10^4. No deviations were found at this level within the wavelength range investigated by *COBE*.

Result number two: The CMB has the dipole distortion in temperature shown in Figure 8.2. That is, although each point on the sky has a blackbody spectrum, in one half of the sky the spectrum is slightly blueshifted to higher temperatures, and in the other half the spectrum is slightly redshifted to lower temperatures.[4] This dipole distortion is a simple Doppler shift, caused by the net

[3] The South Pole is nearly 3 kilometers above sea level; one of the major challenges facing the Amundsen and Scott expeditions was the arduous climb from the Ross Ice Shelf to the central Antarctic plateau.

[4] The stretched "yin-yang" pattern in Figure 8.2 represents the darker, cooler (yin?) hemisphere of the sky and the hotter, brighter (yang?) hemisphere, distorted by the map projection.

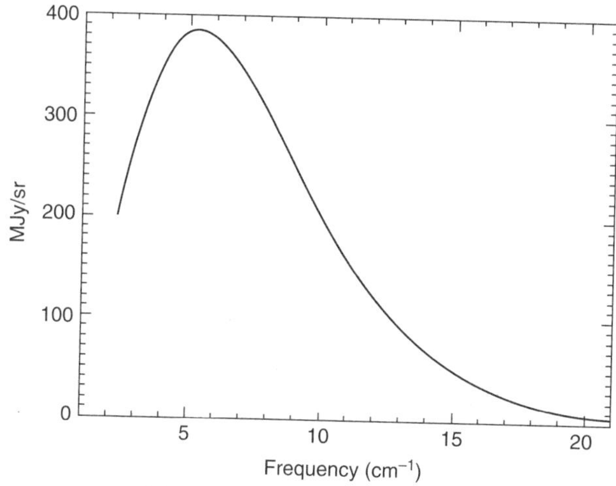

Figure 8.1 The spectrum of the cosmic microwave background, as measured by *COBE*. The uncertainties in the measurement are smaller than the thickness of the line. [Fixsen *et al.* 1996 *ApJ*, **473**, 576]

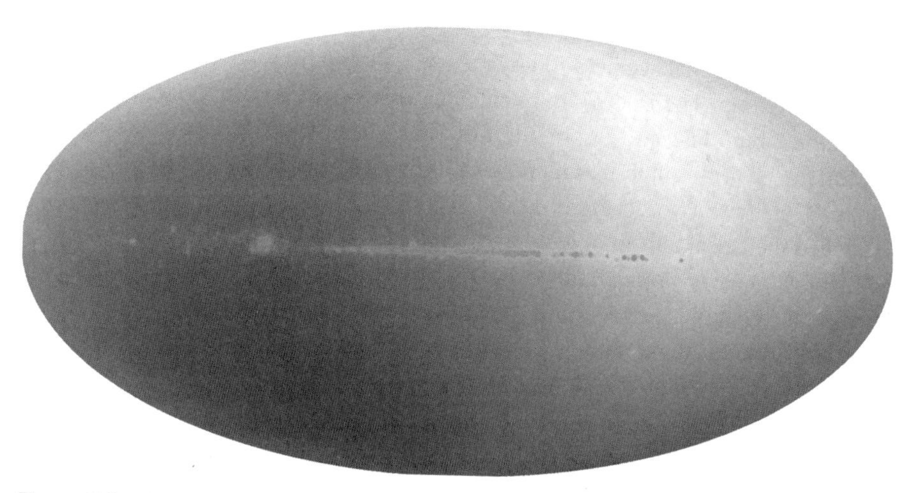

Figure 8.2 The dipole fluctuation in the temperature of the CMB, as measured by *WMAP*. The horizontal band across the middle is non-thermal emission from gas in our own galaxy. [NASA/*WMAP* Science Team]

motion of *WMAP* relative to a frame of reference in which the CMB is isotropic. After correcting for the orbital motion of *WMAP* around the Sun ($v \sim 30\,\mathrm{km\,s^{-1}}$), for the orbital motion of the Sun around the galactic center ($v \sim 235\,\mathrm{km\,s^{-1}}$), and for the orbital motion of our galaxy relative to the center of mass of the Local Group ($v \sim 80\,\mathrm{km\,s^{-1}}$), it is found that the Local Group is moving in the general direction of the constellation Hydra, with a speed $v_{\mathrm{LG}} = 630 \pm 20\,\mathrm{km\,s^{-1}} = 0.0021c$. This peculiar velocity for the Local Group is what you'd expect as the

Figure 8.3 The fluctuations in temperature remaining in the CMB once the dipole fluctuation and the non-thermal foreground emission from our own galaxy are subtracted. [*Planck*/ESA]

result of gravitational acceleration by the largest lumps of matter in the vicinity of the Local Group. The Local Group is being accelerated toward the Virgo cluster, the nearest big cluster to us. In addition, the Virgo cluster is being accelerated toward the Hydra-Centaurus supercluster, the nearest supercluster to us. The combination of these two accelerations, working over the age of the universe, has launched the Local Group in the direction of Hydra, at 0.2% the speed of light.

Result number three: After the dipole distortion of the CMB is subtracted away, the remaining temperature fluctuations, shown in Figure 8.3, are small in amplitude. Let the temperature of the CMB, at a given point on the sky, be $T(\theta, \phi)$. The mean temperature, averaging over all locations, is

$$\langle T \rangle = \frac{1}{4\pi} \int T(\theta, \phi) \sin\theta \, d\theta \, d\phi = 2.7255\,\text{K}. \tag{8.6}$$

The dimensionless temperature fluctuation at a given point (θ, ϕ) on the sky is

$$\frac{\delta T}{T}(\theta, \phi) \equiv \frac{T(\theta, \phi) - \langle T \rangle}{\langle T \rangle}. \tag{8.7}$$

After subtraction of the Doppler dipole, the root mean square temperature fluctuation found by *COBE* was

$$\left\langle \left(\frac{\delta T}{T} \right)^2 \right\rangle^{1/2} = 1.1 \times 10^{-5}. \tag{8.8}$$

(Given the limited angular resolution of the *COBE* satellite, this excludes the temperature fluctuations on an angular scale $< 10^\circ$.) Even taking into account the blurring from *COBE*'s low resolution, the fact that the CMB temperature varies by only 30 microKelvin across the sky represents a remarkably close approach to isotropy.

The observations that the CMB has a nearly perfect blackbody spectrum and that it is nearly isotropic (once the Doppler dipole is removed) provide strong support for the Hot Big Bang model of the universe. A background of nearly isotropic blackbody radiation is natural if the universe was once hot, dense, opaque, and nearly homogeneous, as it was in the Hot Big Bang scenario. If the universe did not go through such a phase, then any explanation of the cosmic microwave background will have to be much more contrived.

8.2 Recombination and Decoupling

To understand in more detail the origin of the cosmic microwave background, we'll have to examine carefully the process by which the baryonic matter goes from being an ionized plasma to a gas of neutral atoms, and the closely related process by which the universe goes from being opaque to being transparent. To avoid muddle, we will distinguish between three closely related (but not identical) moments in the history of the universe. First, the epoch of *recombination* is the time at which the baryonic component of the universe goes from being ionized to being neutral. Numerically, we define it as the instant in time when the number density of ions is equal to the number density of neutral atoms.[5] Second, the epoch of *photon decoupling* is the time when the rate at which photons scatter from electrons becomes smaller than the Hubble parameter (which tells us the rate at which the universe expands). When photons decouple, they cease to interact with the electrons, and the universe becomes transparent. Third, the epoch of *last scattering* is the time at which a typical CMB photon underwent its last scattering from an electron. Surrounding every observer in the universe is a *last scattering surface*, illustrated in Figure 8.4, from which the CMB photons have been streaming freely, with no further scattering by electrons. The probability that a photon will scatter from an electron is small once the expansion rate of the universe is faster than the scattering rate; thus, the epoch of last scattering is very close to the epoch of photon decoupling.

To keep things from getting too complicated, we will assume that the baryonic component of the universe consisted entirely of hydrogen at the epoch of recombination. This is not, however, a strictly accurate assumption. Even at the time of recombination, before stars had a chance to pollute the universe with

[5] Cosmologists sometimes grumble that this should really be called the epoch of "combination" rather than the epoch of "recombination," since this is the very first time when electrons and ions combined to form stable atoms.

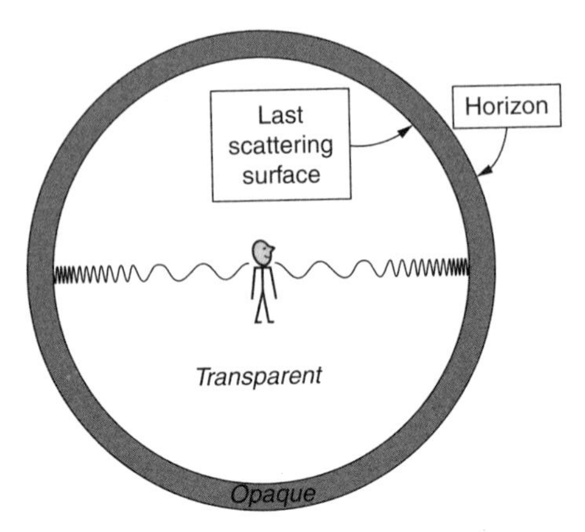

Figure 8.4 An observer is surrounded by a spherical last scattering surface. The photons of the CMB travel straight to us from the last scattering surface, being continuously redshifted.

heavy elements, there was a significant amount of helium present. (In the next chapter, we will examine how this helium was formed in the early universe.) However, the presence of helium is merely a complicating factor. All the significant physics of recombination can be studied in a simplified universe containing no elements other than hydrogen. The hydrogen can take the form of a neutral atom (designated by the letter H), or of a naked hydrogen nucleus, otherwise known as a proton (designated by the letter p). To maintain charge neutrality in this hydrogen-only universe, the number density of free electrons must be equal to that of free protons: $n_e = n_p$. The degree to which the baryonic content of the universe is ionized can be expressed as the fractional ionization X, defined as

$$X \equiv \frac{n_p}{n_p + n_{\rm H}} = \frac{n_p}{n_{\rm bary}} = \frac{n_e}{n_{\rm bary}}. \tag{8.9}$$

The value of X ranges from $X = 1$, when the baryonic content is fully ionized, to $X = 0$, when it consists entirely of neutral atoms.

One useful consequence of assuming that hydrogen is the only element is that there is now a single relevant energy scale in the problem: the ionization energy of hydrogen, $Q = 13.6\,\text{eV}$. A photon with an energy $hf > Q$ is capable of photoionizing a hydrogen atom:

$$\text{H} + \gamma \rightarrow p + e^-. \tag{8.10}$$

This reaction can run in the opposite direction, as well; a proton and an electron can undergo *radiative recombination*, forming a bound hydrogen atom while a photon carries away the excess energy:

$$p + e^- \rightarrow \text{H} + \gamma. \tag{8.11}$$

In a universe containing protons, electrons, and photons, the fractional ionization X will depend on the balance between photoionization and radiative recombination.

Let's travel back in time to a period before the epoch of recombination. For concreteness, let's choose the moment when $a = 10^{-5}$, corresponding to a redshift $z = 10^5$. (In the Benchmark Model, this scale factor was reached when the universe was seventy years old.) The temperature of the background radiation at this time was $T \approx 3 \times 10^5$ K, and the average photon energy was $hf_{\rm mean} \approx 2.7kT \approx 60$ eV, in the extreme ultraviolet. With such a high energy per photon, and with 1.6 billion photons for every baryon, any hydrogen atoms that happened to form by radiative recombination were very short-lived; almost immediately, they were blasted apart into their component electron and proton by a high-energy photon. At early times, then, the fractional ionization of the universe was very close to $X = 1$.

When the universe was fully ionized, photons interacted primarily with electrons, and the main interaction mechanism was Thomson scattering:

$$\gamma + e^- \to \gamma + e^-. \tag{8.12}$$

The scattering interaction is accompanied by a transfer of energy and momentum between the photon and electron. The cross-section for Thomson scattering is $\sigma_e = 6.65 \times 10^{-29}\,{\rm m}^2$. The mean free path of a photon – that is, the mean distance it travels before scattering from an electron – is

$$\lambda = \frac{1}{n_e \sigma_e}. \tag{8.13}$$

Since photons travel with a speed c, the rate at which a photon undergoes scattering interactions is

$$\Gamma = \frac{c}{\lambda} = n_e \sigma_e c. \tag{8.14}$$

When the baryonic component of the universe is fully ionized, $n_e = n_p = n_{\rm bary}$. Currently, the number density of baryons is $n_{\rm bary,0} = 0.25\,{\rm m}^{-3}$. The number density of conserved particles, such as baryons, goes as $1/a^3$, so when the early universe was fully ionized, the free electron density was

$$n_e = n_{\rm bary} = \frac{n_{\rm bary,0}}{a^3}, \tag{8.15}$$

and the scattering rate for photons was

$$\Gamma = \frac{n_{\rm bary,0}\sigma_e c}{a^3} = \frac{5.0 \times 10^{-21}\,{\rm s}^{-1}}{a^3}. \tag{8.16}$$

This means, for instance, that at $a = 10^{-5}$, photons scattered from electrons at a rate $\Gamma = 5.0 \times 10^{-6}\,{\rm s}^{-1}$, about three times a week.

The photons remain coupled to the electrons as long as their scattering rate, Γ, is larger than H, the rate at which the universe expands; this is equivalent to saying that their mean free path λ is shorter than the Hubble distance c/H. As long as photons scatter frequently from electrons, the photons and electrons remain at the same temperature T (and thanks to the electrons' interactions with protons, the protons also have the same temperature). When the photon scattering rate Γ drops below H, then the electrons are being diluted by expansion more rapidly than the photons can interact with them. The photons then decouple from the electrons and the universe becomes transparent. Once the photons are decoupled from the electrons and protons, the baryonic portion of the universe is no longer compelled to have the same temperature as the cosmic microwave background. During the early stages of the universe ($a < a_{rm} \approx 2.9 \times 10^{-4}$) the universe was radiation dominated, and the Friedmann equation was

$$\frac{H^2}{H_0^2} = \frac{\Omega_{r,0}}{a^4}. \tag{8.17}$$

Thus, the Hubble parameter was

$$H = \frac{H_0 \Omega_{r,0}^{1/2}}{a^2} = \frac{2.1 \times 10^{-20}\,\mathrm{s}^{-1}}{a^2}. \tag{8.18}$$

This means, for instance, that at $a = 10^{-5}$, the Hubble parameter was $H = 2.1 \times 10^{-10}\,\mathrm{s}^{-1}$. Since this is much smaller than the scattering rate $\Gamma = 5.0 \times 10^{-6}\,\mathrm{s}^{-1}$ at the same scale factor, the photons were well coupled to the electrons and protons.

If hydrogen had remained ionized (and note the qualifying *if*), then photons would have remained coupled to the electrons and protons until a relatively recent time. Taking into account the transition from a radiation-dominated to a matter-dominated universe, and the resulting change in the expansion rate, we can compute that *if* hydrogen had remained fully ionized, then decoupling would have taken place at a scale factor $a \approx 0.0254$, corresponding to a redshift $z \approx 38$ and a CMB temperature $T \approx 110\,\mathrm{K}$. However, at such a low temperature, the CMB photons are too low in energy to keep the hydrogen ionized. Thus, the decoupling of photons is not a gradual process, caused by the continuous lowering of free electron density as the universe expands. Rather, it is a relatively sudden process, caused by the plummeting of free electron density during the epoch of recombination, as electrons combined with protons to form hydrogen atoms.

8.3 The Physics of Recombination

When does recombination, and the consequent photon decoupling, take place? It's easy to do a quick and dirty approximation of the recombination temperature. Recombination, one could argue, must take place when the mean energy per photon of the cosmic microwave background falls below the ionization energy of hydrogen, $Q = 13.6\,\text{eV}$. When this happens, the average CMB photon is no longer able to photoionize hydrogen. Since the mean CMB photon energy is $\sim 2.7kT$, this line of argument would indicate a recombination temperature of

$$T_{\text{rec}} \sim \frac{Q}{2.7k} \sim \frac{13.6\,\text{eV}}{2.7(8.6 \times 10^{-5}\,\text{eV}\,\text{K}^{-1})} \sim 60\,000\,\text{K}. \tag{8.19}$$

Alas, this crude approximation is a little *too* crude to be useful. It doesn't take into account the fact that CMB photons are not of uniform energy – a blackbody spectrum has an exponential tail (see Figure 2.7) trailing off to high energies. Although the mean photon energy is $2.7kT$, about one photon in 500 will have $E > 10kT$, one in 3 million will have $E > 20kT$, and one in 30 billion will have $E > 30kT$. Although extremely high energy photons make up only a tiny fraction of the CMB photons, the total number of CMB photons is enormous, with 1.6 billion photons for every baryon. The vast swarms of photons that surround every newly formed hydrogen atom greatly increase the probability that the atom will collide with a photon from the high-energy tail of the blackbody spectrum, and be photoionized.

Thus, we expect the recombination temperature to depend on the baryon-to-photon ratio η as well as on the ionization energy Q. An exact calculation of the fractional ionization X, as a function of η and T, requires a smattering of statistical mechanics. Let's start with the reaction that determines the value of X in the early universe:

$$\text{H} + \gamma \rightleftharpoons p + e^-. \tag{8.20}$$

Our calculations will be simplified by the fact that at the time of recombination, the photons, electrons, protons, and hydrogen atoms are in a state of *thermal equilibrium*. Saying that several types of particle are in thermal equilibrium with each other is equivalent to saying that they all have the same temperature T. For instance, in the air around you, the N_2 molecules and O_2 molecules are in thermal equilibrium with each other because of their frequent collisions, and have the same temperature, $T \approx 300\,\text{K}$.[6]

Further simplification comes from the fact that each particle type (photon, electron, proton, or hydrogen atom) is in a state of *kinetic* equilibrium. Saying that a particular type of particle is in kinetic equilibrium is equivalent to saying that the distribution of particle momentum p and energy E is given either by a Fermi–Dirac distribution (if the particles are fermions, with half-integral spin) or by a Bose–Einstein distribution (if the particles are bosons, with integral spin). Suppose, for instance, that particles of type x have a mass m_x. Let $n_x(p)dp$ be the number density of x particles with momentum in the range $p \to p + dp$. If the particles are in kinetic equilibrium at temperature T, then

[6] Slightly cooler if you are using this book for recreational reading as you ski across the Antarctic plateau.

$$n_x(p)dp = g_x \frac{4\pi}{h^3} \frac{p^2 dp}{\exp([E - \mu_x]/kT) \pm 1}, \tag{8.21}$$

where the + sign is chosen for a Fermi–Dirac distribution, and the − sign is chosen for a Bose–Einstein distribution. In Equation 8.21, the factor g_x is the *statistical weight* of the particle; for instance, photons, electrons, protons, and neutrons all have $g = 2$, corresponding to their two different spin states. The factor μ_x is the *chemical potential* for particles of type x. As with other forms of potential energy, such as the gravitational potential, we are particularly interested in the difference in chemical potential between two possible states. For instance, in the reaction given by Equation 8.20, if $\mu_H + \mu_\gamma > \mu_p + \mu_e$, then the reaction runs preferentially from the higher energy state (H + γ) to the lower energy state ($p + e^-$), and photoionizations outnumber radiative recombinations. Conversely, if $\mu_H + \mu_\gamma < \mu_p + \mu_e$, the reaction runs preferentially in the opposite direction, and radiative recombinations outnumber photoionizations.

For photons, the relation between energy, momentum, and frequency is very simple: $E = pc = hf$. Photons are bosons with a statistical weight $g_\gamma = 2$ and a chemical potential $\mu_\gamma = 0$. This means that the number density of photons as a function of frequency f is

$$n_\gamma(f)df = \frac{8\pi}{c^3} \frac{f^2 df}{\exp(hf/kT) - 1}, \tag{8.22}$$

which we have already encountered as the blackbody formula of Equation 2.30. Integrating Equation 8.22 over all photon frequencies gives the total number density of photons:

$$n_\gamma = \frac{2.4041}{\pi^2} \left(\frac{kT}{\hbar c}\right)^3 = 0.2436 \left(\frac{kT}{\hbar c}\right)^3. \tag{8.23}$$

At the time of recombination, electrons, protons, and hydrogen atoms all had $mc^2 \gg kT$, and thus had highly nonrelativistic thermal speeds. When particles of type x are highly nonrelativistic, we can safely use the approximation $p \approx m_x v$ and

$$E \approx m_x c^2 + \frac{1}{2} m_x v^2 \approx m_x c^2 + \frac{p^2}{2m_x}. \tag{8.24}$$

Substituting these values into Equation 8.21, we find that for particles with $m_x c^2 - \mu_x \gg kT$,

$$n_x(p)dp = g_x \frac{4\pi}{h^3 c^3} \exp\left(\frac{-m_x c^2 + \mu_x}{kT}\right) \exp\left(-\frac{p^2}{2m_x kT}\right) p^2 dp, \tag{8.25}$$

representing a Maxwell–Boltzmann distribution of particle speeds. Integrated over all particle momenta, the total number density of the nonrelativistic x particles is

$$n_x = g_x \left(\frac{m_x kT}{2\pi\hbar^2}\right)^{3/2} \exp\left(\frac{-m_x c^2 + \mu_x}{kT}\right), \tag{8.26}$$

regardless of whether they are bosons or fermions.

In general, the chemical potential μ_x for particles other than photons will be nonzero. However, at the time of recombination, we can make the further simplifying assumption that the reaction

$$\text{H} + \gamma \rightleftharpoons p + e^- \tag{8.27}$$

was in *chemical* equilibrium. Saying that a reaction is in chemical equilibrium is equivalent to saying that the reaction rate going from left to right balances the reaction rate going from right to left.[7] In the case of Equation 8.27, for instance, this means that within a given volume of hydrogen gas there will be one radiative recombination, on average, for every photoionization. When chemical equilibrium holds true, the sum of the chemical potentials must be equal on both sides of the equation. For Equation 8.27, given that $\mu_\gamma = 0$, this means that $\mu_\text{H} = \mu_p + \mu_e$.

Using Equation 8.26 to find n_H, n_p, and n_e, and assuming that $\mu_\text{H} = \mu_p + \mu_e$, we find an equation that relates the number density of hydrogen atoms, free protons, and free electrons, as long as photoionization remains in equilibrium with radiative recombination:

$$\frac{n_\text{H}}{n_p n_e} = \frac{g_\text{H}}{g_p g_e}\left(\frac{m_\text{H}}{m_p m_e}\right)^{3/2}\left(\frac{kT}{2\pi\hbar^2}\right)^{-3/2} \exp\left(\frac{[m_p + m_e - m_\text{H}]c^2}{kT}\right). \tag{8.28}$$

Equation 8.28 can be simplified further. First, since the mass of an electron is small compared to that of a proton, we can set $m_\text{H}/m_p = 1$. Second, the binding energy $Q = 13.6\,\text{eV}$ is given by the formula $(m_p + m_e - m_\text{H})c^2 = Q$. The statistical weights of the proton and electron are $g_p = g_e = 2$, while the statistical weight of a hydrogen atom is $g_\text{H} = 4$. Thus, the factor $g_\text{H}/(g_p g_e)$ can be set equal to one. The resulting equation,

$$\frac{n_\text{H}}{n_p n_e} = \left(\frac{m_e kT}{2\pi\hbar^2}\right)^{-3/2} \exp\left(\frac{Q}{kT}\right), \tag{8.29}$$

is called the *Saha equation*, after the astrophysicist Meghnad Saha, who derived it while studying ionization in stellar atmospheres.

Our next job is to convert the Saha equation into a relation between X, T, and η. From the definition of X (Equation 8.9), we can make the substitution

$$n_\text{H} = \frac{1 - X}{X} n_p, \tag{8.30}$$

[7] Although this type of equilibrium is conventionally called "chemical" equilibrium, it can apply to nuclear reactions (as we'll see in the next chapter) as well as to chemical reactions.

and from the requirement of charge neutrality, we can make the substitution $n_e = n_p$. This yields

$$\frac{1-X}{X} = n_p \left(\frac{m_e kT}{2\pi\hbar^2}\right)^{-3/2} \exp\left(\frac{Q}{kT}\right). \tag{8.31}$$

To eliminate n_p from the above equation, we recall that $\eta \equiv n_{\text{bary}}/n_\gamma$. In a universe where hydrogen is the only element, and a fraction X of the hydrogen is in the form of naked protons, we may write

$$\eta = \frac{n_p}{X n_\gamma}. \tag{8.32}$$

Since the photons have a blackbody spectrum, with a photon number density n_γ given by Equation 8.23, we can combine Equations 8.32 and 8.23 to find

$$n_p = 0.2436 X\eta \left(\frac{kT}{\hbar c}\right)^3. \tag{8.33}$$

Substituting Equation 8.33 back into Equation 8.31, we finally find the desired equation for X in terms of T and η:

$$\frac{1-X}{X^2} = 3.84\eta \left(\frac{kT}{m_e c^2}\right)^{3/2} \exp\left(\frac{Q}{kT}\right). \tag{8.34}$$

This is a quadratic equation in X, whose positive root is

$$X = \frac{-1+\sqrt{1+4S}}{2S}, \tag{8.35}$$

where

$$S(T,\eta) = 3.84\eta \left(\frac{kT}{m_e c^2}\right)^{3/2} \exp\left(\frac{Q}{kT}\right). \tag{8.36}$$

If we define the moment of recombination as the instant when $X = \frac{1}{2}$, then (assuming $\eta = 6.1 \times 10^{-10}$) the recombination temperature is

$$kT_{\text{rec}} = 0.324\,\text{eV} = \frac{Q}{42}. \tag{8.37}$$

Because of the exponential dependence of S upon the temperature, the exact value of η doesn't strongly affect the value of T_{rec}. On the Kelvin scale, $kT_{\text{rec}} = 0.324\,\text{eV}$ corresponds to a temperature $T_{\text{rec}} = 3760\,\text{K}$, slightly higher than the melting point of tungsten.[8] The temperature of the universe had a value $T = T_{\text{rec}} = 3760\,\text{K}$ at a redshift $z_{\text{rec}} = 1380$, when the age of the universe, in the Benchmark Model, was $t_{\text{rec}} = 250\,000\,\text{yr}$.

Recombination was not an instantaneous process; it happened sufficiently gradually that at any given instant, the assumption of kinetic and chemical

[8] Not that there was any tungsten around back then to be melted.

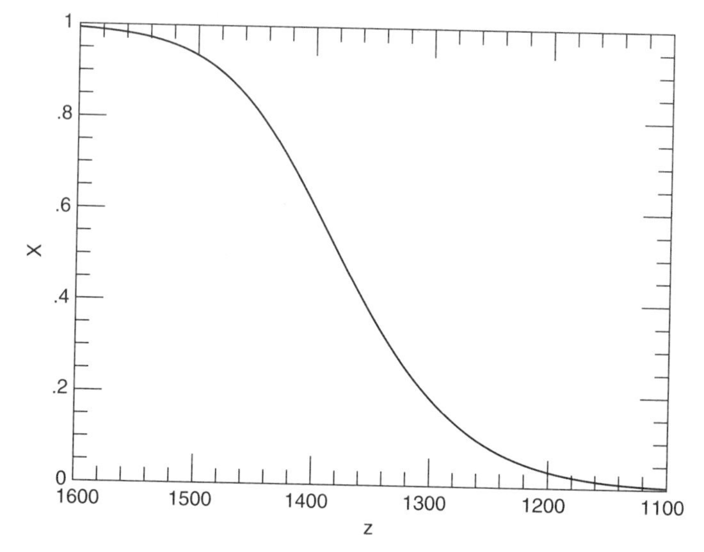

Figure 8.5 Fractional ionization X as a function of redshift during the epoch of recombination. A baryon-to-photon ratio $\eta = 6.1 \times 10^{-10}$ is assumed. Redshift decreases, and thus time increases, from left to right.

equilibrium is a reasonable approximation. However, as shown in Figure 8.5, it did proceed fairly rapidly by cosmological standards. The fractional ionization went from $X = 0.9$ at a redshift $z = 1480$ to $X = 0.1$ at a redshift $z = 1260$. In the Benchmark Model, the time that elapses from $X = 0.9$ to $X = 0.1$ is $\Delta t \approx 70\,000\,\text{yr} \approx 0.28 t_{\text{rec}}$.

Since the number density of free electrons drops rapidly during the epoch of recombination, the time of photon decoupling comes soon after the time of recombination. The rate of photon scattering, when the hydrogen is partially ionized, is

$$\Gamma(z) = n_e(z)\sigma_e c = X(z)(1+z)^3 n_{\text{bary},0}\sigma_e c. \tag{8.38}$$

Using $\Omega_{\text{bary},0} = 0.048$, the numerical value of the scattering rate is

$$\Gamma(z) = 5.0 \times 10^{-21}\,\text{s}^{-1} X(z)(1+z)^3. \tag{8.39}$$

While recombination is taking place, the universe is matter-dominated, so the Hubble parameter is given by the relation

$$\frac{H^2}{H_0^2} = \frac{\Omega_{m,0}}{a^3} = \Omega_{m,0}(1+z)^3. \tag{8.40}$$

Using $\Omega_{m,0} = 0.31$, the numerical value of the Hubble parameter during the epoch of recombination is

$$H(z) = 1.23 \times 10^{-18}\,\text{s}^{-1}(1+z)^{3/2}. \tag{8.41}$$

The redshift of photon decoupling is found by setting $\Gamma = H$, or (combining Equations 8.39 and 8.41),

$$1 + z_{\text{dec}} = \frac{39.3}{X(z_{\text{dec}})^{2/3}}. \tag{8.42}$$

Using the value of $X(z)$ given by the Saha equation (shown in Figure 8.5), the redshift of photon decoupling is found to be $z_{\text{dec}} = 1120$. In truth, the exact redshift of photon decoupling is somewhat smaller than this value. The Saha equation assumes that the reaction $H + \gamma \rightleftharpoons p + e^-$ is in equilibrium. However, when Γ starts to drop below H, the photoionization reaction is no longer in equilibrium. As a consequence, at redshifts smaller than ~ 1200, the fractional ionization X is larger than would be predicted by the Saha equation, and the decoupling of photons is therefore delayed. Without going into the details of the nonequilibrium physics, let's content ourselves by quoting the result $z_{\text{dec}} = 1090$, corresponding to a temperature $T_{\text{dec}} = 2970$ K, when the age of the universe was $t_{\text{dec}} = 371\,000$ yr in the Benchmark Model.

When we examine the CMB with our microwave antennas, the photons we collect have been traveling straight toward us since the last time they scattered from a free electron. During a brief time interval $t \to t + dt$, the probability that a photon undergoes a scattering is $dP = \Gamma(t)dt$, where $\Gamma(t)$ is the scattering rate at time t. Thus, if we detect a CMB photon at time t_0, the expected number of scatterings it has undergone since an earlier time t is

$$\tau(t) = \int_t^{t_0} \Gamma(t)dt. \tag{8.43}$$

The dimensionless number τ is the *optical depth*. The time t for which $\tau = 1$ is the *time of last scattering*, and represents the time that has elapsed since a typical CMB photon last scattered from a free electron. If we change the variable of integration in Equation 8.43 from t to a, we find that

$$\tau(a) = \int_a^1 \Gamma(a)\frac{da}{\dot{a}} = \int_a^1 \frac{\Gamma(a)}{H(a)}\frac{da}{a}, \tag{8.44}$$

using the fact that $H = \dot{a}/a$. Alternatively, we can find the optical depth as a function of redshift by making the substitution $1 + z = 1/a$:

$$\tau(z) = \int_0^z \frac{\Gamma(z)}{H(z)}\frac{dz}{1+z} = 0.0041 \int_0^z X(z)(1+z)^{1/2}dz. \tag{8.45}$$

Here, we have made use of Equations 8.39 and 8.41. As it turns out, the last scattering of a typical CMB photon occurs after the photoionization reaction $H + \gamma \rightleftharpoons p + e^-$ falls out of equilibrium, so the Saha equation doesn't strictly apply. To sufficient accuracy for our purposes, we can state that the redshift of last scattering was comparable to the redshift of photon decoupling: $z_{\text{ls}} \approx z_{\text{dec}} \approx 1090$. Not all the CMB photons underwent their last scattering

Table 8.1 Events in the early universe.

Event	*Redshift*	*Temperature (K)*	*Time (Myr)*
Radiation–matter equality	3440	9390	0.050
Recombination	1380	3760	0.25
Photon decoupling	1090	2970	0.37
Last scattering	1090	2970	0.37

simultaneously; the universe doesn't choreograph its microphysics that well. If we scoop up two photons from the CMB, one may have undergone its last scattering at $z = 1140$, while the other may have scattered more recently, at $z = 1040$. Thus, the "last scattering surface" is really more of a "last scattering layer"; just as we can see a little way into a fog bank here on Earth, we can see a little way into the "electron fog" that hides the early universe from our direct view.

The relevant times of various events around the time of recombination are shown in Table 8.1. For purposes of comparison, the table also contains the time of radiation–matter equality, emphasizing the fact that recombination, photon decoupling, and last scattering took place when the universe was matter-dominated. When we look at the cosmic microwave background, we are getting an intriguing glimpse of the universe as it was when it was only one part in 37 000 of its present age.

8.4 Temperature Fluctuations

The dipole distortion of the cosmic microwave background, shown in Figure 8.2, results from the fact that the universe is not perfectly homogeneous today ($z = 0$). Because we are gravitationally accelerated toward the nearest large lumps of matter, we see a Doppler shift in the radiation of the CMB. The distortions on a smaller angular scale, shown in Figure 8.3, tell us that the universe was not perfectly homogeneous at the time of last scattering ($z \approx 1090$). The angular size of the temperature fluctuations reflects in part the physical size of the density and velocity fluctuations at $z \approx 1090$.

The angular size $\delta\theta$ of a temperature fluctuation in the CMB is related to a physical size ℓ on the last scattering surface by the relation

$$d_A = \frac{\ell}{\delta\theta}, \tag{8.46}$$

where d_A is the angular-diameter distance to the last scattering surface. Since the last scattering surface is at a redshift $z_{\rm ls} = 1090 \gg 1$, a good approximation to d_A is given by Equation 6.40:

$$d_A \approx \frac{d_{\rm hor}(t_0)}{z_{\rm ls}}. \tag{8.47}$$

In the Benchmark Model, the current horizon distance is $d_{\rm hor}(t_0) \approx 14\,000$ Mpc, so the angular-diameter distance to the surface of last scattering is

$$d_A \approx \frac{14\,000\,{\rm Mpc}}{1090} \approx 12.8\,{\rm Mpc}. \tag{8.48}$$

Thus, fluctuations on the last scattering surface with an observed angular size $\delta\theta$ had a physical size

$$\ell = d_A \cdot \delta\theta = 12.8\,{\rm Mpc}\left(\frac{\delta\theta}{1\,{\rm rad}}\right) = 3.7\,{\rm kpc}\left(\frac{\delta\theta}{1\,{\rm arcmin}}\right) \tag{8.49}$$

at the time of last scattering. The smallest fluctuations resolved by the *Planck* satellite (Figure 8.3) have an angular size $\delta\theta \approx 5$ arcmin. This corresponds to a physical size of $\ell \approx 18$ kpc at the time of last scattering, or $\ell(1 + z_{\rm ls}) \approx 20$ Mpc today. A sphere with this radius has a baryon mass $M_{\rm bary} \approx 5 \times 10^{13}\,{\rm M}_\odot$, about that of a cluster of galaxies. (The Coma cluster has $M_{\rm bary} \sim 2 \times 10^{14}\,{\rm M}_\odot$, but it is a very rich cluster.)

Consider the density fluctuations $\delta T/T$ of the cosmic microwave background, as shown in Figure 8.3. Since $\delta T/T$ is defined on the surface of a sphere – the celestial sphere in this case – it is useful to expand it in spherical harmonics:

$$\frac{\delta T}{T}(\theta, \phi) = \sum_{l=0}^{\infty} \sum_{m=-l}^{l} a_{lm} Y_{lm}(\theta, \phi), \tag{8.50}$$

where $Y_{lm}(\theta, \phi)$ are the usual spherical harmonic functions. What concerns cosmologists is not the exact pattern of hot spots and cold spots on the sky, but their statistical properties. The most important statistical property of $\delta T/T$ is the correlation function $C(\theta)$. Consider two points on the last scattering surface. Relative to an observer, they are in the directions $\hat{n}$ and $\hat{n}'$, and are separated by an angle θ given by the relation $\cos\theta = \hat{n} \cdot \hat{n}'$. To find the correlation function $C(\theta)$, multiply together the values of $\delta T/T$ at the two points, then average the product over all points separated by the angle θ:

$$C(\theta) = \left\langle \frac{\delta T}{T}(\hat{n}) \frac{\delta T}{T}(\hat{n}') \right\rangle_{\hat{n}\cdot\hat{n}' = \cos\theta}. \tag{8.51}$$

Using the expansion of $\delta T/T$ in spherical harmonics, the correlation function can be written in the form

$$C(\theta) = \frac{1}{4\pi} \sum_{l=0}^{\infty} (2l + 1) C_l P_l(\cos\theta), \tag{8.52}$$

where P_l are the usual Legendre polynomials:

$$\begin{aligned} P_0(x) &= 1 \\ P_1(x) &= x \\ P_2(x) &= \frac{1}{2}(3x^2 - 1) \end{aligned} \tag{8.53}$$

and so forth. In this way, a measured correlation function $C(\theta)$ can be broken down into its multipole moments C_l. The $l = 0$ (monopole) term of the correlation function vanishes if we've defined the mean temperature correctly. The $l = 1$ (dipole) term results primarily from the Doppler shift due to our motion through space. For larger values of l, the term C_l is a measure of temperature fluctuations on an angular scale $\theta \sim 180°/l$. Thus, the multipole l is interchangeable, for all practical purposes, with the angular scale θ. The moments with $l \geq 2$ are of the most interest to astronomers, since they tell us about the fluctuations present at the time of last scattering.

In presenting the results of CMB observations, it is customary to plot the function

$$\Delta_T \equiv \left(\frac{l(l+1)}{2\pi}C_l\right)^{1/2}\langle T\rangle\,, \tag{8.54}$$

since this function tells us the contribution per logarithmic interval in l to the total temperature fluctuation δT of the cosmic microwave background. Figure 8.6, which shows results from the *Planck* satellite, is a plot of Δ_T as a function of l. The detailed shape of the Δ_T versus l curve contains a wealth of information about the universe at the time of photon decoupling. In the next section we will examine, very briefly, the physics behind the temperature fluctuations, and how we can extract cosmological information from the temperature anisotropy of the cosmic microwave background.

8.5 What Causes the Fluctuations?

At the time of last scattering, a particularly interesting length scale, cosmologically speaking, is the horizon distance,

$$d_{\rm hor}(t_{\rm ls}) = a(t_{\rm ls})c\int_0^{t_{\rm ls}}\frac{dt}{a(t)}. \tag{8.55}$$

Since last scattering takes place long before the cosmological constant plays a significant role in the expansion, we can use the scale factor appropriate for a universe containing just radiation and matter, as given in Equation 5.110. This gives a value for the horizon distance of

$$d_{\rm hor}(t_{\rm ls}) = 2.24ct_{\rm ls} = 0.251\,{\rm Mpc}. \tag{8.56}$$

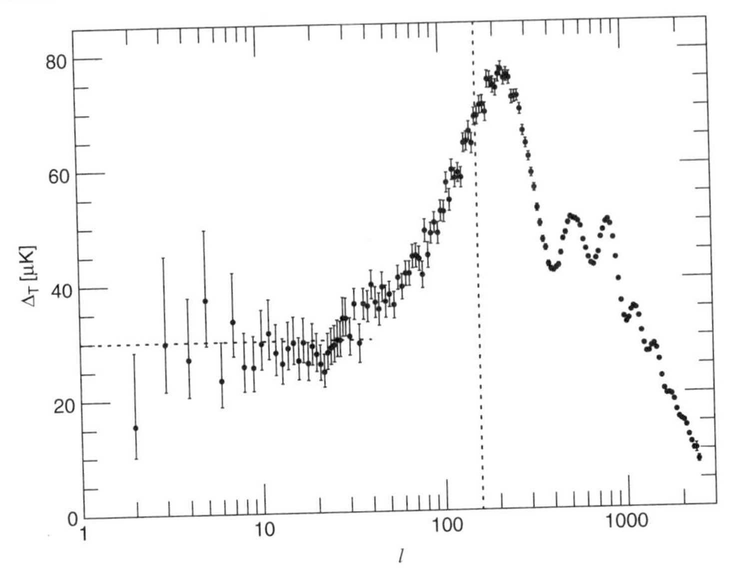

Figure 8.6 Temperature fluctuations Δ_T of the CMB, as observed by *Planck*, expressed as a function of the multipole l. The vertical dotted line shows $l_{\rm hor}$, the multipole corresponding to the horizon size at last scattering. The horizontal dotted line shows the value $\Delta_T \approx 30\,\mu$K at which Δ_T levels off at small l. [data courtesy of *Planck*/ESA]

A patch of the last scattering surface with this physical size will have an angular size, as seen from Earth, of

$$\theta_{\rm hor} = \frac{d_{\rm hor}(t_{\rm ls})}{d_A} = \frac{0.251\,{\rm Mpc}}{12.8\,{\rm Mpc}} \approx 0.020\,{\rm rad} \approx 1.1^\circ. \tag{8.57}$$

This angle corresponds to a multipole $l_{\rm hor} \approx 160$, indicated as the vertical dotted line in Figure 8.6.

The behavior of the Δ_T versus l curve is notably different on large angular scales than on smaller scales. On large angular scales, the value of Δ_T levels off at a nearly constant value $\Delta_T \approx 30\,\mu$K in the limit that $\theta > 4\theta_{\rm hor}$, or $l < 40$. By contrast, on angular scales smaller than $\theta_{\rm hor}$, the Δ_T curve shows a landscape of peaks and valleys rather than a level plateau. The highest peak, at $l \approx 220$, is called the "first peak." The second and third peaks, at $l \approx 520$ and $l \approx 800$, are lower in amplitude. The difference in behavior between the plateau at small l and the peaks and valleys at larger l results from the fact that fluctuations on large angular scales, with $\theta > \theta_{\rm hor}$, have a different physical cause than the fluctuations with $\theta < \theta_{\rm hor}$.

Consider first the large scale fluctuations – those with angular size $\theta > \theta_{\rm hor}$. These temperature fluctuations arise from the gravitational effect of density fluctuations in the distribution of nonbaryonic dark matter. The density of nonbaryonic dark matter at the time of last scattering, since $\varepsilon_{\rm dm} \propto a^{-3} \propto (1+z)^3$, was

$$\varepsilon_{\rm dm}(z_{\rm ls}) = \Omega_{\rm dm,0}\varepsilon_{c,0}(1+z_{\rm ls})^3 \approx 1.7 \times 10^{12}\,{\rm MeV\,m^{-3}}, \tag{8.58}$$

equivalent to a mass density $\rho \sim 3 \times 10^{-18}\,{\rm kg\,m^{-3}}$. The density of baryonic matter at the time of last scattering was

$$\varepsilon_{\rm bary}(z_{\rm ls}) = \Omega_{\rm bary,0}\varepsilon_{c,0}(1+z_{\rm ls})^3 \approx 3.1 \times 10^{11}\,{\rm MeV\,m^{-3}}. \tag{8.59}$$

The density of photons at the time of last scattering, since $\varepsilon_\gamma \propto a^{-4} \propto (1+z)^4$, was

$$\varepsilon_\gamma(z_{\rm ls}) = \Omega_{\gamma,0}\varepsilon_{c,0}(1+z_{\rm ls})^4 \approx 3.9 \times 10^{11}\,{\rm MeV\,m^{-3}}. \tag{8.60}$$

Thus, at the time of last scattering, $\varepsilon_{\rm dm} > \varepsilon_\gamma > \varepsilon_{\rm bary}$, with dark matter, photons, and baryons having energy densities in the ratio 5.5 : 1.24 : 1. The nonbaryonic dark matter dominated the energy density ε, and hence the gravitational potential, of the universe at the time of last scattering.

Suppose that the density of the nonbaryonic dark matter at the time of last scattering was not perfectly homogeneous, but varied as a function of position. Then we could write the energy density of the dark matter as

$$\varepsilon(\vec{r}) = \bar{\varepsilon} + \delta\varepsilon(\vec{r}), \tag{8.61}$$

where $\bar{\varepsilon}$ is the spatially averaged energy density of the nonbaryonic dark matter, and $\delta\varepsilon$ is the local deviation from the mean. In the Newtonian approximation, the spatially varying component of the energy density, $\delta\varepsilon$, gives rise to a spatially varying gravitational potential $\delta\Phi$. The link between $\delta\varepsilon$ and $\delta\Phi$ is Poisson's equation:

$$\nabla^2(\delta\Phi) = \frac{4\pi G}{c^2}\delta\varepsilon. \tag{8.62}$$

Unless the distribution of dark matter were perfectly smooth at the time of last scattering, the fluctuations in its density would necessarily have given rise to fluctuations in the gravitational potential.

Consider the fate of a CMB photon that happens to be at a local minimum of the potential at the time of last scattering. (Minima in the gravitational potential are known colloquially as "potential wells.") In climbing out of the potential well, it loses energy, and consequently is redshifted. Conversely, a photon that happens to be at a potential maximum when the universe became transparent gains energy as it falls down the "potential hill," and thus is blueshifted. When we look at the last scattering surface on large angular scales, cool (redshifted) spots correspond to minima in $\delta\Phi$; hot (blueshifted) spots correspond to maxima in $\delta\Phi$. A detailed general relativistic calculation, first performed by Sachs and Wolfe in 1967, tells us that

$$\frac{\delta T}{T} = \frac{1}{3}\frac{\delta\Phi}{c^2}. \tag{8.63}$$

Thus, the temperature fluctuations on large angular scales ($\theta > \theta_{\rm hor} \approx 1.1^\circ$) give us a map of the potential fluctuations $\delta\Phi$ present at the time of last scattering. In particular, the fact that the observed moments Δ_T are constant over a wide range of angular scales, from $l \sim 2$ to $l \sim 40$, tells us that the potential fluctuations $\delta\Phi$ were constant over a wide range of physical scales. The creation of temperature fluctuations by variations in the gravitational potential is known generally as the *Sachs–Wolfe effect*, and the region of the Δ_T curve where the temperature fluctuations are nearly constant ($l < 40$) is known as the Sachs–Wolfe plateau.

On smaller scales ($\theta < \theta_{\rm hor}$), the origin of the temperature fluctuations in the CMB is complicated by the behavior of the photons and baryons. Consider the situation immediately prior to photon decoupling. The photons, electrons, and protons together make a single photon-baryon fluid, whose energy density is only 40 percent that of the dark matter. Thus, the photon-baryon fluid moves primarily under the gravitational influence of the dark matter, rather than under its own self-gravity. If the photon-baryon fluid finds itself in a potential well of the dark matter, it will start to fall toward the center of the well. However, as the photon-baryon fluid is compressed by gravity, its pressure starts to rise along with its increasing density. Eventually, the pressure is sufficient to cause the fluid to expand outward. As the expansion continues, the pressure drops until gravity causes the photon-baryon fluid to fall inward again. As the cycle continues, the inward and outward oscillations of the photon-baryon fluid are called *acoustic oscillations*, since they represent a standing sound wave.

If the photon-baryon fluid within a potential well is at maximum compression at the time of photon decoupling, its density will be higher than average, and since $T \propto \varepsilon_\gamma^{1/4}$, the liberated photons will be hotter than average. Conversely, if the photon-baryon fluid within a potential well is at maximum expansion at the time of decoupling, the liberated photons will be slightly cooler than average. If the photon-baryon fluid is in the process of expanding or contracting at the time of decoupling, the Doppler effect will cause the liberated photons to be cooler or hotter than average, depending on whether the photon-baryon fluid was moving away from our location or toward it at the time of photon decoupling. Computing the exact shape of the Δ_T versus l curve expected in a particular model universe is a rather complicated chore. Generally speaking, however, the first peak in the Δ_T curve (at $l \approx 220$ or $\theta \approx 0.8^\circ$) represents the potential wells within which the photon-baryon fluid had just reached maximum compression at the time of last scattering. These potential wells have a size comparable to the *sound horizon distance* for the photon-baryon fluid at the time of last scattering. The sound horizon distance d_s is the maximum proper distance that a sound wave in the photon-baryon fluid can have traveled since the Big Bang. By analogy with the usual horizon distance (Equation 8.55), the sound horizon distance at the time of last scattering is

$$d_s(t_{\rm ls}) = a(t_{\rm ls}) \int_0^{t_{\rm ls}} \frac{c_s(t)dt}{a(t)}. \tag{8.64}$$

For most of the time prior to last scattering, the baryons were an insignificant contaminant in the photon-baryon fluid. We can thus make the approximation that the sound speed in the photon-baryon fluid was $c_s \approx c/\sqrt{3}$, the same as that of a pure photon gas. With this approximation, the sound horizon distance at last scattering was

$$d_s(t_{\rm ls}) \approx \frac{1}{\sqrt{3}} d_{\rm hor}(t_{\rm ls}) \approx 0.145\,{\rm Mpc}. \tag{8.65}$$

Hot spots with this physical size have an angular size, as viewed by us today, of

$$\theta_s \approx \frac{d_s(t_{\rm ls})}{d_A} \approx \frac{0.145\,{\rm Mpc}}{12.8\,{\rm Mpc}} \approx 0.011\,{\rm rad} \approx 0.7^\circ. \tag{8.66}$$

The location and the amplitude of the first peak in Figure 8.6 provide very useful cosmological information. The angular size $\theta_{\rm peak}$ at which the peak occurs must be similar to the angle θ_s subtended by the sound horizon distance. However, the estimate for θ_s given in Equation 8.66 assumes a flat universe. In a negatively curved universe ($\kappa = -1$), the angular size θ of an object of known physical size at a known redshift is *smaller* than it is in a positively curved universe ($\kappa = +1$). Thus, if the universe were negatively curved, the first peak in Δ_T would be shifted to a smaller angle; if the universe were positively curved, the peak would be shifted to a larger angle. The observed position of the first acoustic peak is consistent with $\kappa = 0$, or $\Omega_0 = 1$. Figure 8.7 shows the values of $\Omega_{m,0}$ and $\Omega_{\Lambda,0}$ permitted by the present CMB data. Note that the contour that shows the best fit to the CMB data (the black ellipse) is roughly perpendicular to the contour permitted by the type Ia supernova results (the gray ellipse). The small overlap region between the two results tells us that a spatially flat universe, with $\Omega_{m,0} \sim 0.3$ and $\Omega_{\Lambda,0} \sim 0.7$, agrees with both the CMB results and the supernova results.

The *amplitude* of the first peak in the Δ_T versus l plot also yields useful cosmological knowledge. The amplitude is quite sensitive to the sound speed c_s of the photon-baryon fluid, with a lower speed giving a higher amplitude. Since the sound speed is $c_s = \sqrt{w_{\rm pb}}\, c$, and the equation-of-state parameter $w_{\rm pb}$ is in turn dependent on the baryon-to-photon ratio, the amplitude of the peak is a useful diagnostic of the baryon density of the universe. Detailed analysis of the Δ_T curve yields a baryon-to-photon ratio

$$\eta = (6.10 \pm 0.06) \times 10^{-10}. \tag{8.67}$$

This value for η can be converted into a value for the current baryon density by the relation

$$n_{\rm bary,0} = \eta n_{\gamma,0} = 0.251 \pm 0.003\,{\rm m}^{-3}. \tag{8.68}$$

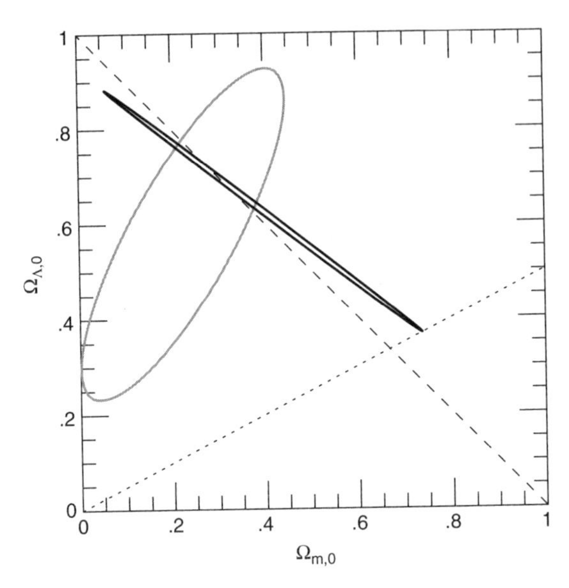

Figure 8.7 The elongated black ellipse represents the 95% confidence interval for the values of $\Omega_{m,0}$ and $\Omega_{\Lambda,0}$ that best fit the *Planck* CMB data. For comparison, the gray ellipse shows the 95% confidence interval from the supernova data, repeated from Figure 6.6. [Anže Slosar & José Alberto Vázquez, BNL]

Since the majority of baryons are protons, we may write, to acceptable accuracy,

$$\varepsilon_{\text{bary},0} = (m_p c^2) n_{\text{bary},0} = 235 \pm 3\,\text{MeV m}^{-3}. \tag{8.69}$$

This translates into a density parameter for baryons of

$$\Omega_{\text{bary},0} = \frac{\varepsilon_{\text{bary},0}}{\varepsilon_{c,0}} = 0.048 \pm 0.003, \tag{8.70}$$

with most of the uncertainty in $\Omega_{\text{bary},0}$ coming from the uncertainty in the critical density $\varepsilon_{c,0}$.

Exercises

8.1 The purpose of this problem is to determine how changing the value of the baryon-to-photon ratio, η, affects the recombination temperature in the early universe. Plot the fractional ionization X as a function of temperature, in the range $3000\,\text{K} < T < 4500\,\text{K}$; first make the plot assuming $\eta = 4 \times 10^{-10}$, then assuming $\eta = 8 \times 10^{-10}$. How much does this change in η affect the computed value of the recombination temperature T_{rec}, if we define T_{rec} as the temperature at which $X = \frac{1}{2}$?

8.2 Assuming a baryon-to-photon ratio $\eta = 6.1 \times 10^{-10}$, at what temperature T will there be one ionizing photon, with $hf > Q = 13.6\,\text{eV}$, per baryon? [Hint: the result of Exercise 2.5 will be useful.] Is the temperature you calculate greater than or less than $T_{\text{rec}} = 3760\,\text{K}$?

8.3 Imagine that at the time of recombination, the baryonic portion of the universe consisted entirely of ^{4}He (that is, helium with two protons and two neutrons in its nucleus). The ionization energy of helium (that is, the energy required to convert neutral He to He$^+$) is $Q_{\text{He}} = 24.6\,\text{eV}$. At what temperature would the fractional ionization of the helium be $X = \frac{1}{2}$? Assume that $\eta = 6 \times 10^{-10}$ and that the number density of He^{++} is negligibly small. [The relevant statistical weight factor for the ionization of helium is $g_{\text{He}}/(g_e g_{\text{He}^+}) = 1/4$.]

8.4 What is the proper distance d_p to the surface of last scattering? What is the luminosity distance d_L to the surface of last scattering? Assume that the Benchmark Model is correct, and that the redshift of the last scattering surface is $z_{\text{ls}} = 1090$.

9

Nucleosynthesis and the Early Universe

The cosmic microwave background tells us a great deal about the state of the universe at the time of last scattering ($t_{\rm ls} \approx 0.37\,{\rm Myr}$). However, the opacity of the early universe prevents us from directly seeing what the universe was like at $t < t_{\rm ls}$. Looking at the last scattering surface is like looking at the surface of a cloud, or the surface of the Sun; our curiosity is piqued, and we wish to find out what conditions are like in the opaque regions so tantalizingly hidden from our direct view.

Theoretically, many properties of the early universe should be quite simple. For instance, when radiation is strongly dominant over matter, at scale factors $a \ll a_{rm} \approx 2.9 \times 10^{-4}$, or times $t \ll t_{rm} \approx 50\,000\,{\rm yr}$, the expansion of the universe has the simple power-law form $a(t) \propto t^{1/2}$. The temperature of the blackbody photons in the early universe, which decreases as $T \propto a^{-1}$ as the universe expands, is given by the convenient relation

$$T(t) \approx 10^{10}\,{\rm K}\left(\frac{t}{1\,{\rm s}}\right)^{-1/2}, \tag{9.1}$$

or equivalently

$$kT(t) \approx 1\,{\rm MeV}\left(\frac{t}{1\,{\rm s}}\right)^{-1/2}. \tag{9.2}$$

Thus the mean energy per photon was

$$E_{\rm mean}(t) \approx 2.7kT(t) \approx 3\,{\rm MeV}\left(\frac{t}{1\,{\rm s}}\right)^{-1/2}. \tag{9.3}$$

The Large Hadron Collider (LHC), on the border between France and Switzerland, accelerates protons to an energy $E = 7 \times 10^{6}\,{\rm MeV} = 7\,{\rm TeV}$, about 7500 times the rest energy of a proton. The LHC is a remarkable piece of engineering; with a main ring 27 kilometers in diameter, it is sometimes called the largest single machine in the world. However, when the universe had an age of $t \approx 2 \times 10^{-13}$ s, the average, run-of-the-mill particle energy was equal to that attained by the LHC. Thus, the early universe is referred to as "the poor man's particle accelerator," since it provided particles of very high energy without running up an enormous electricity bill or requiring billions of euros in funding.

9.1 Nuclear Physics and Cosmology

As the universe has expanded and cooled, the mean energy per photon has dropped from $E_{\text{mean}}(t_P) \sim E_P \sim 10^{28}\,\text{eV}$ at the Planck time to $E_{\text{mean}}(t_0) \sim 10^{-3}\,\text{eV}$ at the present day. Thus, by studying the universe as it expands, we sample 31 orders of magnitude in particle energy. Within this wide energy range, some energies are of more interest than others to physicists. For instance, to physicists studying recombination and photoionization, the most interesting energy scale is the ionization energy of an atom. The ionization energy of hydrogen is $Q = 13.6\,\text{eV}$, as we have already noted. The ionization energies of other elements (that is, the energy required to remove the most loosely bound electron in the neutral atom) are roughly comparable; they range from 24.6 eV for helium to 4 eV for heavy alkali metals like cesium. Thus, atomic physicists, when considering the ionization of atoms, typically deal with energies of $\sim 10\,\text{eV}$, in round numbers.

Nuclear physicists are concerned not with ionization and recombination (removing or adding electrons to an atom), but with the much higher energy processes of fission and fusion (splitting or merging atomic nuclei). An atomic nucleus contains Z protons and N neutrons, where $Z \geq 1$ and $N \geq 0$. Protons and neutrons are collectively called *nucleons*. The total number of nucleons within an atomic nucleus is called the *mass number*, and is given by the formula $A = Z+N$. The proton number Z of a nucleus determines the atomic element to which that nucleus belongs. For instance, hydrogen (H) nuclei all have $Z = 1$, helium (He) nuclei have $Z = 2$, lithium (Li) nuclei have $Z = 3$, beryllium (Be) nuclei have $Z = 4$, and so on, through the complete periodic table. Although all atoms of a given element have the same number of protons in their nuclei, different isotopes of an element can have different numbers of neutrons. A particular isotope of an element is designated by prefixing the mass number A to the symbol for that element. For instance, a standard hydrogen nucleus, with one proton and no neutrons, is symbolized as ^{1}H. (Since an ordinary hydrogen nucleus is nothing more than a proton, we may also write p in place of ^{1}H when considering nuclear reactions.) Heavy hydrogen, or *deuterium*, contains one proton and one neutron, and is symbolized as ^{2}H. (Since the deuterium nucleus is mentioned frequently in the context of nuclear fusion, it has its own name, the "deuteron," and its own special symbol, D.) Ordinary helium contains two protons and two neutrons, and is symbolized as ^{4}He.

The binding energy B of a nucleus is the energy required to pull it apart into its component protons and neutrons. Equivalently, it is the energy released when a nucleus is fused together from individual protons and neutrons. For instance, when a neutron and a proton are bound together to form a deuteron, an energy of $B_D = 2.22\,\text{MeV}$ is released:

$$p + n \rightleftharpoons \text{D} + 2.22\,\text{MeV}. \tag{9.4}$$

The deuteron is not very tightly bound, compared to other atomic nuclei. Figure 9.1 plots the binding energy per nucleon (B/A) for atomic nuclei with different mass numbers. Note that ^{4}He, with a total binding energy of $B = 28.30\,\text{MeV}$, and a binding energy per nucleon of $B/A = 7.07\,\text{MeV}$, is relatively tightly bound, compared to other light nuclei (that is, nuclei with $A \le 10$). The most tightly bound nuclei are those of ^{56}Fe, ^{58}Fe, and ^{62}Ni, which all have $B/A \approx 8.79\,\text{MeV}$. Thus, nuclei more massive than iron or nickel can release energy by fission – splitting into lighter nuclei. Nuclei less massive than iron or nickel can release energy by fusion – merging into heavier nuclei.

Just as studies of ionization and recombination deal with an energy scale of $\sim 10\,\text{eV}$ (a typical ionization energy), so studies of nuclear fusion and fission deal with an energy scale of $\sim 8\,\text{MeV}$ (a typical binding energy per nucleon). Moreover, just as electrons and protons combined to form neutral hydrogen atoms when the temperature dropped sufficiently far below the ionization energy of hydrogen ($Q = 13.6\,\text{eV}$), so protons and neutrons must have fused to form deuterons when the temperature dropped sufficiently far below the binding energy

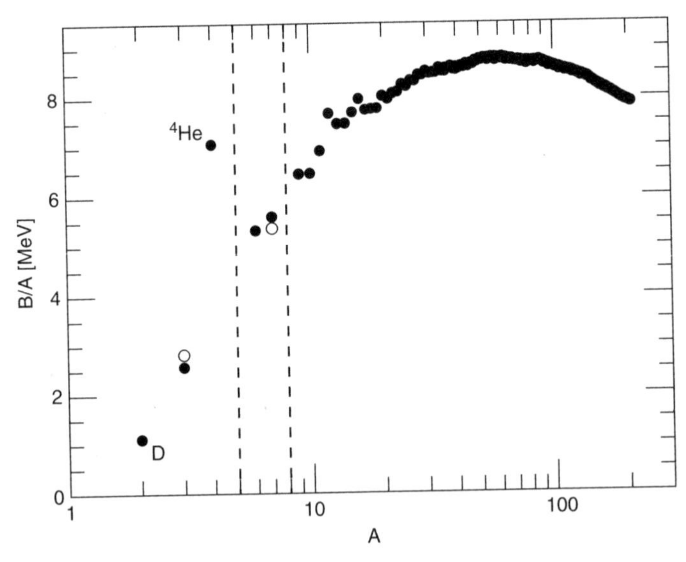

Figure 9.1 Binding energy per nucleon (B/A) as a function of the number of nucleons (A). Stable isotopes are shown as solid dots; the open dots represent the isotopes ^{3}H and ^{7}Be. [data from AME2012 atomic mass evaluation]

of deuterium ($B_D = 2.22\,\text{MeV}$). The epoch of recombination was thus preceded by an epoch of nuclear fusion, commonly called the epoch of Big Bang nucleosynthesis (BBN). Nucleosynthesis in the early universe starts by the fusion of neutrons and protons to form deuterons, then proceeds to form heavier nuclei by successive acts of fusion. Since the binding energy of deuterium is larger than the ionization energy of hydrogen by a factor $B_D/Q = 1.6 \times 10^5$, we would expect, as a rough estimate, the synthesis of deuterium to occur at a temperature 1.6×10^5 times higher than the recombination temperature $T_{\text{rec}} = 3760\,\text{K}$. That is, deuterium synthesis occurred at a temperature $T_{\text{nuc}} \approx 1.6 \times 10^5(3760\,\text{K}) \approx 6 \times 10^8\,\text{K}$, corresponding to a time $t_{\text{nuc}} \approx 300\,\text{s}$. This estimate, as we'll see when we do the detailed calculations, gives a temperature slightly too low, but it certainly gives the right order of magnitude. As indicated in the title of Steven Weinberg's classic book, *The First Three Minutes*, the entire saga of Big Bang nucleosynthesis takes place when the universe is only a few minutes old.

One thing we can say about Big Bang nucleosynthesis, after taking a look at the present-day universe, is that it was shockingly inefficient. From an energy viewpoint, the preferred universe would be one in which the baryonic matter consisted of an iron-nickel alloy. Obviously, we do not live in such a universe. Currently, three-fourths of the baryonic component (by mass) is still in the form of unbound protons, or ^{1}H. Moreover, when we look for nuclei heavier than ^{1}H, we find that they are primarily ^{4}He, a relatively lightweight nucleus; iron and nickel provide only 0.15% of the baryonic mass of our galaxy. The primordial helium fraction of the universe (that is, the helium fraction before nucleosynthesis begins in stars) is usually expressed as the dimensionless number

$$Y_p \equiv \frac{\rho(^4\text{He})}{\rho_{\text{bary}}}. \tag{9.5}$$

That is, Y_p is the mass density of ^{4}He divided by the mass density of all the baryonic matter. The outer regions of the Sun have a helium mass fraction $Y = 0.27$. However, the Sun is made of recycled interstellar gas, which was contaminated by helium formed in earlier generations of stars. When we look at astronomical objects of different sorts, we find a minimum value of $Y = 0.24$. That is, baryonic objects such as stars and gas clouds are all at least 24 percent helium by mass.[1]

9.2 Neutrons and Protons

The basic building blocks for nucleosynthesis are neutrons and protons. The rest energy of a neutron is greater than that of a proton by an amount

$$Q_n = m_n c^2 - m_p c^2 = 1.29\,\text{MeV}. \tag{9.6}$$

[1] Condensed objects that have undergone chemical or physical fractionation can be much lower in helium than this value. For instance, your helium fraction is $\ll 24\%$.

A free neutron is unstable. It decays through the emission of an electron and an electron antineutrino,

$$n \rightarrow p + e^- + \bar{\nu}_e, \tag{9.7}$$

with a decay time $\tau_n = 880\,\text{s}$. That is, if you start with a population of free neutrons, after a time t, a fraction $f = \exp(-t/\tau_n)$ will remain.[2] Since the energy Q_n released by the decay of a neutron into a proton is greater than the rest energy of an electron ($m_e c^2 = 0.51\,\text{MeV}$), the remainder of the energy is carried away by the kinetic energy of the electron and the energy of the electron antineutrino. With a decay time of only fifteen minutes, the existence of a free neutron is as fleeting as fame; once the universe was several hours old, it contained essentially no free neutrons. However, a neutron bound into a stable atomic nucleus is preserved against decay. Neutrons are still around today because they've been tied up in deuterium, helium, and other atoms.

Let's consider the state of the universe when its age was $t = 0.1\,\text{s}$. At that time, the temperature was $T \approx 3 \times 10^{10}\,\text{K}$, and the mean energy per photon was $E_{\text{mean}} \approx 10\,\text{MeV}$. This energy is much greater than the rest energy of an electron or positron, so there were positrons as well as electrons present at $t = 0.1\,\text{s}$, created by pair production:

$$\gamma + \gamma \rightleftharpoons e^- + e^+. \tag{9.8}$$

At $t = 0.1\,\text{s}$, neutrinos were still coupled to the baryonic matter, and neutrons and protons could convert freely back and forth through the interactions

$$n + \nu_e \rightleftharpoons p + e^- \tag{9.9}$$

and

$$n + e^+ \rightleftharpoons p + \bar{\nu}_e. \tag{9.10}$$

At this early time, all particles, including protons and neutrons, were in kinetic equilibrium, at a temperature $kT \approx 3\,\text{MeV} \ll m_p c^2$. Thus, the number density of neutrons and protons can be found from Equation 8.26, which gives the correct number density for nonrelativistic particles in kinetic equilibrium. For neutrons,

$$n_n = g_n \left(\frac{m_n kT}{2\pi\hbar^2} \right)^{3/2} \exp\left(-\frac{m_n c^2}{kT} \right), \tag{9.11}$$

and for protons,[3]

[2] The half-life, the time it takes for half the neutrons to decay, is related to the decay time by the relation $t_{1/2} = \tau_n \ln 2 = 610\,\text{s}$.

[3] I have left out the chemical potential terms, μ_n and μ_p, in Equations 9.11 and 9.12. At the high energies present in the early universe, as it turns out, chemical potentials are small enough to be safely neglected.

$$n_p = g_p \left(\frac{m_p kT}{2\pi\hbar^2}\right)^{3/2} \exp\left(-\frac{m_p c^2}{kT}\right). \tag{9.12}$$

Since the statistical weights of protons and neutrons are equal, with $g_p = g_n = 2$, the neutron-to-proton ratio, from Equations 9.11 and 9.12, is

$$\frac{n_n}{n_p} = \left(\frac{m_n}{m_p}\right)^{3/2} \exp\left(-\frac{(m_n - m_p)c^2}{kT}\right). \tag{9.13}$$

The above equation can be simplified. First, $(m_n/m_p)^{3/2} = 1.002$; there will be no great loss in accuracy if we set this factor equal to one. Second, the difference in rest energy of the neutron and proton is $(m_n - m_p)c^2 = Q_n = 1.29\,\text{MeV}$. Thus, the equilibrium neutron-to-proton ratio has the particularly simple form

$$\frac{n_n}{n_p} = \exp\left(-\frac{Q_n}{kT}\right), \tag{9.14}$$

illustrated as the solid line in Figure 9.2. At temperatures $kT \gg Q_n = 1.29\,\text{MeV}$, corresponding to $T \gg 1.5 \times 10^{10}\,\text{K}$ and $t \ll 1\,\text{s}$, the number of neutrons is nearly equal to the number of protons. However, as the temperature starts to drop below $1.5 \times 10^{10}\,\text{K}$, protons begin to be strongly favored, and the neutron-to-proton ratio plummets exponentially.

If the neutrons and protons remained in equilibrium, then by the time the universe was six minutes old, there would be only one neutron for every million protons. However, neutrons and protons do not remain in equilibrium for nearly

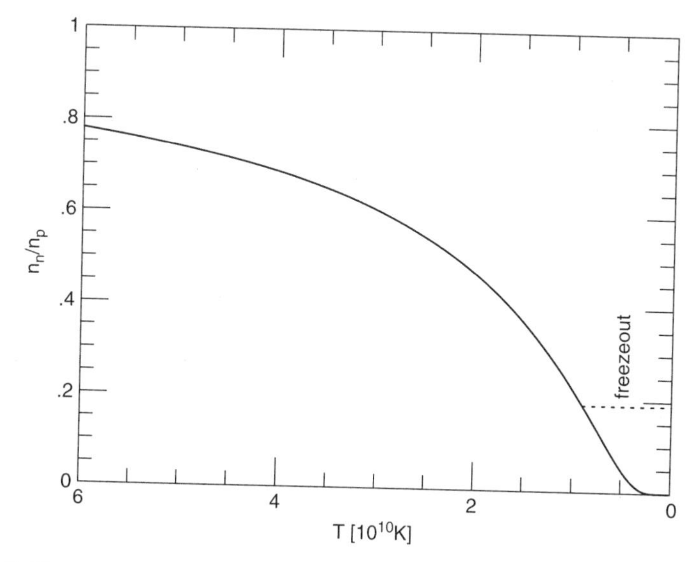

Figure 9.2 Neutron-to-proton ratio in the early universe. The solid line assumes equilibrium; the dotted line gives the value after freezeout. Temperature decreases, and thus time increases, from left to right.

that long. The interactions that mediate between neutrons and protons in the early universe, shown in Equations 9.9 and 9.10, involve the interaction of a baryon with a neutrino (or antineutrino). Neutrinos interact with baryons via the weak nuclear force. At the temperatures we are considering, the cross-sections for weak interactions are small. At temperatures $kT \sim 1\,\mathrm{MeV}$, the cross-section for the interaction of a neutrino with a proton or neutron is

$$\sigma_w \sim 10^{-47}\,\mathrm{m}^2 \left(\frac{kT}{1\,\mathrm{MeV}}\right)^2 . \tag{9.15}$$

(Compare this to the Thomson cross-section for the interaction of electrons via the electromagnetic force: $\sigma_e = 6.65 \times 10^{-29}\,\mathrm{m}^2$.) In the radiation-dominated universe, the temperature falls at the rate $T \propto a(t)^{-1} \propto t^{-1/2}$, and thus the cross-sections for weak interactions diminish at the rate $\sigma_w \propto t^{-1}$. The number density of neutrinos falls at the rate $n_\nu \propto a(t)^{-3} \propto t^{-3/2}$, and hence the rate Γ with which neutrons and protons interact with neutrinos via the weak force falls as a steep power of time:

$$\Gamma = n_\nu c \sigma_w \propto t^{-5/2}. \tag{9.16}$$

Meanwhile, the Hubble parameter is decreasing only at the rate $H \propto t^{-1}$. When $\Gamma \approx H$, the neutrinos decouple from the neutrons and protons, and the ratio of neutrons to protons is "frozen" (at least until the neutrons start to decay, at times $t \sim \tau_n$). An exact calculation of the temperature T_{freeze} at which $\Gamma = H$ requires a knowledge of the exact cross-section of the proton and neutron for weak interactions. Using the best available laboratory information, the "freezeout temperature" turns out to be $kT_{\mathrm{freeze}} = 0.8\,\mathrm{MeV}$, or $T_{\mathrm{freeze}} = 9 \times 10^9\,\mathrm{K}$. The universe reaches this temperature when its age is $t_{\mathrm{freeze}} \sim 1\,\mathrm{s}$. The neutron-to-proton ratio, once the temperature drops below T_{freeze}, is frozen at the value

$$\frac{n_n}{n_p} = \exp\left(-\frac{Q_n}{kT_{\mathrm{freeze}}}\right) \approx \exp\left(-\frac{1.29\,\mathrm{MeV}}{0.8\,\mathrm{MeV}}\right) \approx 0.2. \tag{9.17}$$

At times $t_{\mathrm{freeze}} < t \ll \tau_n$, there was one neutron for every five protons in the universe.

The scarcity of neutrons relative to protons explains why Big Bang nucleosynthesis was so incomplete, leaving three-fourths of the baryons in the form of unfused protons. A neutron will fuse with a proton much more readily than a proton will fuse with another proton. When a proton and neutron fuse to form a deuteron, the reaction is straightforward:

$$p + n \rightleftharpoons \mathrm{D} + \gamma. \tag{9.18}$$

There is no Coulomb barrier between the proton and neutron, and the reaction is mediated by the strong nuclear force. Thus, it has a large cross-section and a fast reaction rate. By contrast, the fusion of two protons to form a deuteron is an inefficient two-step process. First, the protons must overcome the Coulomb barrier between them to form a *diproton*, otherwise known as helium-2:

$$p + p \rightleftharpoons {}^2\text{He}. \tag{9.19}$$

A diproton is wildly unstable, and splits back to a pair of free protons with a lifetime $\tau_{\text{split}} \sim 10^{-23}$ s. There is, however, another possible decay mode for the diproton. It can decay to a deuteron through the reaction

$${}^2\text{He} \rightarrow \text{D} + e^+ + \nu_e, \tag{9.20}$$

which is the required second step of proton–proton fusion. The presence of a neutrino in Equation 9.20 tells us that, like the decay of a free neutron (Equation 9.7), it involves the weak nuclear force. The lifetime for decay through the reaction in Equation 9.20 has not been directly measured. However, decays that happen via the weak nuclear force all have relatively long lifetimes, with $\tau_{\text{weak}} > 10^{-2}$ s. Given the huge disparity between the lifetimes of the two possible decay modes for the diproton, the probability that a diproton will decay to a deuteron rather than back to a pair of free protons is tiny: $P \approx \tau_{\text{split}}/\tau_{\text{weak}} < 10^{-21}$.

It's possible, given enough time, to coax protons into fusing with each other. It's happening in the Sun, for instance, even as you read this sentence. However, fusion in the Sun is a very slow process. If you pick out any particular proton in the Sun's core, it has only one chance in ten billion of being fused into a deuteron during the next year, despite forming short-lived diprotons millions of times per second. The core of the Sun, though, is a stable environment; it's in hydrostatic equilibrium, and its temperature and density change only slowly with time. In the early universe, by strong contrast, the temperature drops as $T \propto t^{-1/2}$ and the density of baryons drops as $n_{\text{bary}} \propto t^{-3/2}$. Big Bang nucleosynthesis is a race against time. After less than an hour, the temperature and density have dropped too low for fusion to occur.

Given the alacrity of neutron–proton fusion when compared to the leisurely rate of proton–proton fusion, we can state, as a lowest order approximation, that BBN proceeds until every free neutron is bonded into an atomic nucleus, with the leftover protons remaining solitary.[4] In this approximation, we can compute the maximum possible value of Y_p, the fraction of the baryon mass in the form of ^{4}He. To compute the maximum possible value of Y_p, suppose that every neutron present after the proton–neutron freezeout is incorporated into a ^{4}He nucleus. Given a neutron-to-proton ratio of $n_n/n_p = 1/5$, we can consider a representative group of 2 neutrons and 10 protons. The 2 neutrons can fuse with 2 of the protons

[4] For the sake of completeness, note that the rate of neutron–neutron fusion is also negligibly small compared to the rate of neutron–proton fusion. A "dineutron," like a diproton, is unstable, and has an overwhelming probability of decaying back to a pair of free neutrons rather than decaying through the weak nuclear force to a deuteron.

to form a single ^{4}He nucleus. The remaining 8 protons, though, will remain unfused. The mass fraction of ^{4}He will then be

$$Y_{\max} = \frac{4}{12} = \frac{1}{3}. \tag{9.21}$$

More generally, if $f \equiv n_n/n_p$, with $0 \leq f \leq 1$, then the maximum possible value of Y_p is $Y_{\max} = 2f/(1+f)$.

If the observed value of $Y_p = 0.24$ were greater than the predicted $Y_{\max}$, that would be a cause for worry; it might mean, for example, that we didn't really understand the process of proton–neutron freezeout. However, the fact that the observed value of Y_p is less than $Y_{\max}$ is not worrisome; various factors act to reduce the actual value of Y_p below its theoretical maximum. First, if nucleosynthesis didn't take place immediately after freezeout at $t \approx 1$ s, then the spontaneous decay of neutrons would inevitably lower the neutron-to-proton ratio, and thus reduce the amount of ^{4}He produced. Next, if some neutrons escape fusion altogether, or end in nuclei lighter than ^{4}He (such as D or ^{3}He), they will not contribute to Y_p. Finally, if nucleosynthesis goes on long enough to produce nuclei heavier than ^{4}He, that too will reduce Y_p.

In order to compute Y_p accurately, as well as the abundances of other isotopes, it will be necessary to consider the process of nuclear fusion in more detail. Fortunately, much of the statistical mechanics we will need is a rehash of what we used when studying recombination.

9.3 Deuterium Synthesis

Let's move on to the next stage of Big Bang nucleosynthesis, just after proton–neutron freezeout is complete. The time is $t \approx 2$ s. The neutron-to-proton ratio is $n_n/n_p = 0.2$. The neutrinos, which ceased to interact with electrons about the same time they stopped interacting with neutrons and protons, are now decoupled from the rest of the universe. The photons, however, are still strongly coupled to the protons and neutrons. Big Bang nucleosynthesis takes place through a series of two-body reactions, building heavier nuclei step by step. The essential first step in BBN is the fusion of a proton and a neutron to form a deuteron:

$$p + n \rightleftharpoons \mathrm{D} + \gamma. \tag{9.22}$$

When a proton and a neutron fuse, the energy released (and carried away by a gamma-ray photon) is the binding energy of a deuteron:

$$B_{\mathrm{D}} = (m_n + m_p - m_{\mathrm{D}})c^2 = 2.22\,\mathrm{MeV}. \tag{9.23}$$

Conversely, a photon with energy $\geq B_{\mathrm{D}}$ can photodissociate a deuteron into its component proton and neutron. The reaction shown in Equation 9.22 should have a haunting familiarity if you've just read Chapter 8; it has the same structural form as the reaction governing the recombination of hydrogen:

$$p + e^- \rightleftharpoons H + \gamma. \tag{9.24}$$

A comparison of Equation 9.22 with Equation 9.24 shows that in each case, two particles become bound together to form a composite object, with the excess energy carried away by a photon. In the case of nucleosynthesis, a proton and neutron are bonded by the strong nuclear force to form a deuteron, with a gamma-ray photon being emitted. In the case of photoionization, a proton and electron are bonded by the electromagnetic force to form a neutral hydrogen atom, with an ultraviolet photon being emitted. A major difference between nucleosynthesis and recombination, of course, is between the energy scales involved.[5]

Despite the difference in energy scales, many of the equations used to analyze recombination can be re-used to analyze deuterium synthesis. Around the time of recombination, for instance, photoionization was in chemical equilibrium with radiative recombination (Equation 9.24). As a consequence, the relative numbers of free protons, free electrons, and neutral hydrogen atoms are given by the Saha equation,

$$\frac{n_H}{n_p n_e} = \left(\frac{m_e kT}{2\pi\hbar^2}\right)^{-3/2} \exp\left(\frac{Q}{kT}\right), \tag{9.25}$$

which tells us that neutral hydrogen is favored in the limit $kT \to 0$, and that ionized hydrogen is favored in the limit $kT \to \infty$. Around the time of deuterium synthesis, neutron–proton fusion was in chemical equilibrium with photodissociation of deuterons (Equation 9.22). As a consequence, the relative numbers of free protons, free neutrons, and deuterons are given by an equation directly analogous to Equation 8.28:

$$\frac{n_{\rm D}}{n_p n_n} = \frac{g_{\rm D}}{g_p g_n}\left(\frac{m_{\rm D}}{m_p m_n}\right)^{3/2}\left(\frac{kT}{2\pi\hbar^2}\right)^{-3/2} \exp\left(\frac{[m_p + m_n - m_{\rm D}]c^2}{kT}\right). \tag{9.26}$$

From Equation 9.23, we can make the substitution $[m_p + m_n - m_{\rm D}]c^2 = B_{\rm D}$. The statistical weight of the deuteron is $g_{\rm D} = 3$, in comparison to $g_p = g_n = 2$ for a proton or neutron. To acceptable accuracy, we may write $m_p = m_n = m_{\rm D}/2$. These substitutions yield the nucleosynthetic equivalent of the Saha equation,

$$\frac{n_{\rm D}}{n_p n_n} = 6\left(\frac{m_n kT}{\pi\hbar^2}\right)^{-3/2} \exp\left(\frac{B_{\rm D}}{kT}\right), \tag{9.27}$$

which tells us that deuterium is favored in the limit $kT \to 0$, and that free protons and neutrons are favored in the limit $kT \to \infty$.

[5] As the makers of bombs have long known, you can release much more energy by fusing atomic nuclei than by simply shuffling electrons around.

To define a precise temperature $T_{\rm nuc}$ at which the nucleosynthesis of deuterium takes place, we need to define what we mean by "the nucleosynthesis of deuterium." Just as recombination takes a finite length of time, so does nucleosynthesis. It is useful, though, to define $T_{\rm nuc}$ as the temperature at which $n_{\rm D}/n_n = 1$; that is, the temperature at which half the free neutrons have been fused into deuterons. As long as Equation 9.27 holds true, the deuteron-to-neutron ratio can be written as

$$\frac{n_{\rm D}}{n_n} = 6n_p \left(\frac{m_n kT}{\pi \hbar^2}\right)^{-3/2} \exp\left(\frac{B_{\rm D}}{kT}\right). \tag{9.28}$$

We can write the deuteron-to-neutron ratio as a function of T and the baryon-to-photon ratio η if we make some simplifying assumptions. Even today, we know that $\sim 75\%$ of all the baryons in the universe are in the form of unbound protons. Before the start of deuterium synthesis, five out of six baryons (or $\sim 83\%$) were in the form of unbound protons. Thus, if we don't want to be fanatical about accuracy, we can write

$$n_p \approx 0.8 n_{\rm bary} = 0.8\eta n_\gamma = 0.8\eta \left[0.2436\left(\frac{kT}{\hbar c}\right)^3\right], \tag{9.29}$$

using Equation 8.23 for the number density n_γ of blackbody photons. By substituting Equation 9.29 into Equation 9.28, we find that the deuteron-to-neutron ratio is a relatively simple function of temperature:

$$\frac{n_{\rm D}}{n_n} \approx 6.5\eta \left(\frac{kT}{m_n c^2}\right)^{3/2} \exp\left(\frac{B_{\rm D}}{kT}\right). \tag{9.30}$$

This function is plotted in Figure 9.3, assuming a baryon-to-photon ratio $\eta = 6.1 \times 10^{-10}$. The temperature $T_{\rm nuc}$ of deuterium nucleosynthesis can be found by solving the equation

$$1 \approx 6.5\eta \left(\frac{kT_{\rm nuc}}{m_n c^2}\right)^{3/2} \exp\left(\frac{B_{\rm D}}{kT_{\rm nuc}}\right). \tag{9.31}$$

With $m_n c^2 = 939.6\,{\rm MeV}$, $B_{\rm D} = 2.22\,{\rm MeV}$, and $\eta = 6.1 \times 10^{-10}$, the temperature of deuterium synthesis is $kT_{\rm nuc} \approx 0.066\,{\rm MeV} \approx B_{\rm D}/34$, corresponding to $T_{\rm nuc} \approx 7.6 \times 10^8\,{\rm K}$. The temperature drops to this value when the age of the universe is $t_{\rm nuc} \approx 200\,{\rm s}$.

Note that the time delay until the start of nucleosynthesis, $t_{\rm nuc} \approx 200\,{\rm s}$, is not negligible compared to the decay time of the neutron, $\tau_n = 880\,{\rm s}$. By the time nucleosynthesis actually gets underway, neutron decay has slightly decreased the neutron-to-proton ratio from $n_n/n_p = 1/5$ to

$$\frac{n_n}{n_p} \approx \frac{\exp(-200/880)}{5 + [1 - \exp(-200/880)]} \approx \frac{0.80}{5.20} \approx 0.15. \tag{9.32}$$

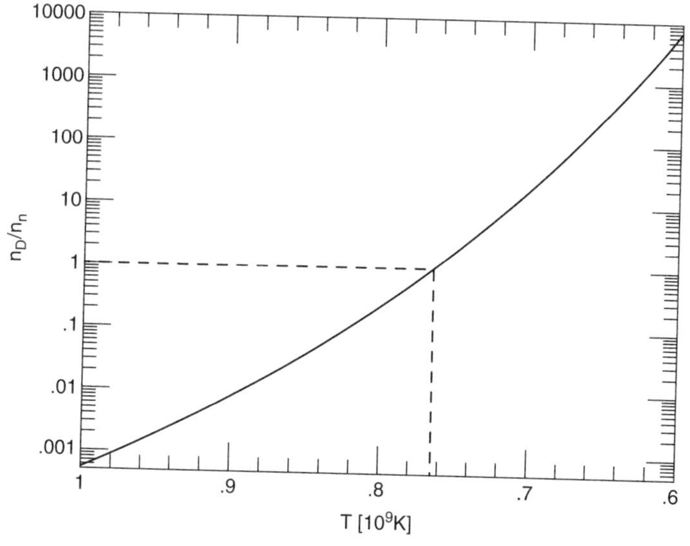

Figure 9.3 The deuteron-to-neutron ratio during the epoch of deuterium synthesis. The nucleosynthetic equivalent of the Saha equation (Equation 9.27) is assumed to hold true. Temperature decreases, and thus time increases, from left to right.

This in turn lowers the maximum possible ^{4}He mass fraction from $Y_{\max} \approx 0.33$ to $Y_{\max} \approx 0.27$.

9.4 Beyond Deuterium

The deuteron-to-neutron ratio $n_{\rm D}/n_n$ does not remain indefinitely at the equilibrium value given by Equation 9.30. Once a significant amount of deuterium forms, many possible nuclear reactions are available. For instance, a deuteron can fuse with a proton to form ^{3}He:

$$\mathrm{D} + p \rightleftharpoons {}^3\mathrm{He} + \gamma. \tag{9.33}$$

Alternatively, it can fuse with a neutron to form ^{3}H, also known as tritium:

$$\mathrm{D} + n \rightleftharpoons {}^3\mathrm{H} + \gamma. \tag{9.34}$$

Tritium is unstable; it spontaneously decays to ^{3}He, emitting an electron and an electron antineutrino in the process. However, the decay time of tritium is approximately 18 years; during the brief time that Big Bang nucleosynthesis lasts, tritium can be regarded as effectively stable.

Deuterons can also fuse with each other to form ^{4}He:

$$\mathrm{D} + \mathrm{D} \rightleftharpoons {}^4\mathrm{He} + \gamma. \tag{9.35}$$

However, it is more likely that the interaction of two deuterons will end in the formation of a tritium nucleus (with the emission of a proton),

$$\mathrm{D}+\mathrm{D} \rightleftharpoons {}^{3}\mathrm{H}+p, \tag{9.36}$$

or the formation of a ^{3}He nucleus (with the emission of a neutron),

$$\mathrm{D}+\mathrm{D} \rightleftharpoons {}^{3}\mathrm{He}+n. \tag{9.37}$$

A large amount of ^{3}H or ^{3}He is never present during the time of nucleosynthesis. Soon after they are formed, they are converted to ^{4}He by reactions such as

$$\begin{aligned} {}^{3}\mathrm{H}+p &\rightleftharpoons {}^{4}\mathrm{He}+\gamma \\ {}^{3}\mathrm{He}+n &\rightleftharpoons {}^{4}\mathrm{He}+\gamma \\ {}^{3}\mathrm{H}+\mathrm{D} &\rightleftharpoons {}^{4}\mathrm{He}+n \\ {}^{3}\mathrm{He}+\mathrm{D} &\rightleftharpoons {}^{4}\mathrm{He}+p. \end{aligned} \tag{9.38}$$

None of the post-deuterium reactions outlined in Equations 9.33 through 9.38 involve neutrinos; they all involve the strong nuclear force, and have large cross-sections and fast reaction rates. Thus, once nucleosynthesis begins, D, ^{3}H, and ^{3}He are all efficiently converted to ^{4}He.

Once ^{4}He is reached, however, the orderly march of nucleosynthesis to heavier and heavier nuclei reaches a roadblock. For such a light nucleus, ^{4}He is exceptionally tightly bound, as illustrated in Figure 9.1. By contrast, there are no stable nuclei with $A = 5$. If you try to fuse a proton or neutron to ^{4}He, it won't work; ^{5}He and ^{5}Li are not stable nuclei. Thus, ^{4}He is resistant to fusion with protons and neutrons. Small amounts of ^{6}Li and ^{7}Li, the two stable isotopes of lithium, are made by reactions such as

$${}^{4}\mathrm{He}+\mathrm{D} \rightleftharpoons {}^{6}\mathrm{Li}+\gamma \tag{9.39}$$

and

$${}^{4}\mathrm{He}+{}^{3}\mathrm{H} \rightleftharpoons {}^{7}\mathrm{Li}+\gamma. \tag{9.40}$$

In addition, small amounts of ^{7}Be are made by reactions such as

$${}^{4}\mathrm{He}+{}^{3}\mathrm{He} \rightleftharpoons {}^{7}\mathrm{Be}+\gamma. \tag{9.41}$$

The synthesis of nuclei with $A > 7$ is hindered by the absence of stable nuclei with $A = 8$. For instance, if ^{8}Be is made by the reaction

$${}^{4}\mathrm{He}+{}^{4}\mathrm{He} \rightarrow {}^{8}\mathrm{Be}, \tag{9.42}$$

then the ^{8}Be nucleus falls back apart into a pair of ^{4}He nuclei with a decay time of only $\tau = 10^{-16}$ s.

The bottom line is that once deuterium begins to be formed, fusion up to the tightly bound ^{4}He nucleus proceeds very rapidly. Fusion of heavier nuclei occurs much less rapidly. The precise yields of the different isotopes involved in BBN are customarily calculated using a fairly complex computer code. The complexity is necessary because of the large number of possible reactions that can occur once deuterium has been formed, all of which have temperature-dependent cross-sections. Thus, there's a good deal of bookkeeping involved. The results of a typical BBN code, which follows the mass fraction of different isotopes as the universe expands and cools, is shown in Figure 9.4. At $t < 10$ s, when $T > 3 \times 10^9$ K, almost all the baryonic matter is in the form of free protons and free neutrons. As the deuterium density climbs upward, however, the point is eventually reached where significant amounts of ^{3}H, ^{3}He, and ^{4}He are formed. By $t \sim 1000$ s, when the temperature has dropped to $T \sim 3 \times 10^8$ K, Big Bang nucleosynthesis is essentially over. Nearly all the baryons are in the form of free protons or ^{4}He nuclei. The small residue of free neutrons decays into protons. Small amounts of D, ^{3}H, and ^{3}He are left over, a tribute to the incomplete nature of Big Bang nucleosynthesis. (^{3}H later decays to ^{3}He.) Very small amounts of ^{6}Li, ^{7}Li, and ^{7}Be are made. (^{7}Be is later converted to ^{7}Li by electron capture: $^7\text{Be} + e^- \rightarrow {}^7\text{Li} + \nu_e$.)

The yields of D, ^{3}He, ^{4}He, ^{6}Li, and ^{7}Li depend on various physical parameters. Most importantly, they depend on the baryon-to-photon ratio η. Figure 9.5 shows the abundance of various elements produced by Big Bang nucleosynthesis, plotted as a function of η. A high baryon-to-photon ratio increases the temperature T_{nuc} at which deuterium synthesis occurs, and hence gives an earlier start to Big Bang nucleosynthesis. Since BBN is a race against the clock as the density and

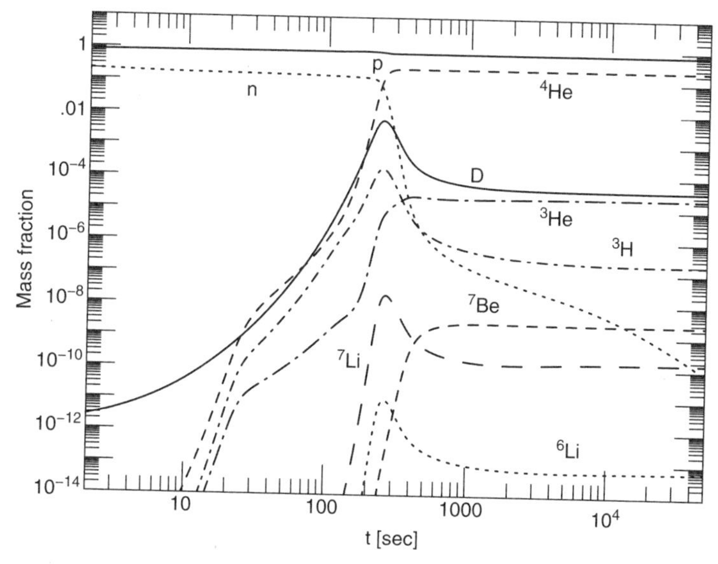

Figure 9.4 Mass fraction of nuclei as a function of time during the epoch of nucleosynthesis. Time increases, and thus temperature decreases, from left to right. [data courtesy of Alain Coc]

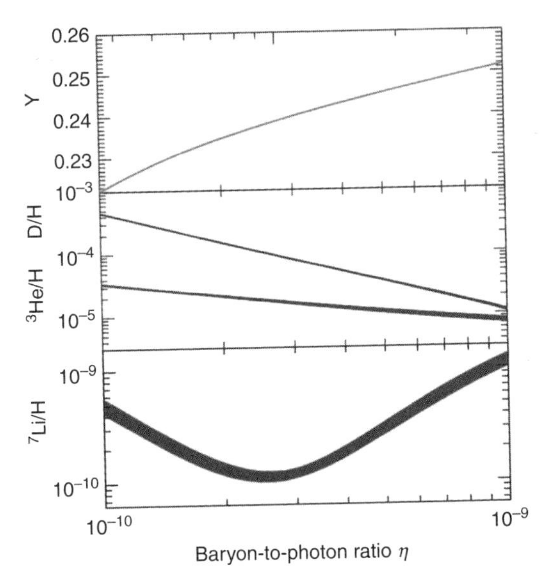

Figure 9.5 The mass fraction of ^{4}He, and the number densities of D, ^{3}He, and ^{7}Li expressed as a fraction of the H number density. The width of each line represents the 1σ confidence interval in the density. [Cyburt *et al.* 2016, *Rev. Mod. Phys.*, **88**, 015004]

temperature of the universe drop, getting an earlier start means that nucleosynthesis is more efficient at producing ^{4}He, leaving less D and ^{3}He as leftovers. The dependence of ^{7}Li on η is more complicated. Within the range of η plotted in Figure 9.5, the direct production of ^{7}Li by the fusion of ^{4}He and ^{3}H is a decreasing function of η, while the indirect production of ^{7}Li by ^{7}Be electron capture is an increasing function of η. The net result is a minimum in the predicted density of ^{7}Li at $\eta \approx 3 \times 10^{-10}$.

To determine the value of η using the predictions of Big Bang nucleosynthesis, it is necessary to make accurate observations of the *primordial* densities of the light elements; that is, the densities before nucleosynthesis in stars started to alter the chemical composition of the universe. In determining the value of η, it is most useful to determine the primordial abundance of deuterium. This is because the deuterium abundance is strongly dependent on η in the range of interest. Thus, determining the deuterium abundance with only modest accuracy will enable us to determine η fairly well. By contrast, the primordial helium fraction, Y_p, has only a weak dependence on η for the range of interest, as shown in Figure 9.5. Thus, determining η with a fair degree of accuracy would require measuring Y_p with fanatic precision.

Deuterium abundances are customarily given as the ratio of the number of deuterium atoms to the number of ordinary hydrogen atoms (D/H). In the local interstellar gas, within $\sim 50\,\text{pc}$ of the Sun, the deuterium-to-hydrogen ratio is of

$D/H \approx 1.6 \times 10^{-5}$. However, deuterium is very easily destroyed in stars. Since the interstellar gas is contaminated by deuterium-depleted gas that has been cycled through stellar interiors, we expect the primordial deuterium-to-hydrogen ratio was $D/H > 1.6 \times 10^{-5}$. Currently, the best way to find the primordial value of D/H is to look at the spectra of distant quasars. In the search for deuterium, we don't care what a quasar actually is, or how much deuterium is inside the quasar itself; instead, we just want to use the very luminous quasar as a flashlight to illuminate the intergalactic gas clouds that lie between it and us. If an intergalactic gas cloud contains no detectable stars, and has very low levels of elements heavier than lithium, we can hope that its D/H value is close to the primordial value, and hasn't been driven downward by the effects of fusion within stars. Neutral hydrogen atoms within these intergalactic clouds absorb photons whose energy corresponds to the Lyman-α transition; that is, the transition of the atom's electron from the ground state ($n = 1$) to the next higher energy level ($n = 2$). In an ordinary hydrogen atom (^{1}H), the Lyman-α transition corresponds to a wavelength $\lambda_{\rm H} = 121.567\,{\rm nm}$. In a deuterium atom, the greater mass of the nucleus causes a small isotopic shift in the electron's energy levels. As a consequence, the Lyman-α transition in deuterium corresponds to a slightly shorter wavelength, $\lambda_{\rm D} = 121.534\,{\rm nm}$. When we look at light from a quasar that has passed through an intergalactic cloud at redshift $z_{\rm cl}$, we see a strong absorption line at $\lambda_{\rm H}(1 + z_{\rm cl})$, due to absorption from ordinary hydrogen, and a much weaker absorption line at $\lambda_{\rm D}(1 + z_{\rm cl})$, due to absorption from deuterium. Detailed studies of the strength of the absorption lines in the spectra of different quasars give the ratio $({\rm D/H}) = (2.53 \pm 0.04) \times 10^{-5}$. Using the results of BBN calculations such as those plotted in Figure 9.5, this translates into a baryon-to-photon ratio $\eta = (6.0 \pm 0.1) \times 10^{-10}$, consistent with the value found from the temperature fluctuations of the cosmic microwave background (Equation 8.67).

9.5 Baryon–Antibaryon Asymmetry

The results of Big Bang nucleosynthesis tell us what the universe was like when it was relatively hot ($T_{\rm nuc} \approx 7.6 \times 10^8\,{\rm K}$) and dense:

$$\varepsilon_{\rm nuc} \approx \alpha T_{\rm nuc}^4 \approx 1.6 \times 10^{33}\,{\rm MeV\,m^{-3}}. \tag{9.43}$$

This energy density corresponds to a mass density $\varepsilon_{\rm nuc}/c^2 \approx 2800\,{\rm kg\,m^{-3}}$, or nearly three times the density of water. Remember, though, that the energy density at the time of BBN was almost entirely in the form of radiation. The mass density of baryons at the time of BBN was

$$\rho_{\rm bary}(t_{\rm nuc}) = \Omega_{\rm bary,0}\rho_{c,0}\left(\frac{T_{\rm nuc}}{T_0}\right)^3 \approx 0.009\,{\rm kg\,m^{-3}}. \tag{9.44}$$

A density of several grams per cubic meter is not outlandishly high, by everyday standards; it's equal to the density of the Earth's stratosphere. A mean photon energy of $2.7kT_{\rm nuc} \approx 0.18\,{\rm MeV}$ is not outlandishly high, by everyday standards; you are bombarded with photons of about a third that energy when you have your teeth X-rayed at the dentist. The physics of Big Bang nucleosynthesis is well understood.

Some of the initial conditions for Big Bang nucleosynthesis, however, are rather puzzling. The baryon-to-photon ratio, $\eta \approx 6 \times 10^{-10}$, is a remarkably small number; the universe seems to have a strong preference for photons over baryons. It's also worthy of remark that the universe seems to have a strong preference for baryons over antibaryons. The laws of physics demand the presence of antiprotons ($\bar{p}$), containing two "anti-up" quarks and one "anti-down" quark apiece, as well as antineutrons ($\bar{n}$), containing one "anti-up" quark and two "anti-down" quarks apiece.[6] In practice, though, we find that the universe has an extremely large excess of protons and neutrons over antiprotons and antineutrons (and hence an excess of quarks over antiquarks). At the time of Big Bang nucleosynthesis, the number density of antibaryons ($\bar{n}$ and $\bar{p}$) was tiny compared to the number density of baryons, which in turn was tiny compared to the number density of photons. This imbalance, $n_{\rm antibary} \ll n_{\rm bary} \ll n_\gamma$, has its origin in the physics of the very early universe.

When the temperature of the early universe was greater than $kT \approx 150\,{\rm MeV}$, the quarks it contained were not confined within baryons and other particles, as they are today, but formed a sea of free quarks (sometimes referred to by the oddly culinary name of "quark soup"). During the first few microseconds of the universe, when the quark soup was piping hot, quarks and antiquarks were constantly being created by pair production and destroyed by mutual annihilation:

$$\gamma + \gamma \rightleftharpoons q + \bar{q}, \tag{9.45}$$

where q and $\bar{q}$ could represent, for instance, an "up" quark and an "anti-up" quark, or a "down" quark and an "anti-down" quark. During this period of quark pair production, the numbers of "up" quarks, "anti-up" quarks, "down" quarks, "anti-down" quarks, and photons were nearly equal to each other. However, suppose there were a very tiny asymmetry between quarks and antiquarks, such that

$$\delta_q \equiv \frac{n_q - n_{\bar{q}}}{n_q + n_{\bar{q}}} \ll 1. \tag{9.46}$$

As the universe expanded and the quark soup cooled, quark–antiquark pairs would no longer be produced. The existing antiquarks would then annihilate with

6 Note that an "anti-up" quark is *not* the same as a "down" quark; nor is "anti-down" equivalent to "up."

the quarks. However, because of the small excess of quarks over antiquarks, there would be a residue of quarks with number density

$$\frac{n_q}{n_\gamma} \sim \delta_q. \tag{9.47}$$

Thus, if there were 800 000 003 quarks for every 800 000 000 antiquarks in the early universe, three lucky quarks would be left over after the others encountered antiquarks and were annihilated. The leftover quarks, however, would be surrounded by 1.6 billion photons, the product of the annihilations. After the three quarks were bound together into a baryon at $kT \approx 150\,\text{MeV}$, the resulting baryon-to-photon ratio would be $\eta \sim 6 \times 10^{-10}$.

Thus, the very strong asymmetry between baryons and antibaryons today and the large number of photons per baryon are both products of a tiny asymmetry between quarks and antiquarks in the early universe. The exact origin of the quark–antiquark asymmetry in the early universe is still not known. The physicist Andrei Sakharov, as far back as 1967, was the first to outline the necessary physical conditions for producing a small asymmetry; however, the precise mechanism by which the quarks first developed their few-parts-per-billion advantage over antiquarks still remains to be found.

Exercises

9.1 Suppose the neutron decay time were $\tau_n = 88\,\text{s}$ instead of $\tau_n = 880\,\text{s}$, with all other physical parameters unchanged. Estimate Y_{max}, the maximum possible mass fraction in ^4He, assuming that all available neutrons are incorporated into ^4He nuclei.

9.2 Suppose the difference in rest energy of the neutron and proton were $Q_n = (m_n - m_p)c^2 = 0.129\,\text{MeV}$ instead of $Q_n = 1.29\,\text{MeV}$, with all other physical parameters unchanged. Estimate Y_{max}, the maximum possible mass fraction in ^4He, assuming that all available neutrons are incorporated into ^4He nuclei.

9.3 The total luminosity of the stars in our galaxy is $L \approx 3 \times 10^{10}\,\text{L}_\odot$. Suppose that the luminosity of our galaxy has been constant for the past 10 Gyr. How much energy has our galaxy emitted in the form of starlight during that time? Most stars are powered by the fusion of H into ^4He, with the release of 28.4 MeV for every helium nucleus formed. How many helium nuclei have been created within stars in our galaxy over the course of the past 10 Gyr, assuming that the fusion of H into ^4He is the only significant energy source? If the baryonic mass of our galaxy is $M \approx 10^{11}\,\text{M}_\odot$, by what amount has the helium fraction Y of our galaxy been increased over its primordial value $Y_4 = 0.24$?

9.4 In Section 9.2, it is asserted that the maximum possible value of the primordial helium fraction is

$$Y_{\text{max}} = \frac{2f}{1+f}, \tag{9.48}$$

where $f = n_n/n_p \leq 1$ is the neutron-to-proton ratio at the time of nucleosynthesis. Prove that this assertion is true.

9.5 The typical energy of a neutrino in the cosmic neutrino background, as pointed out in Chapter 5, is $E_\nu \sim kT_\nu \sim 5 \times 10^{-4}\,\text{eV}$. What is the approximate interaction cross-section σ_w for one of these cosmic neutrinos? Suppose you had a large lump of ^{56}Fe (with density $\rho = 7900\,\text{kg}\,\text{m}^{-3}$). What is the number density of protons, neutrons, and electrons within the lump of iron? How far, on average, would a cosmic neutrino travel through the iron before interacting with a proton, neutron, or electron? (Assume that the cross-section for interaction is simply σ_w, regardless of the type of particle the neutrino interacts with.)

10

Inflation and the Very Early Universe

The observed properties of galaxies, quasars, and supernovae at relatively small redshift ($z < 10$) tell us about the universe at times $t > 0.5\,\mathrm{Gyr}$. The properties of the cosmic microwave background tell us about the universe at the time of last scattering ($z_{\mathrm{ls}} = 1090$, $t_{\mathrm{ls}} = 0.37\,\mathrm{Myr}$). The abundances of light elements such as deuterium and helium tell us about the universe at the time of Big Bang nucleosynthesis ($z_{\mathrm{nuc}} \approx 3 \times 10^8$, $t_{\mathrm{nuc}} \approx 3\,\mathrm{min}$). In fact, the observation that primordial gas clouds are roughly one-fourth helium by mass, rather than being all helium or all hydrogen, tells us that we have a fair understanding of what was happening at the time of neutron–proton freezeout ($z_{\mathrm{freeze}} \approx 4 \times 10^9$, $t_{\mathrm{freeze}} \approx 1\,\mathrm{s}$).

So far, I've been emphasizing the successes of the Hot Big Bang model for the universe. These successes are indeed impressive; understanding the universe when it was just one second old is nothing to scorn. However, the standard Hot Big Bang model, in which the early universe was radiation-dominated, is not without its flaws. In particular, after the discovery of the cosmic microwave background led to the widespread embrace of the Big Bang, it was realized that the standard Hot Big Bang scenario had three underlying problems. These nagging problems were called the *flatness problem*, the *horizon problem*, and the *monopole problem*. The flatness problem can be summarized by the statement, "The universe is nearly flat today, and was even flatter in the past." The horizon problem can be summarized by the statement, "The universe is nearly isotropic and homogeneous today, and was even more so in the past." The monopole problem can be summarized by the statement, "The universe is apparently free of magnetic monopoles." To see why these simple statements pose a problem to the standard Hot Big Bang scenario, it is necessary to go a little deeper into the physics of the expanding universe.

10.1 The Flatness Problem

Let's start by examining the flatness problem. The spatial curvature of the universe is related to the density parameter Ω by the Friedmann equation:

$$1 - \Omega(t) = -\kappa \left(\frac{c/H(t)}{a(t)R_0} \right)^2 . \tag{10.1}$$

(Here we use the Friedmann equation in the form given by Equation 4.34.) At the present moment, the density parameter and curvature are linked by the equation

$$1 - \Omega_0 = -\kappa \left(\frac{c/H_0}{R_0} \right)^2 . \tag{10.2}$$

The CMB and type Ia supernova results (Figure 8.7), when combined with other observations, give the constraint

$$|1 - \Omega_0| \leq 0.005, \tag{10.3}$$

implying $R_0 \geq 14c/H_0$. Why should the value of Ω_0 be so close to one today? It could have had, for instance, the value $\Omega_0 = 10^{-6}$ or $\Omega_0 = 10^6$ without violating any laws of physics. We could of course invoke coincidence by saying that the initial conditions for the universe just happened to produce $\Omega_0 \approx 1$ today. However, when you extrapolate the value of $\Omega(t)$ backward into the past, the closeness of Ω to unity becomes more and more difficult to dismiss as a coincidence.

By combining Equations 10.1 and 10.2, we find the equation that gives the density parameter as a function of time:

$$1 - \Omega(t) = \frac{H_0^2(1 - \Omega_0)}{H(t)^2 a(t)^2} . \tag{10.4}$$

When the universe was dominated by radiation and matter, at times $t \ll t_{m\Lambda} \approx$ 10 Gyr, the Hubble parameter was given by Equation 5.108:

$$\frac{H(t)^2}{H_0^2} = \frac{\Omega_{r,0}}{a^4} + \frac{\Omega_{m,0}}{a^3} . \tag{10.5}$$

Thus, the density parameter evolved at the rate

$$1 - \Omega(t) = \frac{(1 - \Omega_0)a^2}{\Omega_{r,0} + a\Omega_{m,0}} . \tag{10.6}$$

During the period when the universe was dominated by radiation and matter, the deviation of Ω from one was constantly growing. During the radiation-dominated phase,

$$|1 - \Omega|_r \propto a^2 \propto t. \tag{10.7}$$

During the later matter-dominated phase,

$$|1 - \Omega|_m \propto a \propto t^{2/3} . \tag{10.8}$$

Suppose, as the available evidence indicates, that the universe is described by a model close to the Benchmark Model, with $\Omega_{r,0} \approx 9.0 \times 10^{-5}$, $\Omega_{m,0} \approx 0.31$, and $\Omega_{\Lambda,0} \approx 0.69$. If the total density parameter today falls within the limits $|1 - \Omega_0| \leq 0.005$, then at the time of radiation–matter equality, $a_{rm} \approx 2.9 \times 10^{-4}$, the density parameter Ω_{rm} was equal to one with an accuracy

$$|1 - \Omega_{rm}| \leq 2 \times 10^{-6}. \tag{10.9}$$

If we extrapolate backward to the time of Big Bang nucleosynthesis, at $a_{\text{nuc}} \approx 3.6 \times 10^{-9}$, the deviation of the density parameter Ω_{nuc} from one was only

$$|1 - \Omega_{\text{nuc}}| \leq 7 \times 10^{-16}. \tag{10.10}$$

At the time that deuterium was forming, the density of the universe was equal to the critical density to an accuracy of better than one part in a quadrillion. Let's push our extrapolation as far back as we dare, to the Planck time at $t_P \sim 5 \times 10^{-44}$ s, when the scale factor was $a_P \sim T_0/T_P \sim 2.7\,\text{K}/1.4 \times 10^{32}\,\text{K} \sim 2 \times 10^{-32}$. At the Planck time, we find that the density parameter Ω_P was extraordinarily close to one:

$$|1 - \Omega_P| \leq 2 \times 10^{-62}. \tag{10.11}$$

The number 2×10^{-62} is, of course, quite tiny. To use an analogy, in order to change the Sun's mass by two parts in 10^{62}, you would have to add or subtract a twentieth of an electron. Our very existence depends on the fanatically close balance between the actual density and the critical density in the early universe. If, for instance, the deviation of Ω from one at the time of nucleosynthesis had been a part per million instead of a part per quadrillion, the universe would have collapsed in a Big Crunch or expanded to a low-density Big Chill after only a few decades. In that case, galaxies, stars, planets, and cosmologists would not have had time to form.

You might try to dismiss the extreme flatness of the early universe as a coincidence, by saying, "Ω_P might have had any value, but it just happened to be $1 \pm 2 \times 10^{-62}$." However, a coincidence at this level is *extremely* far-fetched. It would be far more satisfactory if we could find a physical mechanism for flattening the universe early in its history, instead of relying on extremely contrived initial conditions at the Planck time.

10.2 The Horizon Problem

The "flatness problem," the remarkable closeness of Ω to one in the early universe, is puzzling. It is accompanied, however, by the "horizon problem," which is, if anything, even more puzzling. The horizon problem is simply the statement that the universe is nearly homogeneous and isotropic on very large scales. Why should we regard this as a problem? So far, we've treated the homogeneity and isotropy of the universe as a blessing rather than a curse. It's the homogeneity and isotropy of the universe, after all, that permit us to describe its curvature by the relatively simple Robertson–Walker metric, and its expansion by the relatively simple Friedmann equation. If the universe were inhomogeneous and anisotropic on large scales, it would be much more difficult to describe mathematically.

The universe, however, is under no obligation to make things simple for cosmologists. To see why the large-scale homogeneity and isotropy of the universe is so unexpected in the standard Hot Big Bang scenario, consider two antipodal points on the last scattering surface, as illustrated in Figure 10.1. The current proper distance to the last scattering surface is

$$d_p(t_0) = c \int_{t_{\rm ls}}^{t_0} \frac{dt}{a(t)}. \tag{10.12}$$

Since the last scattering of the CMB photons occurred a long time ago ($t_{\rm ls} \ll t_0$), the current proper distance to the last scattering surface is only slightly smaller than the current horizon distance. For the Benchmark Model, the current proper distance to the last scattering surface is $d_p(t_0) = 0.98 d_{\rm hor}(t_0)$. Thus, two antipodal points on the last scattering surface, separated by 180° as seen by an observer on Earth, are currently separated by a proper distance of $1.96 d_{\rm hor}(t_0)$. Since the two points are farther apart than the horizon distance, they are causally disconnected. That is, they haven't had time to send messages to each other, and in particular, haven't had time to come into thermal equilibrium with each other. Nevertheless, the two points have the same temperature (once the dipole distortion is subtracted) to within one part in 10^5. Why? How can two points that haven't had time to swap information be so nearly identical in their properties?

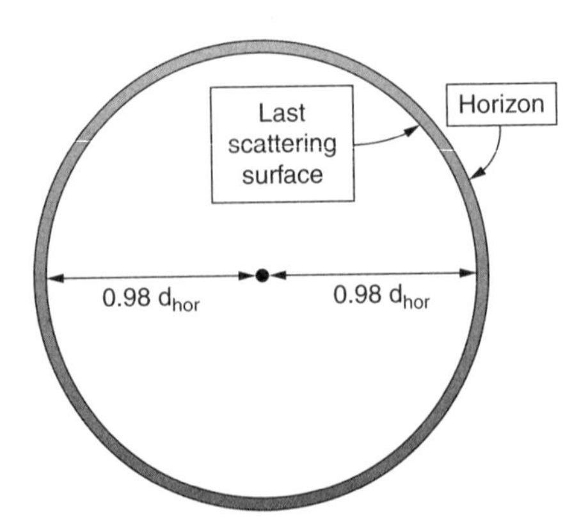

Figure 10.1 In the standard Hot Big Bang scenario, the current proper distance to the last scattering surface is 98 percent of the current horizon distance.

The near-isotropy of the cosmic microwave background is still more remarkable when we recall that the temperature fluctuations in the CMB result from the density and velocity fluctuations that existed at the time of last scattering. In the standard Hot Big Bang scenario, the horizon distance at the time of last scattering was, from Equation 8.56, $d_{\rm hor}(t_{\rm ls}) = 0.251\,{\rm Mpc}$. Thus, points more than 0.251 megaparsecs apart at the time of last scattering were not in causal contact in the standard Hot Big Bang scenario. The angular-diameter distance to the last scattering surface is $d_A \approx 12.8\,{\rm Mpc}$, as computed in Section 8.4. Thus, points on the last scattering surface that were separated by a horizon distance will have an angular separation equal to

$$\theta_{\rm hor} = \frac{d_{\rm hor}(t_{\rm ls})}{d_A} \approx \frac{0.251\,{\rm Mpc}}{12.8\,{\rm Mpc}} \approx 0.020\,{\rm rad} \approx 1.1^\circ \tag{10.13}$$

as seen from the Earth today. However, as Figure 8.6 shows, points on the last scattering surface separated by distances $\theta > 1.1^\circ$, corresponding to $l < 160$, have temperatures that are the same to within $30\,\mu{\rm K}$, just one part in 10^5 of the mean temperature, $T_0 = 2.7255\,{\rm K}$.

Why should regions that were out of causal contact with each other at $t_{\rm ls}$ have been so nearly homogeneous in their properties? Invoking coincidence ("The different patches just happened to have the same temperature") requires a great stretch of the imagination. The surface of last scattering can be divided into some 40 000 patches, each 1.1° across. In the standard Hot Big Bang scenario, the center of each of these patches was out of touch with the other patches at the time of last scattering. Now, if you invite two people to a potluck dinner, and they both bring potato salad, you can dismiss that as coincidence, even if they had 10^5 different dishes to choose from. However, if you invite 40 000 people to a potluck dinner, and they all bring potato salad, it starts to dawn on you that they must have been in contact with each other: "Psst ... let's all bring potato salad. Pass it on." Similarly, it starts to dawn on you that the different patches of the last scattering surface, in order to be so nearly equal in temperature, must have been in contact with each other: "Psst ... let's all be at $T = 2.7255\,{\rm K}$ when the universe is 13.7 gigayears old. Pass it on."

10.3 The Monopole Problem

The monopole problem – that is, the apparent lack of magnetic monopoles in the universe – is not a purely cosmological problem, but one that results from combining the Hot Big Bang scenario with the particle physics concept of a Grand Unified Theory. In particle physics, a Grand Unified Theory, or GUT, is a field theory that attempts to unify the electromagnetic force, the weak nuclear force, and the strong nuclear force. Unification of forces has been a goal of scientists since the 1870s, when James Clerk Maxwell demonstrated that electricity and magnetism are both manifestations of a single underlying electromagnetic field. Currently, it is customary to speak of the four fundamental forces of nature: gravitational, electromagnetic, weak, and strong. In the view of many physicists, though, four forces are three too many; they've spent much time and effort to show that two or more of the "fundamental forces" are actually different aspects of a single underlying force. About a century after Maxwell, Steven Weinberg, Abdus Salam, and Sheldon Glashow successfully devised an electroweak theory. They demonstrated that at particle energies greater than $E_{\rm ew} \sim 1\,{\rm TeV}$, the electromagnetic force and the weak force unite to form a single "electroweak" force. The electroweak energy of $E_{\rm ew} \sim 1\,{\rm TeV}$ corresponds to a temperature $T_{\rm ew} \sim E_{\rm ew}/k \sim 10^{16}\,{\rm K}$; the universe had this temperature when its age was $t_{\rm ew} \sim 10^{-12}\,{\rm s}$. Thus, when the universe was less than a picosecond old, there were only three fundamental forces: the gravitational, strong, and electroweak force. When the predictions of the electroweak energy were confirmed experimentally, Weinberg, Salam, and Glashow toted home their Nobel Prizes, and physicists braced themselves for the next step: unifying the electroweak force with the strong force.

By extrapolating the known properties of the strong and electroweak forces to higher particle energies, physicists estimate that at an energy $E_{\rm GUT} \sim 10^{12}\,{\rm TeV}$, the strong and electroweak forces should be unified as a single Grand Unified Force. The GUT energy $E_{\rm GUT} \sim 10^{12}\,{\rm TeV}$ corresponds to a temperature $T_{\rm GUT} \sim 10^{28}\,{\rm K}$ and an age for the universe of $t_{\rm GUT} \sim 10^{-36}\,{\rm s}$. The GUT energy is about four orders of magnitude smaller than the Planck energy, $E_P \sim 10^{16}\,{\rm TeV}$. Physicists are searching for a Theory of Everything (TOE) that describes how the Grand Unified Force and the force of gravity ultimately unite to form a single unified force at the Planck scale. The different unification energy scales, and the corresponding temperatures and times in the early universe, are shown in Figure 10.2.

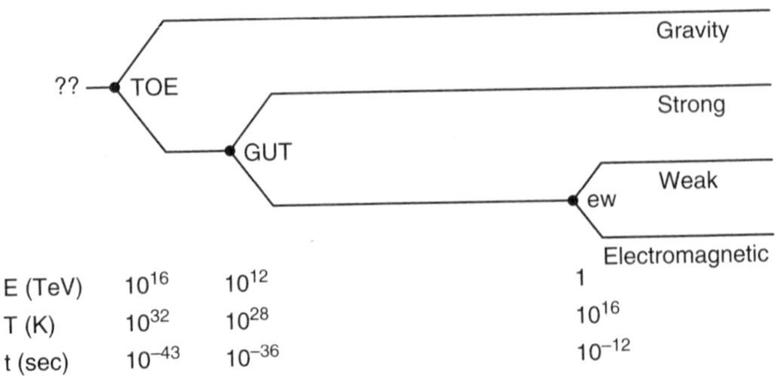

Figure 10.2 The energy, temperature, and time scales at which the different force unifications occur.

One of the predictions of Grand Unified Theories is that the universe underwent a *phase transition* as the temperature dropped below the GUT temperature. Generally speaking, phase transitions are associated with a spontaneous loss of symmetry as the temperature of a system is lowered. Take, as an example, the phase transition known as "freezing water." At temperatures $T > 273\,\text{K}$, water is liquid. Individual water molecules are randomly oriented, and the liquid water thus has rotational symmetry about any point; in other words, it is isotropic. However, when the temperature drops below $T = 273\,\text{K}$, the water undergoes a phase transition, from liquid to solid, and the rotational symmetry of the water is lost. The water molecules are locked into a crystalline structure, and the ice no longer has rotational symmetry about an arbitrary point. In other words, the ice crystal is anisotropic, with preferred directions corresponding to the crystal's axes of symmetry.[1] In a broadly similar vein, there is a loss of symmetry when the universe undergoes the GUT phase transition at $t_{\text{GUT}} \sim 10^{-36}\,\text{s}$. At $T > T_{\text{GUT}}$, there was a symmetry between the strong and electroweak forces. At $T < T_{\text{GUT}}$, the symmetry is spontaneously lost; the strong and electroweak forces begin to behave quite differently from each other.

In general, phase transitions associated with a loss of symmetry give rise to flaws known as *topological defects*. To see how topological defects form, consider a large tub of water cooled below $T = 273\,\text{K}$. Usually, the freezing of the water will start at two or more widely separated nucleation sites. The crystal that forms about any given nucleation site is very regular, with well-defined axes of symmetry. However, the axes of symmetry of two adjacent ice crystals will be misaligned. At the boundary of two adjacent crystals, there will be a two-dimensional topological defect, called a *domain wall*, where the axes of symmetry fail to line up. Other types of phase transitions give rise to one-dimensional, or line-like, topological defects (in a cosmological context, these linear defects are known as *cosmic strings*). Still other types of phase transitions give rise to zero-dimensional, or point-like, topological defects. Grand Unified Theories predict that the GUT phase transition creates point-like topological defects that act as *magnetic monopoles*. That is, they act as the isolated north pole or south pole of a magnet. The rest energy of the magnetic monopoles created in the GUT phase transition is predicted to be $m_{\text{MC}}c^2 \sim E_{\text{GUT}} \sim 10^{12}\,\text{TeV}$. This corresponds to a mass of over a nanogram (comparable to that of a bacterium), which is a lot of mass for a single particle to be carrying around. At the time of the GUT phase transition, points further apart than the horizon size will be out of causal contact with each other. Thus, we expect roughly one topological defect per horizon volume, due to the mismatch of fields that are not causally linked. The number density of magnetic monopoles, at the time of their creation, would be

[1] Suppose, for instance, that the water freezes in the familiar six-pointed form of a snowflake. It is now only symmetric with respect to rotations of 60° (or integral multiples of that angle) about the snowflake's center.

$$n_{\rm M}(t_{\rm GUT}) \sim \frac{1}{(2ct_{\rm GUT})^3} \sim 10^{82}\,{\rm m}^{-3}, \tag{10.14}$$

and their energy density would be

$$\varepsilon_{\rm M}(t_{\rm GUT}) \sim (m_{\rm M}c^2)n_{\rm M} \sim 10^{94}\,{\rm TeV\,m}^{-3}. \tag{10.15}$$

This is a large energy density, but it is smaller by ten orders of magnitude than the energy density of radiation at the time of the GUT phase transition:

$$\varepsilon_\gamma(t_{\rm GUT}) \approx \alpha T_{\rm GUT}^4 \sim 10^{104}\,{\rm TeV\,m}^{-3}. \tag{10.16}$$

Thus, the magnetic monopoles wouldn't have kept the universe from being radiation-dominated at the time of the GUT phase transition. However, the magnetic monopoles, being so massive, would soon have become highly non-relativistic, with energy density $\varepsilon_{\rm M} \propto a^{-3}$. The energy density in radiation, though, was falling off at the rate $\varepsilon_\gamma \propto a^{-4}$. Thus, the magnetic monopoles would have dominated the energy density of the universe when the scale factor had grown by a factor $\sim 10^{10}$; that is, when the temperature had fallen to $T \sim 10^{-10}T_{\rm GUT} \sim 10^{18}$ K, and the age of the universe was only $t \sim 10^{-16}$ s. It is obvious from observations, however, that the universe is *not* dominated by magnetic monopoles today. In fact, there is no strong evidence that they exist at all; monopole searches have placed an upper limit on the number density of monopoles, with $n_{\rm M,0} < 10^{-29}n_{\rm bary,0}$. Even with a hefty monopole mass $m_{\rm M}c^2 \sim 10^{12}$ TeV $\sim 10^{15}m_pc^2$, this implies that the density parameter in monopoles today is $\Omega_{\rm M,0} < 5 \times 10^{-16}$, a far cry indeed from monopole domination.

The monopole problem can be rephrased as the question, "Where have all the magnetic monopoles gone?" Now, you can always answer the question by saying, "There were never any monopoles to begin with." There is not yet a single, definitive Grand Unified Theory, and in some variants on the GUT theme, magnetic monopoles are not produced. However, the flatness and horizon problems are not so readily dismissed. When the physicist Alan Guth first proposed the idea of *inflation* in 1981, he introduced it as a way of resolving the flatness problem, the horizon problem, and the monopole problem with a single cosmological mechanism.

10.4 The Inflation Solution

What is inflation? In a cosmological context, inflation can most generally be defined as the hypothesis that there was a period, early in the history of our universe, when the expansion was accelerating outward; that is, an epoch when $\ddot{a} > 0$. The acceleration equation,

$$\frac{\ddot{a}}{a} = -\frac{4\pi G}{3c^2}(\varepsilon + 3P), \tag{10.17}$$

tells us that $\ddot{a} > 0$ when $P < -\varepsilon/3$. Thus, inflation would have taken place if the universe were temporarily dominated by a component with equation-of-state parameter $w < -1/3$. The simplest implementation of inflation states that the universe was temporarily dominated by a positive cosmological constant Λ_i (with $w = -1$), and thus had an acceleration equation that could be written in the form

$$\frac{\ddot{a}}{a} = \frac{\Lambda_i}{3} > 0. \tag{10.18}$$

In an inflationary phase when the energy density was dominated by a cosmological constant, the Friedmann equation was

$$\left(\frac{\dot{a}}{a}\right)^2 = \frac{\Lambda_i}{3}. \tag{10.19}$$

The Hubble constant H_i during the inflationary phase was thus constant, with the value $H_i = (\Lambda_i/3)^{1/2}$, and the scale factor grew exponentially with time:

$$a(t) \propto e^{H_i t}. \tag{10.20}$$

To see how a period of exponential growth can resolve the flatness, horizon, and monopole problems, suppose that the universe had a period of exponential expansion sometime in the midst of its early, radiation-dominated phase. For simplicity, suppose the exponential growth was switched on instantaneously at a time t_i, and lasted until some later time t_f, when the exponential growth was switched off instantaneously, and the universe reverted to its former state of radiation-dominated expansion. In this simple case, we can write the scale factor as

$$a(t) = \begin{cases} a_i(t/t_i)^{1/2} & t < t_i \\ a_i e^{H_i(t-t_i)} & t_i < t < t_f \\ a_i e^{H_i(t_f-t_i)}(t/t_f)^{1/2} & t > t_f. \end{cases} \tag{10.21}$$

Thus, between the time t_i, when the exponential inflation began, and the time t_f, when the inflation stopped, the scale factor increased by a factor

$$\frac{a(t_f)}{a(t_i)} = e^N, \tag{10.22}$$

where N, the number of e-foldings of inflation, was

$$N \equiv H_i(t_f - t_i). \tag{10.23}$$

If the duration of inflation, $t_f - t_i$, was long compared to the Hubble time during inflation, H_i^{-1}, then N was large, and the growth in scale factor during inflation was enormous.

For concreteness, let's take one possible model for inflation. This model states that exponential inflation started around the GUT time, $t_i \approx t_{\rm GUT} \approx 10^{-36}$ s, with a Hubble parameter $H_i \approx t_{\rm GUT}^{-1} \approx 10^{36}\,{\rm s}^{-1}$, and lasted for N e-foldings, ending at $t_f \approx (N+1)t_{\rm GUT}$. Note that the cosmological constant Λ_i present at the time of inflation in this model was very large compared to the cosmological constant that is present today. Currently, the evidence is consistent with an energy density in Λ of $\varepsilon_{\Lambda,0} \approx 0.69\varepsilon_{c,0} \approx 0.0034\,{\rm TeV\,m^{-3}}$. To produce exponential expansion with a Hubble parameter $H_i \approx 10^{36}\,{\rm s}^{-1}$, the cosmological constant during inflation would have had an energy density (see Equation 4.69)

$$\varepsilon_{\Lambda_i} = \frac{c^2}{8\pi G}\Lambda_i = \frac{3c^2}{8\pi G}H_i^2 \sim 10^{105}\,{\rm TeV\,m^{-3}}, \tag{10.24}$$

over 107 orders of magnitude larger.

How does inflation resolve the flatness problem? Equation 10.1, which gives Ω as a function of time, can be written in the form

$$|1-\Omega(t)| = \frac{c^2}{R_0^2 a(t)^2 H(t)^2} \tag{10.25}$$

for any universe not perfectly flat. During the exponential expansion of the inflationary epoch, $H(t) = H_i =$ const and $a(t) \propto \exp(H_i t)$, leading to

$$|1-\Omega(t)| \propto e^{-2H_i t}. \tag{10.26}$$

The difference between Ω and one plummets exponentially with time. If we compare the density parameter at the beginning of exponential inflation ($t = t_i$) with the density parameter at the end of inflation ($t = t_f = [N+1]t_i$), we find

$$|1-\Omega(t_f)| = e^{-2N}|1-\Omega(t_i)|. \tag{10.27}$$

Suppose that prior to inflation, the universe was actually strongly curved, with

$$|1-\Omega(t_i)| \sim 1. \tag{10.28}$$

After N e-foldings of inflation, the deviation of Ω from one would be

$$|1-\Omega(t_f)| \sim e^{-2N}. \tag{10.29}$$

During the subsequent radiation-dominated and matter-dominated eras, the difference of Ω from one would increase again. However, starting from the fact that $|1-\Omega_0| \leq 0.005$ today, we can use Equation 10.6 to extrapolate back to the time $t_f = (N+1)t_i \approx (N+1)\cdot 10^{-36}$ s, when the scale factor was

$$a(t_f) \approx 2\times 10^{-28}\sqrt{N+1}. \tag{10.30}$$

At this time, immediately after the end of inflation, the deviation of Ω from one had the value

$$|1-\Omega(t_f)| \leq 2\times 10^{-54}(N+1). \tag{10.31}$$

Equation 10.29 tells us the amount of flattening produced by exponential inflation; Equation 10.31 tells us the amount of flattening required by observations of the universe today. Comparing the two equations, we find that the minimum number of e-foldings of inflation needed to match the observations is $N \approx 60$. The minimum required increase in the scale factor during inflation is thus $a(t_f)/a(t_i) \sim e^{60} \sim 10^{26}$. (Keep in mind, however, that N may have been much greater than 60, since the observational data are consistent with $|1 - \Omega_0| \ll 1$.)

How does inflation resolve the horizon problem? At any time t, the horizon distance $d_{\rm hor}(t)$ is given by the relation

$$d_{\rm hor}(t) = a(t)c \int_0^t \frac{dt}{a(t)}. \tag{10.32}$$

Prior to the inflationary period, the universe was radiation-dominated. Thus, the horizon distance at the beginning of inflation was

$$d_{\rm hor}(t_i) = a_i c \int_0^{t_i} \frac{dt}{a_i (t/t_i)^{1/2}} = 2ct_i. \tag{10.33}$$

The horizon size at the end of inflation was

$$d_{\rm hor}(t_f) = a_i e^N c \left(\int_0^{t_i} \frac{dt}{a_i (t/t_i)^{1/2}} + \int_{t_i}^{t_f} \frac{dt}{a_i \exp[H_i (t - t_i)]} \right). \tag{10.34}$$

If N, the number of e-foldings of inflation, is large, then the horizon size at the end of inflation was

$$d_{\rm hor}(t_f) = e^N c(2t_i + H_i^{-1}). \tag{10.35}$$

An epoch of exponential inflation causes the horizon size to grow exponentially. If inflation started at $t_i \approx 10^{-36}$ s, then the horizon size immediately before inflation was

$$d_{\rm hor}(t_i) = 2ct_i \approx 6 \times 10^{-28}\,{\rm m}. \tag{10.36}$$

For concreteness, let's assume $N = 65$ e-foldings of inflation, just a bit more than the minimum of 60 e-foldings required to explain the flatness of today's universe. In this fairly minimal model, if we take $H_i \approx t_i^{-1}$, the horizon size immediately after inflation was

$$d_{\rm hor}(t_f) \approx e^N 3ct_i \sim 15\,{\rm m}. \tag{10.37}$$

During the brief period of $\sim 10^{-34}$ s that inflation lasts in this model, the horizon size is boosted exponentially from submicroscopic scales to something the size of a whale.

The exponential increase in the horizon size during inflation is illustrated by the solid line in Figure 10.3. In the post-inflation era, when the universe reverts

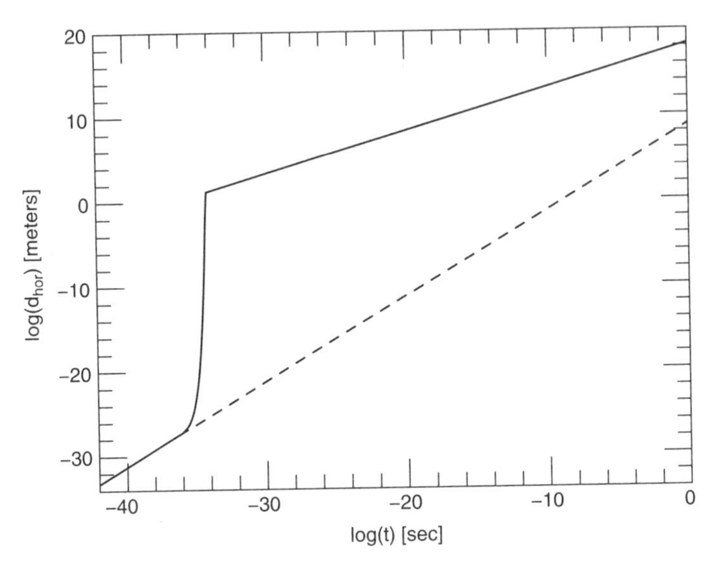

Figure 10.3 The solid line shows the growth of the horizon distance in a universe where exponential inflation begins at $t = 10^{-36}$ s and lasts for $N = 65$ e-foldings. The dashed line, for comparison, shows the horizon distance in a radiation-dominated universe without an inflationary epoch.

to being radiation-dominated, the horizon size grows at the rate $d_{\rm hor} \propto a \propto t^{1/2}$, as points that were separated by a distance $d_{\rm hor}(t_f)$ at the end of inflation continue to be carried apart from each other by the expansion of the universe. In the model we've adopted, where inflation started around the GUT time and lasted for $N = 65$ e-foldings, the scale factor was $a(t_f) \sim 2 \times 10^{-27}$ at the end of inflation, estimated from Equation 10.30. At the time of last scattering, the scale factor was $a(t_{\rm ls}) \approx 1/1090 \approx 9.1 \times 10^{-4}$. Thus, in our model, the horizon distance grew from $d_{\rm hor}(t_f) \sim 15$ m at the end of inflation to $d_{\rm hor}(t_{\rm ls}) \sim 200$ Mpc at the time of last scattering. This is 800 times bigger than the horizon size $d_{\rm hor}(t_{\rm ls}) \approx 0.25$ Mpc that we calculated in the absence of inflation, and is large enough that antipodal points on the last scattering surface are causally connected.

To look at the resolution of the horizon problem from a slightly different viewpoint, consider the entire universe directly visible to us today; that is, the region bounded by the surface of last scattering. Currently, the proper distance to the surface of last scattering is $d_p(t_0) \approx 14\,000$ Mpc. In our model, where inflation started at the GUT time and ended after $N = 65$ e-foldings, the scale factor at the end of inflation was $a(t_f) \sim 2 \times 10^{-27}$. This implies that immediately after inflation, the portion of the universe currently visible to us was crammed into a sphere of physical radius

$$d_p(t_f) = a_f d_p(t_0) \sim 3 \times 10^{-23}\,\text{Mpc} \sim 0.9\,\text{m}. \tag{10.38}$$

Just after inflation, in this model, all the mass-energy destined to become the hundreds of billions of galaxies we see today was enclosed within a sphere six feet across.[2]

If your mind has not yet been blown by the idea of packing the visible universe into a sphere of radius $\sim 0.9\,\text{m}$, then hang on to your skull. If there were $N = 65$ e-foldings of inflation, then immediately before the inflationary epoch, the currently visible universe was enclosed within a sphere of proper radius

$$d_p(t_i) = \mathrm{e}^{-N} d_p(t_f) \sim 4 \times 10^{-29}\,\text{m}. \tag{10.39}$$

Note that this distance is smaller than the horizon size immediately prior to inflation: $d_{\text{hor}}(t_i) \sim 2ct_i \sim 6 \times 10^{-28}\,\text{m}$. Thus, the portion of the universe we can see today, even given the minimum number of e-foldings, had time to exchange information and achieve uniformity before inflation began.[3]

How does inflation resolve the monopole problem? If magnetic monopoles were created before or during inflation, then the number density of monopoles was diluted to an undetectably low level. During a period when the universe was expanding exponentially ($a \propto \mathrm{e}^{H_i t}$), the number density of monopoles, if they were neither created nor destroyed, was decreasing exponentially ($n_{\text{M}} \propto \mathrm{e}^{-3H_i t}$). For instance, if inflation started around the GUT time, when the number density of magnetic monopoles was $n_{\text{M}}(t_{\text{GUT}}) \approx 10^{82}\,\text{m}^{-3}$, then after 65 e-foldings of inflation, the number density would have been $n_{\text{M}}(t_f) = \mathrm{e}^{-195} n_{\text{M}}(t_{\text{GUT}}) \approx 0.002\,\text{m}^{-3}$. The number density today, after the additional expansion from $a(t_f) \sim 2 \times 10^{-27}$ to $a_0 = 1$, would then be $n_{\text{M}}(t_0) \approx 2 \times 10^{-83}\,\text{m}^{-3} \approx 5 \times 10^{-16}\,\text{Mpc}^{-3}$. With this tiny density, there probably would not be any monopoles at all within the last scattering surface.

10.5 The Physics of Inflation

Inflation explains some otherwise puzzling aspects of our universe, by flattening it, ensuring its homogeneity over large scales, and driving down the number density of magnetic monopoles it contains. However, we have not yet answered many crucial questions about the inflationary epoch. What triggers inflation at $t = t_i$, and (just as important) what turns it off at $t = t_f$? If inflation reduces the number density of monopoles to undetectably low levels, why doesn't it reduce the number density of photons to undetectably low levels? If inflation

[2] If just after inflation the horizon distance was the size of a whale, then the currently visible universe was the size of Jonah.

[3] Because of the boost given to the horizon distance by an exponential period of inflation, the "horizon distances" that we computed in previous chapters must be slightly redefined as "post-inflationary horizon distances." That is, instead of being the distance traveled by light since the Big Bang at $t = 0$, they are the distance traveled by light since the end of inflation at $t = t_f \sim 10^{-34}\,\text{s}$.

is so efficient at flattening the global curvature of the universe, why doesn't it also flatten out the local curvature due to fluctuations in the energy density? We know that the universe wasn't *perfectly* homogeneous after inflation, because the cosmic microwave background isn't perfectly isotropic.

To answer these questions, we will have to examine, at least in broad outline, the physics behind inflation. At present, there is not a consensus among cosmologists about the exact mechanism driving inflation. We will restrict ourselves to speaking in general terms about one plausible mechanism for bringing about an inflationary epoch.

Suppose the universe contains a scalar field $\phi(\vec{r}, t)$ whose value can vary as a function of position and time. Some early implementations of inflation associated the scalar field ϕ with the Higgs field, which mediates interactions between particles at energies higher than the GUT energy; however, to keep the discussion general, let's just call the field $\phi(\vec{r}, t)$ the *inflaton field*.[4] Generally speaking, a scalar field can have an associated potential energy $V(\phi)$. (As a simple illustrative example, suppose that the scalar field ϕ is the elevation above sea level at a given point on the Earth's surface. The associated potential energy, in this case, is the gravitational potential $V = g\phi$, where $g = 9.8\,\mathrm{m\,s^{-2}}$.)

If ϕ has units of energy, and its potential V has units of energy density, then the energy density of the inflaton field is

$$\varepsilon_\phi = \frac{1}{2}\frac{1}{\hbar c^3}\dot{\phi}^2 + V(\phi) \tag{10.40}$$

in a region of space where ϕ is homogeneous. The pressure of the inflaton field is

$$P_\phi = \frac{1}{2}\frac{1}{\hbar c^3}\dot{\phi}^2 - V(\phi). \tag{10.41}$$

If the inflaton field changes only very slowly as a function of time, with

$$\dot{\phi}^2 \ll \hbar c^3 V(\phi), \tag{10.42}$$

then the inflaton field acts like a cosmological constant, with

$$\varepsilon_\phi \approx -P_\phi \approx V(\phi). \tag{10.43}$$

Thus, an inflaton field can drive exponential inflation if there is a temporary period when its rate of change $\dot{\phi}$ is small (satisfying Equation 10.42), and its potential $V(\phi)$ is large enough to dominate the energy density of the universe.

Under what circumstances are the conditions for inflation (small $\dot{\phi}$ and large V) met in the early universe? To determine the value of $\dot{\phi}$, start with the fluid equation for the energy density of the inflaton field,

$$\dot{\varepsilon}_\phi + 3H(t)(\varepsilon_\phi + P_\phi) = 0, \tag{10.44}$$

[4] This is not a typo: the *inflaton* field drives *inflation*.

where $H(t) = \dot{a}/a$. Substituting from Equations 10.40 and 10.41, we find the equation that governs the rate of change of ϕ:

$$\ddot{\phi} + 3H(t)\dot{\phi} = -\hbar c^3 \frac{dV}{d\phi}. \quad (10.45)$$

Note that Equation 10.45 mimics the equation of motion for a particle being accelerated by a force proportional to $-dV/d\phi$ and being impeded by a frictional force proportional to the particle's speed. Thus, the expansion of the universe provides a "Hubble friction" term, $3H\dot{\phi}$, which slows the transition of the inflaton field to a value that will minimize the potential V. Just as a skydiver reaches terminal velocity when the downward force of gravity is balanced by the upward force of air resistance, so the inflaton field can reach "terminal velocity" (with $\ddot{\phi} = 0$) when

$$3H\dot{\phi} = -\hbar c^3 \frac{dV}{d\phi}, \quad (10.46)$$

or

$$\dot{\phi} = -\frac{\hbar c^3}{3H}\frac{dV}{d\phi}. \quad (10.47)$$

If the inflaton field has reached this terminal velocity, then the requirement that $\dot{\phi}^2 \ll \hbar c^3 V$, necessary if the inflaton field is to play the role of a cosmological constant, translates into

$$\left(\frac{dV}{d\phi}\right)^2 \ll \frac{9H^2 V}{\hbar c^3}. \quad (10.48)$$

If the universe is undergoing exponential inflation driven by the potential energy of the inflaton field, this means that the Hubble parameter is

$$H = \left(\frac{8\pi G\varepsilon_\phi}{3c^2}\right)^{1/2} = \left(\frac{8\pi GV}{3c^2}\right)^{1/2}, \quad (10.49)$$

and Equation 10.48 becomes

$$\left(\frac{dV}{d\phi}\right)^2 \ll \frac{24\pi GV^2}{\hbar c^5}, \quad (10.50)$$

which can also be written as

$$\left(\frac{E_P}{V}\frac{dV}{d\phi}\right)^2 \ll 1, \quad (10.51)$$

where E_P is the Planck energy. If the slope of the inflaton's potential is sufficiently shallow, satisfying Equation 10.51, and if the amplitude of the potential is sufficiently large to dominate the energy density of the universe, then the inflaton field is capable of giving rise to exponential expansion.

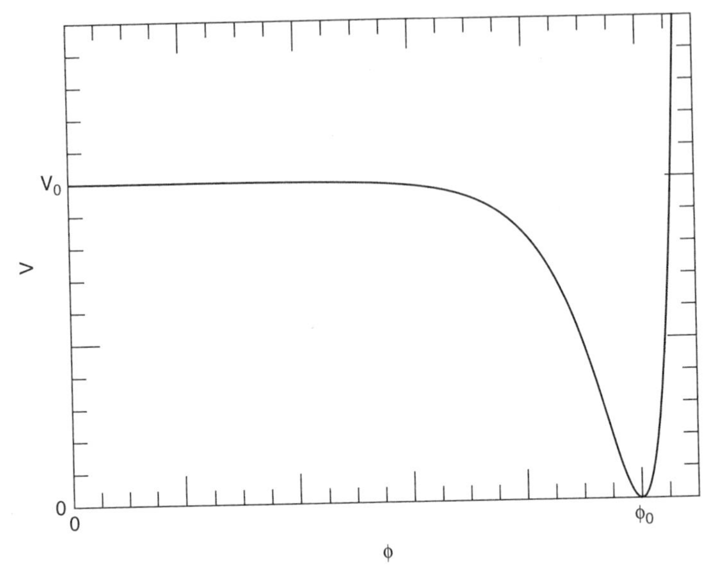

Figure 10.4 A potential that can give rise to an inflationary epoch. The global minimum in V (or "true vacuum") is at $\phi = \phi_0$. If the scalar field starts at $\phi = 0$, it is in a "false vacuum" state.

As a concrete example of a potential $V(\phi)$ which can give rise to inflation, consider the potential shown in Figure 10.4. The global minimum in the potential occurs when the value of the inflaton field is $\phi = \phi_0$. Suppose, however, that the inflaton field starts at $\phi \approx 0$, where the potential is $V(\phi) \approx V_0$. If

$$\left(\frac{dV}{d\phi}\right)^2 \ll \frac{V_0^2}{E_P^2} \tag{10.52}$$

on the "plateau" where $V \approx V_0$, then while ϕ is slowly rolling toward ϕ_0, the inflaton field contributes an energy density $\varepsilon_\phi \approx V_0 \approx$ constant to the universe.

When an inflaton field has a potential similar to that of Figure 10.4, it is referred to as being in a *metastable false vacuum state* when it is near the maximum at $\phi = 0$. Such a state is not truly stable; if the inflaton field is nudged from $\phi = 0$ to $\phi = +d\phi$, it will continue to slowly roll toward the *true vacuum state* at $\phi = \phi_0$ and $V = 0$. However, if the plateau is sufficiently broad as well as sufficiently shallow in slope, it can take many Hubble times for the inflaton field to roll down to the true vacuum state. Whether the inflaton field is dynamically significant during its transition from the false vacuum to the true vacuum depends on the value of V_0. As long as $\varepsilon_\phi \approx V_0$ is tiny compared to the energy density of radiation, $\varepsilon_r \sim \alpha T^4$, the contribution of the inflaton field to the Friedmann equation can be ignored. Exponential inflation, driven by the energy density of the inflaton field, will begin at a temperature

$$T_i \approx \left(\frac{V_0}{\alpha}\right)^{1/4} \approx 2 \times 10^{28}\,\mathrm{K}\left(\frac{V_0}{10^{105}\,\mathrm{TeV\,m^{-3}}}\right)^{1/4} \tag{10.53}$$

or

$$kT_i \approx (\hbar^3 c^3 V_0)^{1/4} \approx 2 \times 10^{12}\,\mathrm{TeV}\left(\frac{V_0}{10^{105}\,\mathrm{TeV\,m^{-3}}}\right)^{1/4}. \tag{10.54}$$

This corresponds to a time

$$t_i \approx \left(\frac{c^2}{GV_0}\right)^{1/2} \approx 3 \times 10^{-36}\,\mathrm{s}\left(\frac{V_0}{10^{105}\,\mathrm{TeV\,m^{-3}}}\right)^{-1/2}. \tag{10.55}$$

While the inflaton field is slowly rolling toward the true vacuum state, it produces exponential expansion, with a Hubble parameter

$$H_i \approx \left(\frac{8\pi G V_0}{3c^2}\right)^{1/2} \approx t_i^{-1}. \tag{10.56}$$

The exponential expansion ends as the inflaton field reaches the true vacuum at $\phi = \phi_0$. The duration of inflation thus depends on the exact shape of the potential $V(\phi)$. The number of e-foldings of inflation, for the potential shown in Figure 10.4, should be

$$N \sim H_i \frac{\phi_0}{\dot{\phi}} \sim \left(\frac{E_P}{V_0}\frac{dV}{d\phi}\right)^{-1}\left(\frac{\phi_0}{E_P}\right). \tag{10.57}$$

Large values of ϕ_0 and V_0 (that is, a broad, high plateau) and small values of $dV/d\phi$ (that is, a shallowly sloped plateau) lead to more e-foldings of inflation.

After rolling off the plateau in Figure 10.4, the inflaton field ϕ oscillates about the minimum at ϕ_0. The amplitude of these oscillations is damped by the "Hubble friction" term proportional to $H\dot{\phi}$ in Equation 10.45. If the inflaton field is coupled to any of the other fields in the universe, however, the oscillations in ϕ are damped more rapidly, with the energy of the inflaton field being carried away by photons or other relativistic particles. These photons *reheat* the universe after the precipitous drop in temperature caused by inflation. The energy lost by the inflaton field after its phase transition from the false vacuum to the true vacuum can be thought of as the latent heat of that transition. When water freezes, to use a low-energy analogy, it loses an energy of $3 \times 10^8\,\mathrm{J\,m^{-3}} \sim 2 \times 10^{15}\,\mathrm{TeV\,m^{-3}}$, which goes to heat its surroundings.[5] Similarly, the transition from false to true vacuum releases an energy V_0, which goes to reheat the universe.

If the scale factor increases by a factor

$$\frac{a(t_f)}{a(t_i)} = e^N \tag{10.58}$$

[5] This is why orange growers spray their trees with water when a hard freeze threatens. The energy released by water as it freezes keeps the delicate leaves warm. (The thin layer of ice also cuts down on convective and radiative heat loss, but the release of latent heat is the largest effect.)

during inflation, then the temperature will drop by a factor e^{-N}. If inflation starts around the GUT time, and lasts for $N = 65$ e-foldings, then the temperature drops from a toasty $T(t_i) \sim T_{\rm GUT} \sim 10^{28}$ K to a chilly $T(t_f) \sim e^{-65} T_{\rm GUT} \sim 0.6$ K. Not only is inflation very effective at driving down the number density of magnetic monopoles, it is also effective at driving down the number density of every other type of particle, including photons. The chilly post-inflationary period didn't last, though. As the energy density associated with the inflaton field was converted to relativistic particles such as photons, the temperature of the universe was restored to its pre-inflationary value T_i.

Inflation successfully explains the flatness, homogeneity, and isotropy of the universe. It ensures that we live in a universe with a negligibly low density of magnetic monopoles, while the inclusion of reheating ensures that we *don't* live in a universe with a negligibly low density of photons. In some ways, though, inflation seems to be too successful. It makes the universe homogeneous and isotropic all right, but it makes it too homogeneous and isotropic. Exponential inflation not only flattens the global curvature of the universe, it also flattens the local curvature due to fluctuations in the energy density. If energy fluctuations prior to inflation were $\delta\varepsilon/\bar{\varepsilon} \sim 1$, a naïve calculation predicts that density fluctuations immediately after 65 e-foldings of inflation would be

$$\frac{\delta\varepsilon}{\bar{\varepsilon}} \sim e^{-65} \sim 10^{-28}. \tag{10.59}$$

This is a very close approach to homogeneity. Even allowing for the growth in amplitude of density fluctuations prior to the time of last scattering, this would leave the cosmic microwave background much smoother than is actually observed.

Remember, however, the saga of how a submicroscopic patch of the universe ($d \sim 4\times10^{-29}$ m) was inflated to macroscopic size ($d \sim 0.9$ m), before growing to the size of the currently visible universe. Inflation excels in taking submicroscopic scales and blowing them up to macroscopic scales. On submicroscopic scales, the vacuum, whether true or false, is full of constantly changing quantum fluctuations, with virtual particles popping into and out of existence. On quantum scales, the universe is intrinsically inhomogeneous. Inflation takes the submicroscopic quantum fluctuations in the inflaton field and expands them to macroscopic scales. The energy fluctuations that result are the origin, in the inflationary scenario, of the inhomogeneities in the current universe. We can replace the old proverb, "Great oaks from tiny acorns grow," with the yet more amazing proverb, "Great superclusters from tiny quantum fluctuations grow."

Exercises

10.1 What upper limit is placed on $\Omega(t_P)$ by the requirement that the universe not end in a Big Crunch between the Planck time, $t_P \approx 5 \times 10^{-44}$ s, and the start of the inflationary epoch at t_i? Compute the maximum permissible value of $\Omega(t_P)$, first assuming $t_i \approx 10^{-36}$ s, then assuming $t_i \approx 10^{-26}$ s. (Hint: prior to inflation, the Friedmann equation will be dominated by the radiation term and the curvature term.)

10.2 Current observational limits on the density of magnetic monopoles tell us that their density parameter is currently $\Omega_{M,0} < 10^{-6}$. If monopoles formed at the GUT time, with one monopole per horizon of mass $m_M = m_{\rm GUT}$, how many e-foldings of inflation would be required to drive the current density of monopoles below the bound $\Omega_{M,0} < 10^{-6}$? Assume that inflation took place immediately after the formation of monopoles.

10.3 It has been speculated that the present-day acceleration of the universe is due to the existence of a false vacuum, which will eventually decay. Suppose that the energy density of the false vacuum is $\varepsilon_\Lambda = 0.69\varepsilon_{c,0} = 3360\,{\rm MeV\,m^{-3}}$, and that the current energy density of matter is $\varepsilon_{m,0} = 0.31\varepsilon_{c,0} = 1510\,{\rm MeV\,m^{-3}}$. What will be the value of the Hubble parameter once the false vacuum becomes strongly dominant? Suppose that the false vacuum is fated to decay instantaneously to radiation at a time $t_f = 50t_0$. (Assume, for simplicity, that the radiation takes the form of blackbody photons.) To what temperature will the universe be reheated at $t = t_f$? What will the energy density of matter be at $t = t_f$? At what time will the universe again be dominated by matter?

11

Structure Formation: Gravitational Instability

The universe can be approximated as being homogeneous and isotropic only if we smooth it with a filter $\sim$ 100 Mpc across. On smaller scales, the universe contains density fluctuations ranging from subatomic quantum fluctuations up to the large superclusters and voids, $\sim$ 50 Mpc across, which characterize the distribution of galaxies in space. If we relax the strict assumption of homogeneity and isotropy that underlies the Robertson–Walker metric and the Friedmann equation, we can ask (and, to some extent, answer) the question, "How do density fluctuations in the universe evolve with time?"

The formation of relatively small objects, such as planets, stars, or even galaxies, involves fairly complicated physics. Because of the greater complexity when baryons are involved in structure formation, we will postpone the discussion of galaxies and stars until the next chapter. In this chapter, we will focus on the formation of structures larger than galaxies – clusters, superclusters, and voids. Cosmologists use the term *large scale structure of the universe* to refer to all structures bigger than individual galaxies. A map of the large scale structure of the universe, as traced by the positions of galaxies, can be made by measuring the redshifts of a sample of galaxies and using the Hubble relation, $d = (c/H_0)z$, to estimate their distances. For instance, Figure 11.1 shows two slices through the universe based on data from the 2dF Galaxy Redshift Survey (2dFGRS). From our location near the midplane of our galaxy's dusty disk, we get our best view of distant galaxies when we look perpendicular to the disk, toward what are conventionally called the "north galactic pole" (in the constellation Coma Berenices) and the antipodal "south galactic pole" (in the constellation Sculptor). The 2dFGRS selected two long narrow stripes on the sky, one near the north galactic pole and the other near the south galactic pole. In each stripe, the redshift was measured for $\sim$ 100 000 galaxies. By plotting the redshift of each galaxy as a function of its angular position along the stripe, a pair of two-dimensional slices through the universe were mapped out.

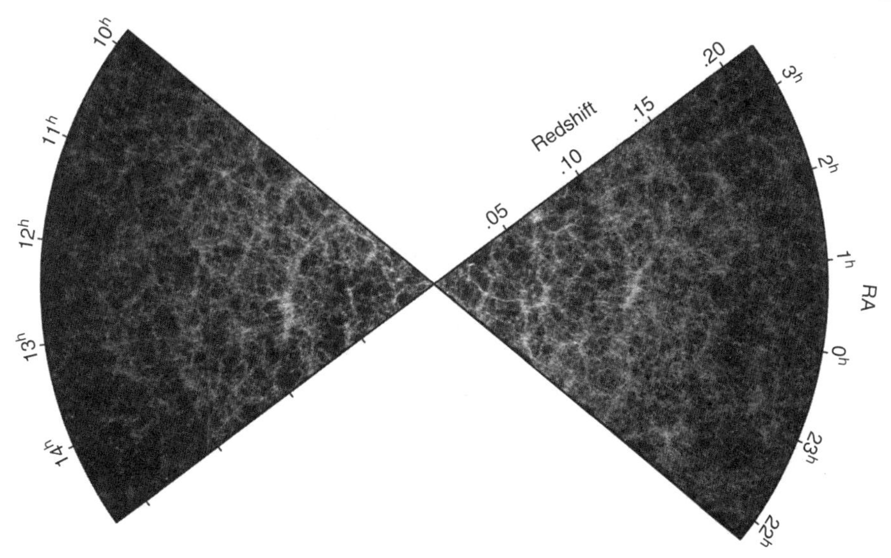

Figure 11.1 Two slices through the universe, based on the redshift distribution of $\sim 200\,000$ galaxies measured by the 2dF Galaxy Redshift Survey; white regions are higher in density, black regions are lower. Left slice: data from the north stripe. Right slice: data from the south stripe. [van de Weygaert & Schaap 2009, *Lect. Notes Phys.*, **665**, 291]

Figure 11.1 shows the two slices that the 2dF Galaxy Redshift Survey made through the universe. They reach as far as a redshift $z \approx 0.22$, corresponding to a proper distance $d_p(t_0) \sim 1000\,\text{Mpc}$. The white regions in the plot represent regions with a high number density of galaxies. The galaxies obviously do not have a random Poisson distribution, in which the position of each galaxy is uncorrelated with the position of other galaxies. Instead, the galaxies are found primarily in long superclusters separated by lower density voids. Superclusters are objects that are just now collapsing under their own self-gravity. Superclusters are typically flattened (roughly planar) or elongated (roughly linear) structures. A supercluster will contain one or more clusters embedded within it; a cluster is a fully collapsed object that has come to equilibrium, more or less, and hence obeys the virial theorem, as discussed in Section 7.3. In comparison to the flattened or elongated superclusters, the underdense voids are roughly spherical in shape. When gazing at the large scale structure of the universe, as traced by the distribution of galaxies, cosmologists are likely to call it "bubbly" or "spongy," with the low-density voids constituting the bubbles, or the holes of the sponge. Alternatively, if they focus on the high-density superclusters rather than the low-density voids, cosmologists refer to a "cosmic web" of filaments and walls.

Figure 11.2 The northeastern United States and southeastern Canada at night, as seen by the *Suomi-NPP* satellite. [NASA Worldview]

11.1 The Matthew Effect

Being able to describe the distribution of galaxies in space doesn't automatically lead to an understanding of the origin of large scale structure. Consider, as an analogy, the distribution of luminous objects shown in Figure 11.2. The distribution of illuminated cities on the Earth's surface is obviously not random. There are "superclusters" of cities, such as the Boswash supercluster stretching from Boston to Washington. There are "voids" such as the Appalachian void. However, the influences that determine the exact location of cities are often far removed from fundamental physics.[1]

Fortunately, the distribution of galaxies in space is more closely tied to fundamental physics than is the distribution of cities on the Earth. The basic mechanism for growing large structures, such as voids and superclusters, is *gravitational instability*. Suppose that at some time in the past, the density of the universe had slight inhomogeneities. We know, for instance, that such density fluctuations existed at the time of last scattering, since they left their stamp on the cosmic microwave background. When the universe is matter-dominated, the overdense regions expand less rapidly than the universe as a whole; if their density is sufficiently great, they will collapse and become gravitationally bound objects such as clusters. The dense clusters will, in addition, draw matter to themselves from the surrounding underdense regions. The effect of gravity on density fluctuations is sometimes referred to as the Matthew effect: "For whosoever hath, to him shall be given, and he shall have more abundance; but whosoever hath not, from him shall be taken away even that he hath" [Matthew 13:12]. In less biblical language, the rich get richer and the poor get poorer.[2]

1 Consider, for instance, the complicated politics that went into determining the location of Washington, DC.

2 The singer-songwriter Billie Holiday may have summed it up best: "Them that's got shall have / Them that's not shall lose."

To put our study of gravitational instability on a more quantitative basis, consider some component of the universe whose energy density $\varepsilon(\vec{r}, t)$ is a function of position as well as time. At a given time t, the spatially averaged energy density is

$$\bar{\varepsilon}(t) = \frac{1}{V}\int_V \varepsilon(\vec{r}, t) d^3r. \tag{11.1}$$

To ensure that we have found the true average, the volume V over which we are averaging must be large compared to the size of the biggest structure in the universe. It is useful to define a dimensionless density fluctuation

$$\delta(\vec{r}, t) \equiv \frac{\varepsilon(\vec{r}, t) - \bar{\varepsilon}(t)}{\bar{\varepsilon}(t)}. \tag{11.2}$$

The value of δ is thus negative in underdense regions and positive in overdense regions. If the density ε is constrained to be non-negative, as it is for matter and radiation, then the minimum possible value of δ is $\delta = -1$, corresponding to $\varepsilon = 0$. There is no such upper limit on δ, which can have a value $\delta \gg +1$.

To get a feel for how initially small density contrasts grow with time, consider a particularly simple case. Start with a static, homogeneous, matter-only universe with uniform mass density $\bar{\rho}$. (At this point, we stumble over the inconvenient fact that there's no such thing as a static, homogeneous, matter-only universe. This is the awkward fact that inspired Einstein to introduce the cosmological constant. However, there are conditions under which we can consider some region of the universe to be approximately static and homogeneous. For instance, the air in a closed room is approximately static and homogeneous; it is stabilized by a pressure gradient with a scale length much greater than the height of the ceiling.) In a region of the universe that is *approximately* static and homogeneous, we add a small amount of mass within a sphere of radius R, as seen in Figure 11.3, so that the density within the sphere is $\bar{\rho}(1 + \delta)$, with $\delta \ll 1$. If the density excess

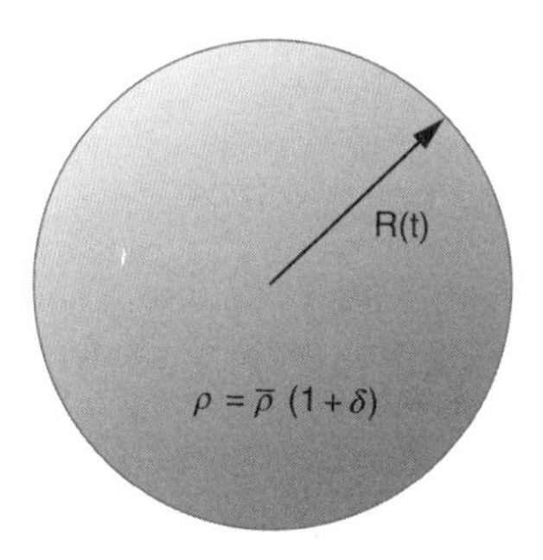

Figure 11.3 A sphere of radius $R(t)$ expanding or contracting under the influence of the density fluctuation $\delta(t)$.

δ is uniform within the sphere, then the gravitational acceleration at the sphere's surface, due to the excess mass, will be

$$\ddot{R} = -\frac{G(\Delta M)}{R^2} = -\frac{G}{R^2}\left(\frac{4\pi}{3}R^3\bar{\rho}\delta\right), \tag{11.3}$$

or

$$\frac{\ddot{R}}{R} = -\frac{4\pi G\bar{\rho}}{3}\delta(t). \tag{11.4}$$

Thus, a mass excess ($\delta > 0$) will cause the sphere to collapse inward ($\ddot{R} < 0$).

Equation 11.4 contains two unknowns, $R(t)$ and $\delta(t)$. If we want to find an explicit solution for $\delta(t)$, we need a second equation involving $R(t)$ and $\delta(t)$. Conservation of mass tells us that the mass of the sphere,

$$M = \frac{4\pi}{3}\bar{\rho}[1+\delta(t)]R(t)^3, \tag{11.5}$$

remains constant during the collapse. Thus we can write another relation between $R(t)$ and $\delta(t)$ that must hold true during the collapse:

$$R(t) = R_0[1+\delta(t)]^{-1/3}, \tag{11.6}$$

where

$$R_0 \equiv \left(\frac{3M}{4\pi\bar{\rho}}\right)^{1/3} = \text{constant}. \tag{11.7}$$

When $\delta \ll 1$, we may make the approximation

$$R(t) \approx R_0\left[1 - \frac{1}{3}\delta(t)\right]. \tag{11.8}$$

Taking the second time derivative yields

$$\ddot{R} \approx -\frac{1}{3}R_0\ddot{\delta} \approx -\frac{1}{3}R\ddot{\delta}. \tag{11.9}$$

Thus, mass conservation tells us that

$$\frac{\ddot{R}}{R} \approx -\frac{1}{3}\ddot{\delta} \tag{11.10}$$

in the limit that $\delta \ll 1$. Combining Equations 11.4 and 11.10, we find a tidy equation that tells us how the small overdensity δ evolves as the sphere collapses:

$$\ddot{\delta} = 4\pi G\bar{\rho}\delta. \tag{11.11}$$

The most general solution of Equation 11.11 has the form

$$\delta(t) = A_1 e^{t/t_{\rm dyn}} + A_2 e^{-t/t_{\rm dyn}}, \tag{11.12}$$

where the dynamical time for collapse is

$$t_{\rm dyn} = \frac{1}{(4\pi G\bar{\rho})^{1/2}} \approx 9.6\,\text{hours}\left(\frac{\bar{\rho}}{1\,\text{kg}\,\text{m}^{-3}}\right)^{-1/2}. \tag{11.13}$$

Note that the dynamical time depends only on $\bar{\rho}$, and not on R. The constants A_1 and A_2 in Equation 11.12 depend on the initial conditions of the sphere. For instance, if the overdense sphere starts at rest, with $\dot{\delta} = 0$ at $t = 0$, then $A_1 = A_2 = \delta(0)/2$. After a few dynamical times, however, only the exponentially growing term of Equation 11.12 is significant. Thus, gravity tends to make small density fluctuations in a static, pressureless medium grow exponentially with time.

11.2 The Jeans Length

The exponential growth of density perturbations is slightly alarming, at first glance. For instance, the density of the air around you is $\bar{\rho} \approx 1\,\text{kg}\,\text{m}^{-3}$, yielding a dynamical time for collapse of $t_{\rm dyn} \sim 10\,\text{hours}$.[3] What keeps small density perturbations in the air from undergoing a runaway collapse over the course of a few days? The answer, of course, is pressure. A nonrelativistic gas, as shown in Section 4.4, has an equation-of-state parameter

$$w \approx \frac{kT}{\mu c^2}, \tag{11.14}$$

where T is the temperature of the gas and μ is the mean mass per gas particle. Thus, the pressure of an ideal gas will never totally vanish, but will only approach zero in the limit that the temperature approaches absolute zero.

When a sphere of gas is compressed by its own gravity, a pressure gradient will build up that tends to counter the effects of gravity. In the universe today, a star is the prime example of a sphere of gas in which the inward force of gravity is balanced by the outward force provided by a pressure gradient; the hot intracluster gas of the Coma cluster is also in a balance between gravity and pressure. However, hydrostatic equilibrium, the state in which gravity is exactly balanced by a pressure gradient, cannot always be attained. Consider an overdense sphere with initial radius R. If pressure were not present, it would collapse on a timescale

$$t_{\rm dyn} \sim \frac{1}{(G\bar{\rho})^{1/2}} \sim \left(\frac{c^2}{G\bar{\varepsilon}}\right)^{1/2}. \tag{11.15}$$

If the pressure is nonzero, the attempted collapse will be countered by a steepening of the pressure gradient within the perturbation. The steepening of the pressure gradient, however, doesn't occur instantaneously. Any change in pressure

[3] Slightly longer if you are using this book for recreational reading as you climb Mount Everest.

travels at the sound speed.[4] Thus, the time it takes for the pressure gradient to build up in a region of radius R will be

$$t_{\text{pre}} \sim \frac{R}{c_s}, \tag{11.16}$$

where c_s is the local sound speed. In a medium with equation-of-state parameter $w > 0$, the sound speed is

$$c_s = c\left(\frac{dP}{d\varepsilon}\right)^{1/2} = \sqrt{w}c. \tag{11.17}$$

For hydrostatic equilibrium to be attained, the pressure gradient must build up before the overdense region collapses, implying

$$t_{\text{pre}} < t_{\text{dyn}}. \tag{11.18}$$

Comparing Equation 11.15 with Equation 11.16, we find that for a density perturbation to be stabilized by pressure against collapse, it must be smaller than some reference size λ_J, given by the relation

$$\lambda_J \sim c_s t_{\text{dyn}} \sim c_s\left(\frac{c^2}{G\bar{\varepsilon}}\right)^{1/2}. \tag{11.19}$$

The length scale λ_J is known as the *Jeans length*, after the astrophysicist James Jeans, who was among the first to study gravitational instability in a cosmological context. Overdense regions larger than the Jeans length collapse under their own gravity. Overdense regions smaller than the Jeans length merely oscillate in density; they constitute stable standing sound waves.

A more precise derivation of the Jeans length, including all the appropriate factors of π, yields the result

$$\lambda_J = c_s\left(\frac{\pi c^2}{G\bar{\varepsilon}}\right)^{1/2} = 2\pi c_s t_{\text{dyn}}. \tag{11.20}$$

The Jeans length of the Earth's atmosphere, for instance, where the sound speed is a third of a kilometer per second and the dynamical time is ten hours, is $\lambda_J \sim 10^5$ km, far longer than the scale height of the Earth's atmosphere. You don't have to worry about density fluctuations in the air undergoing a catastrophic collapse.

To consider the behavior of density fluctuations on cosmological scales, consider a spatially flat universe in which the mean density is $\bar{\varepsilon}$, but which contains density fluctuations with amplitude $|\delta| \ll 1$. The characteristic time for expansion of such a universe is the Hubble time,

$$H^{-1} = \left(\frac{3c^2}{8\pi G\bar{\varepsilon}}\right)^{1/2}. \tag{11.21}$$

[4] What is sound, after all, but a traveling change in pressure?

Comparison of Equation 11.13 with Equation 11.21 reveals that the Hubble time is comparable in magnitude to the dynamical time $t_{\rm dyn}$ for the collapse of an overdense region:

$$H^{-1} = \left(\frac{3}{2}\right)^{1/2} t_{\rm dyn} \approx 1.22 t_{\rm dyn}. \tag{11.22}$$

The Jeans length in an expanding flat universe will then be

$$\lambda_J = 2\pi c_s t_{\rm dyn} = 2\pi \left(\frac{2}{3}\right)^{1/2} \frac{c_s}{H}. \tag{11.23}$$

If we focus on one particular component of the universe, with equation-of-state parameter w and sound speed $c_s = \sqrt{w}c$, the Jeans length for that component will be

$$\lambda_J = 2\pi \left(\frac{2}{3}\right)^{1/2} \sqrt{w}\frac{c}{H}. \tag{11.24}$$

Consider, for instance, the "radiation" component of the universe. With $w = 1/3$, the sound speed in a gas of photons or other relativistic particles is $c_s = c/\sqrt{3}$. The Jeans length for radiation in an expanding universe is then

$$\lambda_J = \frac{2\pi\sqrt{2}}{3}\frac{c}{H} \approx 3.0\frac{c}{H}. \tag{11.25}$$

Density fluctuations in the radiative component will be pressure-supported if they are smaller than three times the Hubble distance. Although a universe containing nothing but radiation can have density perturbations smaller than $\lambda_J \sim 3c/H$, they will be stable sound waves, and will not collapse under their own gravity.

In order for a universe to have gravitationally collapsed structures much smaller than the Hubble distance, it must have a nonrelativistic component, with $\sqrt{w} \ll 1$. The gravitational collapse of the *baryonic* component of the universe is complicated by the fact that it was coupled to photons until the time of decoupling at a redshift $z_{\rm dec} \approx z_{\rm ls} \approx 1090$. Before this time, the photons, electrons, and baryons were all coupled together to form a single photon–baryon fluid. Since photons were still dominant over baryons at the time of decoupling, with $\varepsilon_\gamma > \varepsilon_{\rm bary}$, we can regard the baryons (with mild exaggeration) as being a dynamically insignificant contaminant in the photon–baryon fluid. Just *before* decoupling, the Jeans length of the photon–baryon fluid was roughly the same as the Jeans length of a pure photon gas:

$$\lambda_J(\text{before}) \approx 3c/H(z_{\rm dec}) \approx 0.66\,\text{Mpc} \approx 2.0 \times 10^{22}\,\text{m}. \tag{11.26}$$

The *baryonic Jeans mass*, M_J, is defined as the mass of baryons contained within a sphere of radius λ_J:

$$M_J \equiv \rho_{\rm bary}\left(\frac{4\pi}{3}\lambda_J^3\right). \tag{11.27}$$

Immediately before decoupling, when the baryon density was $\rho_{\rm bary} \approx 5.6 \times 10^{-19}\,{\rm kg\,m^{-3}}$, the baryonic Jeans mass was

$$M_J({\rm before}) = \rho_{\rm bary}\frac{4\pi}{3}\lambda_J({\rm before})^3 \approx 2\times 10^{49}\,{\rm kg} \approx 10^{19}\,{\rm M}_\odot. \tag{11.28}$$

This is very large compared to the baryonic mass of the Coma cluster ($M_{\rm bary} \sim 2\times 10^{14}\,{\rm M}_\odot$); it is even large when compared to the baryonic mass of a supercluster ($M_{\rm bary} \sim 10^{16}\,{\rm M}_\odot$).

Now consider what happens to the baryonic Jeans mass immediately after decoupling. Once the photons are decoupled, the photons and baryons form two separate gases, instead of a single photon–baryon fluid. The sound speed in the photon gas is

$$c_s({\rm photon}) = c/\sqrt{3} \approx 0.58c. \tag{11.29}$$

The sound speed in the baryonic gas, by contrast, is

$$c_s({\rm baryon}) = \left(\frac{kT}{mc^2}\right)^{1/2} c. \tag{11.30}$$

At the time of decoupling, the thermal energy per particle was $kT_{\rm dec} \approx 0.26\,{\rm eV}$, and the mean rest energy of the atoms in the baryonic gas was $mc^2 = 1.22 m_p c^2 \approx 1140\,{\rm MeV}$, taking into account the helium mass fraction of $Y_p = 0.24$. Thus, the sound speed of the baryonic gas immediately after decoupling was

$$c_s({\rm baryon}) \approx \left(\frac{0.26\,{\rm eV}}{1140\times 10^6\,{\rm eV}}\right)^{1/2} c \approx 1.5\times 10^{-5}c, \tag{11.31}$$

only 5 kilometers per second. Thus, once the baryons were decoupled from the photons, their associated Jeans length decreased by a factor

$$F = \frac{c_s({\rm baryon})}{c_s({\rm photon})} \approx \frac{1.5\times 10^{-5}}{0.58} \approx 2.6\times 10^{-5}. \tag{11.32}$$

Decoupling causes the baryonic Jeans mass to decrease by a factor $F^3 \approx 1.8 \times 10^{-14}$, plummeting from $M_J({\rm before}) \approx 10^{19}\,{\rm M}_\odot$ to

$$M_J({\rm after}) = F^3 M_J({\rm before}) \approx 2\times 10^5\,{\rm M}_\odot. \tag{11.33}$$

This is very small compared to the baryonic mass of our own galaxy ($M_{\rm bary} \sim 10^{11}\,{\rm M}_\odot$); it is even small when compared to a more modest galaxy like the Small Magellanic Cloud ($M_{\rm bary} \sim 10^9\,{\rm M}_\odot$).

The abrupt decrease of the baryonic Jeans mass at the time of decoupling marks an important epoch in the history of structure formation. Perturbations in the baryon density, from supercluster scales down to the size of the smallest dwarf galaxies, couldn't grow in amplitude until the time of photon decoupling, when the universe had reached the ripe old age of $t_{\rm dec} \approx 0.37\,\rm Myr$. After decoupling, the growth of density perturbations in the baryonic component was off and running.

11.3 Instability in an Expanding Universe

Density perturbations smaller than the Hubble distance can grow in amplitude only when they are no longer pressure-supported. However, if the pressure (and hence the Jeans length) of some component becomes negligibly small, this doesn't necessarily imply that the amplitude of density fluctuations is free to grow exponentially with time. The analysis of Section 11.1, which yielded $\delta \propto \exp(t/t_{\rm dyn})$, assumed that the universe was *static* as well as pressureless. In an expanding Friedmann universe, the timescale for the growth of a density perturbation by self-gravity, $t_{\rm dyn} \sim (c^2/G\bar{\varepsilon})^{1/2}$, is comparable to the timescale for expansion, $H^{-1} \sim (c^2/G\bar{\varepsilon})^{1/2}$. Self-gravity, in the absence of global expansion, causes overdense regions to become *more dense* with time. The global expansion of the universe, in the absence of self-gravity, causes overdense regions to become *less dense* with time. Because the timescales for these two competing processes are similar, they must both be taken into account when computing the time evolution of a density perturbation.

To see how small density perturbations in an expanding universe evolve with time, let's do a Newtonian analysis of the problem, similar in spirit to the Newtonian derivation of the Friedmann equation given in Chapter 4. Suppose you are in a universe containing only pressureless matter with mass density $\bar{\rho}(t)$. As the universe expands, the density decreases at the rate $\bar{\rho}(t) \propto a(t)^{-3}$. Within a spherical region of radius R, a small amount of matter is added, or removed, so that the density within the sphere is

$$\rho(t) = \bar{\rho}(t)[1 + \delta(t)], \tag{11.34}$$

with $|\delta| \ll 1$. The total gravitational acceleration at the surface of the sphere will be

$$\ddot{R} = -\frac{GM}{R^2} = -\frac{G}{R^2}\left(\frac{4\pi}{3}\rho R^3\right) = -\frac{4\pi}{3}G\bar{\rho}R - \frac{4\pi}{3}G(\bar{\rho}\delta)R. \tag{11.35}$$

The equation of motion for a point at the surface of the sphere can then be written in the form

$$\frac{\ddot{R}}{R} = -\frac{4\pi}{3}G\bar{\rho} - \frac{4\pi}{3}G\bar{\rho}\delta. \tag{11.36}$$

Mass conservation tells us that the mass inside the sphere,

$$M = \frac{4\pi}{3}\bar{\rho}(t)[1+\delta(t)]R(t)^3, \tag{11.37}$$

remains constant as the sphere expands. Thus,

$$R(t) \propto \bar{\rho}(t)^{-1/3}[1+\delta(t)]^{-1/3}, \tag{11.38}$$

or, since $\bar{\rho} \propto a^{-3}$,

$$R(t) \propto a(t)[1+\delta(t)]^{-1/3}. \tag{11.39}$$

That is, if the sphere is slightly overdense, its radius will grow slightly less rapidly than the scale factor $a(t)$. If the sphere is slightly underdense, it will grow slightly more rapidly than the scale factor.

Taking two time derivatives of Equation 11.39 yields

$$\frac{\ddot{R}}{R} = \frac{\ddot{a}}{a} - \frac{1}{3}\ddot{\delta} - \frac{2}{3}\frac{\dot{a}}{a}\dot{\delta}, \tag{11.40}$$

when $|\delta| \ll 1$. Combining Equations 11.36 and 11.40, we find

$$\frac{\ddot{a}}{a} - \frac{1}{3}\ddot{\delta} - \frac{2}{3}\frac{\dot{a}}{a}\dot{\delta} = -\frac{4\pi}{3}G\bar{\rho} - \frac{4\pi}{3}G\bar{\rho}\delta. \tag{11.41}$$

When $\delta = 0$, Equation 11.41 reduces to

$$\frac{\ddot{a}}{a} = -\frac{4\pi}{3}G\bar{\rho}, \tag{11.42}$$

which is the correct acceleration equation for a homogeneous, isotropic universe containing nothing but pressureless matter (compare to Equation 4.49). By subtracting Equation 11.42 from Equation 11.41 to leave only the terms linear in the perturbation δ, we find the equation that governs the growth of small amplitude perturbations:

$$-\frac{1}{3}\ddot{\delta} - \frac{2}{3}\frac{\dot{a}}{a}\dot{\delta} = -\frac{4\pi}{3}G\bar{\rho}\delta, \tag{11.43}$$

or

$$\ddot{\delta} + 2H\dot{\delta} = 4\pi G\bar{\rho}\delta, \tag{11.44}$$

remembering that $H \equiv \dot{a}/a$. In a static universe, with $H = 0$, Equation 11.44 reduces to the result we already found in Equation 11.11:

$$\ddot{\delta} = 4\pi G\bar{\rho}\delta. \tag{11.45}$$

The additional term, $\propto H\dot{\delta}$, found in an expanding universe, is a "Hubble friction" term; it acts to slow the growth of density perturbations in an expanding universe.

A fully relativistic calculation for the growth of density perturbations yields the more general result

$$\ddot{\delta} + 2H\dot{\delta} = \frac{4\pi G}{c^2}\bar{\varepsilon}_m \delta. \quad (11.46)$$

This form of the equation can be applied to a universe that contains components with non-negligible pressure, such as radiation ($w = 1/3$) or a cosmological constant ($w = -1$). In multiple-component universes, however, it should be remembered that δ represents the fluctuation in the density of *matter* alone. That is,

$$\delta = \frac{\varepsilon_m - \bar{\varepsilon}_m}{\bar{\varepsilon}_m}, \quad (11.47)$$

where $\bar{\varepsilon}_m(t)$, the average matter density, might be only a small fraction of $\bar{\varepsilon}(t)$, the average total density. Rewritten in terms of the density parameter for matter,

$$\Omega_m = \frac{\bar{\varepsilon}_m}{\varepsilon_c} = \frac{8\pi G\bar{\varepsilon}_m}{3c^2H^2}, \quad (11.48)$$

Equation 11.46 takes the form

$$\ddot{\delta} + 2H\dot{\delta} - \frac{3}{2}\Omega_m H^2 \delta = 0. \quad (11.49)$$

During epochs when the universe is not dominated by matter, density perturbations in the matter do not grow rapidly in amplitude. Take, for instance, the early radiation-dominated phase of the universe. During this epoch, $\Omega_m \ll 1$ and $H = 1/(2t)$, meaning that Equation 11.49 takes the form

$$\ddot{\delta} + \frac{1}{t}\dot{\delta} \approx 0, \quad (11.50)$$

which has a solution of the form

$$\delta(t) \approx B_1 + B_2 \ln t. \quad (11.51)$$

During the radiation-dominated epoch, density fluctuations in the dark matter grew only at a logarithmic rate. In the far future, if the universe is indeed dominated by a cosmological constant, the density parameter for matter will again be negligibly small, the Hubble parameter will have the constant value $H = H_\Lambda$, and Equation 11.49 will take the form

$$\ddot{\delta} + 2H_\Lambda\dot{\delta} \approx 0, \quad (11.52)$$

which has a solution of the form

$$\delta(t) \approx C_1 + C_2 e^{-2H_\Lambda t}. \quad (11.53)$$

In a lambda-dominated phase, therefore, fluctuations in the matter density reach a constant fractional amplitude, while the average matter density plummets at the rate $\bar{\varepsilon}_m \propto e^{-3H_\Lambda t}$.

Only when matter dominates the energy density can fluctuations in the matter density grow at a significant rate. In a flat, matter-dominated universe, $\Omega_m = 1$, $H = 2/(3t)$, and Equation 11.49 takes the form

$$\ddot{\delta} + \frac{4}{3t}\dot{\delta} - \frac{2}{3t^2}\delta = 0. \tag{11.54}$$

If we guess that the solution to the above equation has the power-law form Dt^n, plugging this guess into the equation yields

$$n(n-1)Dt^{n-2} + \frac{4}{3t}nDt^{n-1} - \frac{2}{3t^2}Dt^n = 0, \tag{11.55}$$

or

$$n(n-1) + \frac{4}{3}n - \frac{2}{3} = 0. \tag{11.56}$$

The two possible solutions for this quadratic equation are $n = -1$ and $n = 2/3$. Thus, the general solution for the time evolution of density perturbations in a spatially flat, matter-only universe is

$$\delta(t) \approx D_1 t^{2/3} + D_2 t^{-1}. \tag{11.57}$$

The values of D_1 and D_2 are determined by the initial conditions for $\delta(t)$. The decaying mode, $\propto t^{-1}$, eventually becomes negligibly small compared to the growing mode, $\propto t^{2/3}$. When the growing mode is the only survivor, the density perturbations in a flat, matter-only universe grow at the rate

$$\delta \propto t^{2/3} \propto a(t) \propto \frac{1}{1+z}, \tag{11.58}$$

as long as $|\delta| \ll 1$.

If baryonic matter were the only type of nonrelativistic matter in the universe, then density perturbations could have started to grow at $z_{\rm dec} = 1090$, and they could have grown in amplitude by a factor ~ 1090 by the present day. However, the dominant form of nonrelativistic matter is dark matter. The density perturbations in the dark matter started to grow effectively at $z_{rm} = 3440$. At the time of decoupling, the baryons fell into the preexisting gravitational wells of the dark matter. The situation is schematically illustrated in Figure 11.4. Having nonbaryonic dark matter allows the universe to get a "head start" on structure formation; perturbations in the matter density can start growing at $z_{rm} = 3440$ rather than $z_{\rm dec} = 1090$, as they would in a universe without dark matter.

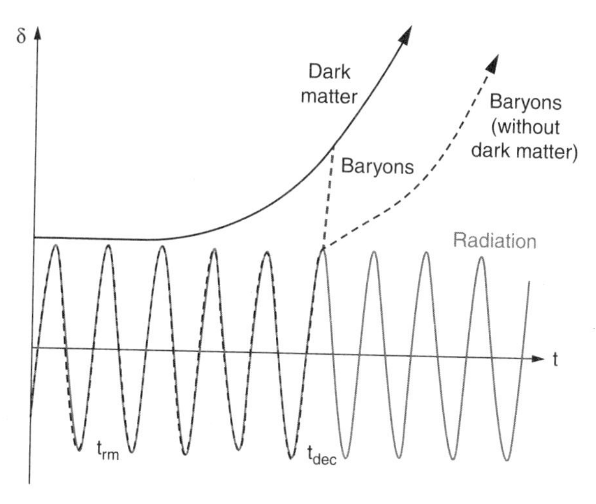

Figure 11.4 A highly schematic drawing of how density fluctuations in different components of the universe evolve with time.

11.4 The Power Spectrum

When deriving Equation 11.44, which determines the growth rate of density perturbations in a Newtonian universe, we assumed that the perturbation was spherically symmetric. In fact, Equation 11.44 and its relativistically correct brother, Equation 11.46, both apply to low-amplitude perturbations of any shape. This is fortunate, since the density perturbations in the real universe are not spherically symmetric. The bubbly structure shown in redshift maps of galaxies, such as Figure 11.1, has grown from the density perturbations that were present when the universe became matter dominated. Great oaks from tiny acorns grow – but then, great pine trees from tiny pinenuts grow. By looking at the current large scale structure (the "tree"), we can deduce the properties of the early, low-amplitude, density fluctuations (the "nut").[5]

Let us consider the properties of the early density fluctuations at some time t_i, when they were still very low in amplitude ($|\delta| \ll 1$). As long as the density fluctuations are small in amplitude, the expansion of the universe is still nearly isotropic, and the geometry of the universe is still well described by the Robertson–Walker metric (Equation 3.41):

$$ds^2 = -c^2dt^2 + a(t)^2[dr^2 + S_\kappa(r)^2 d\Omega^2]. \tag{11.59}$$

Under these circumstances, it is useful to set up a comoving coordinate system. Choose some point as the origin. In a universe described by the Robertson–Walker

[5] At the risk of carrying the arboreal analogy too far, I should mention that the temperature fluctuations of the cosmic microwave background, as shown in Figure 8.3, offer us a look at the "sapling."

metric, as shown in Section 3.6, the proper distance of any point from the origin can be written in the form

$$d_p(t_i) = a(t_i)r, \tag{11.60}$$

where the comoving distance r is what the proper distance would be at the present day ($a = 1$) if the expansion continued to be perfectly isotropic. If we label each bit of matter in the universe with its comoving coordinate position $\vec{r} = (r, \theta, \phi)$, then $\vec{r}$ will remain very nearly constant as long as $|\delta| \ll 1$. Thus, when considering the regime where density fluctuations are small, cosmologists typically consider $\delta(\vec{r})$, the density fluctuation at a comoving location $\vec{r}$, at some time t_i. (The exact value of t_i doesn't matter, as long as it's a time after the density perturbations are in place, but before they reach an amplitude $|\delta| \sim 1$. Switching to a different value of t_i, under these restrictions, simply changes the amplitude of $\delta(\vec{r})$, and not its shape.)

When discussing the temperature fluctuations of the cosmic microwave background in Chapter 8, I pointed out that cosmologists weren't interested in the exact pattern of hot and cold spots on the last scattering surface, but rather in the statistical properties of the field $\delta T/T(\theta, \phi)$. Similarly, cosmologists are not interested in the exact locations of the density maxima and minima in the early universe, but rather in the statistical properties of the field $\delta(\vec{r})$. When studying the temperature fluctuations of the CMB, it is useful to expand $\delta T/T(\phi, \theta)$ in spherical harmonics. A similar decomposition of $\delta(\vec{r})$ is also useful. Since δ is defined in three-dimensional space rather than on the surface of a sphere, a useful expansion of δ is in terms of Fourier components.

Within a large box expanding along with the universe, of comoving volume V, the density fluctuation field $\delta(\vec{r})$ can be expressed as

$$\delta(\vec{r}) = \frac{V}{(2\pi)^3} \int \delta_{\vec{k}} e^{-i\vec{k}\cdot\vec{r}} d^3k, \tag{11.61}$$

where the individual Fourier components $\delta_{\vec{k}}$ are found by performing the integral

$$\delta_{\vec{k}} = \frac{1}{V} \int \delta(\vec{r}) e^{i\vec{k}\cdot\vec{r}} d^3r. \tag{11.62}$$

In performing the Fourier transform, we are breaking up the function $\delta(\vec{r})$ into an infinite number of sine waves, each with comoving wavenumber $\vec{k}$ and comoving wavelength $\lambda = 2\pi/k$. If we have complete, uncensored knowledge of $\delta(\vec{r})$, we can compute all the Fourier components $\delta_{\vec{k}}$; conversely, if we know all the Fourier components, we can reconstruct the density field $\delta(\vec{r})$.

Each Fourier component is a complex number, which can be written in the form

$$\delta_{\vec{k}} = |\delta_{\vec{k}}| e^{i\phi_{\vec{k}}}. \tag{11.63}$$

When $|\delta_{\vec{k}}| \ll 1$, each Fourier component obeys Equation 11.49,

$$\ddot{\delta}_{\vec{k}} + 2H\dot{\delta}_{\vec{k}} - \frac{3}{2}\Omega_m H^2 \delta_{\vec{k}} = 0, \qquad (11.64)$$

as long as the proper wavelength, $a(t)2\pi/k$, is large compared to the Jeans length and small compared to the Hubble distance c/H. The phase $\phi_{\vec{k}}$ remains constant as long as the amplitude $|\delta_{\vec{k}}|$ remains small. Even after fluctuations with a short proper wavelength have reached $|\delta_{\vec{k}}| \sim 1$ and collapsed under their own gravity, the growth of the longer wavelength perturbations is still described by Equation 11.64. This means, helpfully enough, that we can use linear perturbation theory to study the growth of very large scale structures even after smaller structures, such as galaxies and clusters of galaxies, have already collapsed.

The mean square amplitude of the Fourier components defines the *power spectrum*:

$$P(k) = \langle |\delta_{\vec{k}}|^2 \rangle, \qquad (11.65)$$

where the average is taken over all possible orientations of the wavenumber $\vec{k}$. (If $\delta(\vec{r})$ is isotropic, then no information is lost, statistically speaking, if we average the power spectrum over all angles.) When the phases $\phi_{\vec{k}}$ of the different Fourier components are uncorrelated with each other, then $\delta(\vec{r})$ is called a *Gaussian field*. If a Gaussian field is homogeneous and isotropic, then all its statistical properties are summed up in the power spectrum $P(k)$. If $\delta(\vec{r})$ is a Gaussian field, then the value of δ at a randomly selected point is drawn from the Gaussian probability distribution

$$p(\delta) = \frac{1}{\sqrt{2\pi}\,\sigma_\delta} \exp\left(-\frac{\delta^2}{2\sigma_\delta^2}\right), \qquad (11.66)$$

where the standard deviation σ_δ can be computed from the power spectrum:

$$\sigma_\delta^2 = \frac{V}{(2\pi)^3}\int P(k)d^3k = \frac{V}{2\pi^2}\int_0^\infty P(k)k^2 dk. \qquad (11.67)$$

The study of Gaussian density fields is of particular interest to cosmologists because most inflationary scenarios predict that the density fluctuations created by inflation (Section 10.5) constitute an isotropic, homogeneous Gaussian field. In addition, it is expected that the power spectrum of inflationary density fluctuations can be well described by a power law, with

$$P(k) \propto k^n. \qquad (11.68)$$

Theoretically, the value of the power-law index favored by inflation is $n \approx 1$. A power spectrum with $n = 1$ *exactly* is often referred to as a Harrison–Zel'dovich spectrum, after Edward Harrison and Yakov Zel'dovich, who independently proposed it as a physically plausible spectrum for density perturbations as early as 1970 (even before the concept of inflation entered cosmology). Observationally, the value of the power-law index n can be deduced from the temperature fluctuations of the cosmic microwave background on large scales ($\theta > 1.1°$). The temperature correlation function measured by the *Planck* satellite (Figure 8.6) is consistent with $n = 0.97 \pm 0.01$. Thus, the observed inflationary power spectrum is only slightly "tilted" relative to a Harrison–Zel'dovich spectrum.

What does a universe with $P(k) \propto k^n$ look like? Imagine going through such a universe and marking out randomly located spheres of comoving radius r. The mean mass of each sphere (considering only the nonrelativistic matter which it contains) will be

$$\langle M \rangle = \frac{4\pi}{3} r^3 \rho_{m,0} = 1.67 \times 10^{11}\,\mathrm{M}_\odot \left(\frac{r}{1\,\mathrm{Mpc}} \right)^3 . \tag{11.69}$$

However, the actual mass of each sphere will vary; some spheres will be slightly underdense, and others will be slightly overdense. The mean square density fluctuation of the mass inside each sphere is a function of the power spectrum and of the size of the sphere:

$$\begin{aligned} \left\langle \left(\frac{M - \langle M \rangle}{\langle M \rangle} \right)^2 \right\rangle &= \frac{V}{2\pi^2} \int P(k) \left[\frac{3 j_1(kr)}{kr} \right]^2 k^2 dk \\ &= \frac{9V}{2\pi^2 r^2} \int P(k) j_1(kr)^2 dk, \end{aligned} \tag{11.70}$$

where $j_1(x) = (\sin x - x \cos x)/x^2$ is a spherical Bessel function of the first kind.

In general, to compute the mean square mass fluctuation, you must know the power spectrum $P(k)$. However, if the power spectrum has the simple form $P(k) \propto k^n$, we can substitute a new variable of integration, $u = kr$, to find

$$\left\langle \left(\frac{M - \langle M \rangle}{\langle M \rangle} \right)^2 \right\rangle = \frac{9V}{2\pi^2} r^{-3-n} \int_0^\infty u^n j_1(u)^2 du. \tag{11.71}$$

Thus, the root mean square mass fluctuation within spheres of comoving radius r will have the dependence

$$\frac{\delta M}{M} \equiv \left\langle \left(\frac{M - \langle M \rangle}{\langle M \rangle} \right)^2 \right\rangle^{1/2} \propto r^{-(3+n)/2}. \tag{11.72}$$

This can also be expressed in the form $\delta M/M \propto M^{-(3+n)/6}$. (If you scattered point masses randomly throughout the universe, so that they formed a Poisson distribution, you would expect mass fluctuations of amplitude $\delta M/M \propto N^{-1/2}$, where N is the expected number of point masses within the sphere. Since the average mass within a sphere is $M \propto N$, a Poisson distribution has $\delta M/M \propto M^{-1/2}$, or $n = 0$. The Harrison–Zel'dovich spectrum, with $n = 1$, thus will produce more power on small length scales than a Poisson distribution of points.) Note that the potential fluctuations associated with mass fluctuations on a length scale r will have an amplitude $\delta\Phi \sim G\delta M/r \propto \delta M/M^{1/3} \propto M^{(1-n)/6}$. Thus, the Harrison–Zel'dovich spectrum, with $n = 1$, is the only power law that prevents the divergence of the potential fluctuations on both large and small scales. For this reason, the Harrison–Zel'dovich spectrum is also referred to as a *scale invariant* power spectrum of density perturbations.

11.5 Hot versus Cold

Immediately after inflation, the power spectrum for density perturbations has the form $P(k) \propto k^n$, with an index $n \approx 1$. However, the shape of this primordial power spectrum will be modified between the end of inflation at t_f and the time of radiation–matter equality at $t_{rm} \approx 5.0 \times 10^4$ yr. The shape of the power spectrum at t_{rm}, when density perturbations start to grow significantly in amplitude, depends on the properties of the dark matter. More specifically, it depends on whether the dark matter is predominantly *cold dark matter* or *hot dark matter.*

Saying that dark matter is "cold" or "hot" is a statement about the thermal velocity of the dark matter particles. More specifically, it's a statement about the thermal velocity at a particular time in the history of the universe; the time when the mass within a Hubble volume (that is, a sphere of proper radius c/H) was equal to the total mass of a large galaxy like our own, $M_{\rm gal} \approx 10^{12}\,{\rm M}_\odot$. In the radiation-dominated universe, the Hubble parameter was, from Equation 5.81,

$$H(a) = H_0 \frac{\sqrt{\Omega_{r,0}}}{a^2}. \tag{11.73}$$

This means the mass within a Hubble volume during the early, radiation-dominated era was

$$M_H = \frac{4\pi}{3}\left(\frac{c}{H_0}\frac{a^2}{\sqrt{\Omega_{r,0}}}\right)^3 \frac{\rho_{m,0}}{a^3} = 1.6 \times 10^{28}\,{\rm M}_\odot\, a^3, \tag{11.74}$$

using the parameters of the Benchmark Model. The mass within a Hubble volume was equal to $M_{\rm gal} \approx 10^{12}\,{\rm M}_\odot$ at a scale factor $a \approx 4 \times 10^{-6}$, corresponding to a temperature $kT \approx 60\,{\rm eV}$ and a cosmic time $t \approx 8 \times 10^{-10} H_0^{-1} \approx 12\,{\rm yr}$. Particles with a rest energy $mc^2 \ll 3kT \sim 180\,{\rm eV}$ were highly relativistic at this time, with a particle energy $E \sim 3kT \sim 180\,{\rm eV}$. Particles with a rest energy $mc^2 \gg 180\,{\rm eV}$ were highly nonrelativistic at this time, with a particle energy E almost entirely contributed by their rest energy mc^2.

Cold dark matter consists of particles that were nonrelativistic at $t \approx 12\,{\rm yr}$, either because they were nonrelativistic at the time they decoupled from photons and baryonic matter, or because they have $mc^2 \gg 3kT \sim 180\,{\rm eV}$, and thus had cooled to nonrelativistic thermal velocities by $t \approx 12\,{\rm yr}$. For instance, WIMPs decoupled at $t \sim 1$ s, when the universe had a mean particle energy $\sim 3kT \sim 3$ MeV. However, WIMPs are predicted to have a mass $mc^2 \sim 100$ GeV. With a thermal energy much smaller than their rest energy at the time of decoupling, WIMPs were already nonrelativistic when they decoupled, and thus qualify as cold dark matter. Axions are a type of elementary particle first proposed by particle physicists for noncosmological purposes. If they exist, however, they would have formed out of equilibrium in the early universe, with very low thermal velocities. Thus, axions would act as cold dark matter as well.

Hot dark matter, by contrast, consists of particles that were *relativistic* at the time they decoupled from the other components of the universe, and that remained relativistic until $t \gg 12$ yr, when the mass inside the Hubble volume was large compared to the mass of a galaxy. For instance, neutrinos, like WIMPs, decoupled at $t \sim 1$ s, when the mean particle energy was ~ 3 MeV. Thus, a neutrino with mass $m_\nu c^2 \ll 3$ MeV was hot enough to be relativistic at the time it decoupled. Moreover, a neutrino with mass $5 \times 10^{-4}\,\text{eV} \ll m_\nu c^2 \ll 180\,\text{eV}$ was relativistic at $t \approx 12$ yr, but is nonrelativistic today, at $t_0 \approx 13.7$ Gyr. Thus, neutrinos in this mass range qualify as hot dark matter.

To see how the existence of hot dark matter modifies the spectrum of density perturbations, consider what would happen in a universe filled with hot dark matter consisting of particles with mass m_h. The initially relativistic hot dark matter particles cool as the universe expands, until their thermal velocities drop below c when $3kT \approx m_h c^2$. This happens at a temperature

$$T_h \approx \frac{m_h c^2}{3k} \approx 12\,000\,\text{K}\left(\frac{m_h c^2}{3\,\text{eV}}\right). \tag{11.75}$$

For a particle mass $m_h c^2 > 2.4$ eV, the particles become nonrelativistic when the universe is still radiation-dominated. In a radiation-dominated universe, the temperature of Equation 11.75 corresponds to a cosmic time

$$t_h \approx 42\,000\,\text{yr}\left(\frac{m_h c^2}{3\,\text{eV}}\right)^{-2}. \tag{11.76}$$

Prior to the time t_h, the hot dark matter particles move freely in random directions with a speed close to that of light. This motion, called *free streaming*, acts to wipe out any density fluctuations present in the hot dark matter. Thus, the net effect of free streaming in the hot dark matter is to wipe out any density fluctuations whose wavelength is smaller than $\sim ct_h$. When the hot dark matter particles become nonrelativistic, there will be no density fluctuations on scales smaller than the physical scale

$$d_{\min} \approx ct_h \approx 13\,\text{kpc}\left(\frac{m_h c^2}{3\,\text{eV}}\right)^{-2}, \tag{11.77}$$

corresponding to a comoving length scale

$$r_{\min} = \frac{d_{\min}}{a(t_h)} \approx \left(\frac{T_h}{2.7255\,\mathrm{K}}\right) d_{\min} \approx 55\,\mathrm{Mpc}\left(\frac{m_h c^2}{3\,\mathrm{eV}}\right)^{-1} . \tag{11.78}$$

The total amount of matter within a sphere of comoving radius $r_{\min}$ is

$$M_{\min} = \frac{4\pi}{3} r_{\min}^3 \rho_{m,0} \approx 2.7 \times 10^{16}\,\mathrm{M}_\odot \left(\frac{m_h c^2}{3\,\mathrm{eV}}\right)^{-3} . \tag{11.79}$$

Hot dark matter particles with $m_h c^2 \sim 50\,\mathrm{eV}$ will smear out fluctuations smaller than the Local Group, particles with $m_h c^2 \sim 8\,\mathrm{eV}$ will rub out fluctuations smaller than the Coma cluster, and particles with $m_h c^2 \sim 3\,\mathrm{eV}$ will wipe away fluctuations smaller than superclusters.

The left panel of Figure 11.5 shows the power spectrum of density fluctuations in hot dark matter, once the hot dark matter particles have cooled enough to become nonrelativistic; the plot assumes that $m_h c^2 = 3\,\mathrm{eV}$ and thus $r_{\min} \approx 55\,\mathrm{Mpc}$. For wavenumbers $k \ll 2\pi/r_{\min} \approx 0.1\,\mathrm{Mpc}^{-1}$, the power spectrum of hot dark matter (shown as the dotted line) is close to the original $P \propto k$ spectrum (shown as the dashed line). However, the free streaming of the hot dark matter results in a severe loss of power for wavenumbers $k \gg 2\pi/r_{\min}$. The right panel of Figure 11.5 shows the root mean square mass fluctuations, $\delta M/M$, calculated using Equation 11.70. For the original $P \propto k$ spectrum (dashed line), the amplitude of density fluctuations is larger on smaller mass scales, with $\delta M/M \propto M^{-2/3}$. However, for the hot dark matter power spectrum (dotted line),

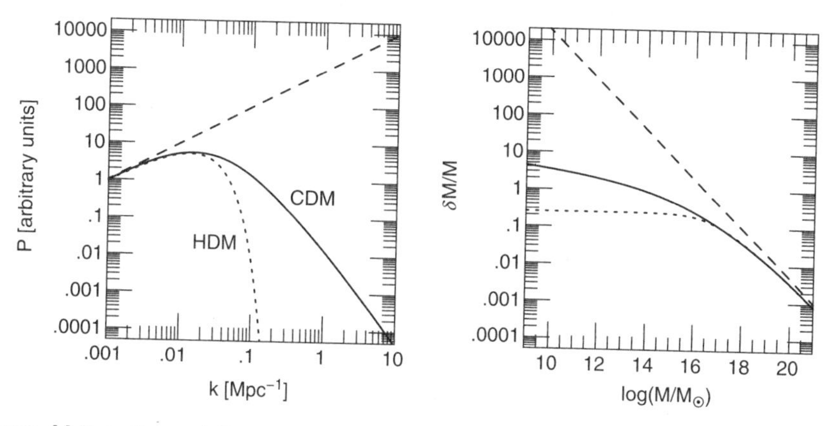

Figure 11.5 Left panel: Power spectrum at the time of radiation–matter equality for cold dark matter (solid line) and for hot dark matter (dotted line). The initial power spectrum produced by inflation (dashed line) is assumed to have the form $P(k) \propto k$. Right panel: The root mean square mass fluctuations $\delta M/M$ are shown as a function of M. Line types are the same as in the left panel.

the value of $\delta M/M$ levels off on scales smaller than $M < M_{\rm min} \sim 3 \times 10^{16}\,{\rm M}_\odot$. In a universe with this hot dark matter power spectrum, the first structures to collapse are regions of mass $M \sim M_{\rm min}$ with nearly uniform overdensity. These regions collapse to form superclusters; smaller structures, such as clusters and galaxies, then form by fragmentation of the superclusters. This scenario, in which the largest observable structures form first, is called the *top-down* scenario.

If most of the dark matter in the universe were hot dark matter with $m_h c^2 \sim 3\,{\rm eV}$, then we would expect the oldest structures in the universe to be superclusters, and that galaxies would be relatively young. In fact, the opposite seems to be true in our universe. Superclusters are just collapsing today, while galaxies have been around since at least $z \sim 10$, when the universe was about half a gigayear old. Thus, most of the dark matter in the universe must be *cold* dark matter, for which free streaming has been negligible.

The evolution of the power spectrum of cold dark matter, given the absence of free streaming, is different from the evolution of the power spectrum for hot dark matter. Remember, when the universe is radiation-dominated, density fluctuations $\delta_{\vec{k}}$ in the dark matter do not grow significantly in amplitude, as long as their proper wavelength $a(t)2\pi/k$ is small compared to the Hubble distance $c/H(t)$. However, when the proper wavelength of a density perturbation is large compared to the Hubble distance, its amplitude will be able to increase, regardless of whether the universe is radiation-dominated or matter-dominated. If the cold dark matter consists of WIMPs, they decouple from the radiation at a time $t_d \sim 1\,{\rm s}$, when the scale factor is $a_d \sim 3 \times 10^{-10}$. At the time of WIMP decoupling, the Hubble distance is $c/H \sim 2ct_d \sim 6 \times 10^8\,{\rm m}$. This corresponds to a comoving length scale

$$r_d = \frac{2ct_d}{a_d} \sim 60\,{\rm pc}, \tag{11.80}$$

a mass scale

$$M_d = \frac{4\pi}{3} r_d^3 \rho_{m,0} \sim 0.05\,{\rm M}_\odot, \tag{11.81}$$

and a comoving wavenumber $k_d \sim 2\pi/r_d \sim 10^5\,{\rm Mpc}^{-1}$. Thus, density fluctuations with a wavenumber $k < k_d$ and a mass $M > M_d$ will have a wavelength greater than the Hubble distance at the time of WIMP decoupling, and will be able to grow freely in amplitude as long as their wavelength remains longer than the Hubble distance.

How the fluctuations grow once their wavelength is smaller than the Hubble distance depends on whether that transition happens before or after the time of radiation–matter equality $t_{rm} = 0.050\,{\rm Myr}$, when the scale factor was $a_{rm} = 2.9 \times 10^{-4}$. At the time of radiation–matter equality, the Hubble distance was $c/H \approx 1.8ct_{rm} \approx 0.027\,{\rm Mpc}$. This corresponds to a comoving length scale

$$r_{rm} = \frac{1.8ct_{rm}}{a_{rm}} \approx 90\,\text{Mpc}, \qquad (11.82)$$

a mass scale $M_{rm} \approx 1.3 \times 10^{17}\,\text{M}_\odot$, and a comoving wavenumber $k_{rm} = 2\pi/r_{rm} \approx 0.07\,\text{Mpc}^{-1}$. Density fluctuations with a wavenumber $k < k_{rm}$ and a mass $M > M_{rm}$ will remain larger than the Hubble distance during the entire radiation-dominated era, and will grow steadily in amplitude during all that time. Thus, for wavenumbers $k < k_{rm} \approx 0.07\,\text{Mpc}^{-1}$, the power spectrum for cold dark matter retains the original $P(k) \propto k$ form that it had immediately after inflation (see the left panel of Figure 11.5). By contrast, cold dark matter density perturbations with a wavenumber $k_d > k > k_{rm}$ will grow in amplitude only until their physical wavelength $a(t)r \propto t^{1/2}$ is smaller than the Hubble distance $c/H(t) \propto t$. At that time, their amplitude is frozen until the time t_{rm}, when matter dominates, and density perturbations smaller than the Hubble distance are free to grow again. Thus, for wavenumbers $k > k_{rm}$, the power spectrum for cold dark matter is suppressed in amplitude, with the suppression being greatest for the largest wavenumbers (corresponding to shorter wavelengths, which become smaller than the Hubble distance at an earlier time).

The left panel of Figure 11.5 shows, as the solid line, the power spectrum for cold dark matter (CDM) at the time of radiation–matter equality. Note the broad maximum in the power spectrum at $k \sim k_{rm} \sim 0.1\,\text{Mpc}^{-1}$. The root mean square mass fluctuations in the cold dark matter, $\delta M/M$, are shown as the solid line in the right panel of Figure 11.5. The amplitude of the fluctuations is normalized so that $\delta M/M = 1$ at $M = 10^{14}\,\text{M}_\odot$. This normalization gives agreement with the observed density fluctuations today on very large scales, where $\delta M/M < 1$, and the growth of density perturbations is still in the linear regime. At the time of radiation–matter equality, the amplitude of the density fluctuations was smaller, with $(\delta M/M)_{rm}$ equal to $\sim a_{rm} \sim 3 \times 10^{-4}$ times the value of $\delta M/M$ shown in Figure 11.5. Notice that the mass fluctuations in the CDM scenario are largest in amplitude for the smallest mass scales. This implies that in a universe filled with cold dark matter, the first objects to form are the smallest, with galaxies forming first, then clusters, then superclusters. This scenario, called the *bottom-up* scenario, is consistent with the observed relative ages of galaxies and superclusters.

Assuming that the dark matter consists of nothing but hot dark matter gives a poor fit to the observed large scale structure of the universe. However, there is strong evidence that neutrinos do have some mass, and thus that the universe contains at least *some* hot dark matter. Cosmologists studying the large scale structure of the universe can adjust the assumed power spectrum of the dark matter, by mixing together hot and cold matter. Comparison of the assumed power spectrum with the observed large scale structure indicates that $\Omega_{\text{hdm},0} \leq 0.007$. If the hot dark matter consists entirely of the three standard flavors of neutrino in the cosmic neutrino background, this implies an upper limit on the sum of the neutrino masses, with (Equation 2.26)

$$[m(\nu_e) + m(\nu_\mu) + m(\nu_\tau)]c^2 \leq 0.3\,\text{eV}. \tag{11.83}$$

If there were more hot dark matter than this amount, free streaming of the hot dark matter particles would make the universe too smooth on small scales. Some like it hot, but most like it cold – the majority of the dark matter in the universe must be *cold* dark matter.

Cosmologists frequently refer to the "ΛCDM model" for the universe. This is the particular variant of the Hot Big Bang model in which the dominant contributors to the energy density today are dark energy in the form of a cosmological constant (this is the "Λ" of ΛCDM) and nonrelativistic matter in the form of cold dark matter (this is the "CDM" of ΛCDM). The basic ΛCDM model is spatially flat. The Greco-Roman abbreviation "ΛCDM" was first used in the mid-1990s; in the years since then, the accumulation of evidence has led the ΛCDM model to be adopted as the standard model of cosmology.

11.6 Baryon Acoustic Oscillations

The cold dark matter power spectrum shown as the solid line in Figure 11.5 assumes that the only density fluctuations are the quantum fluctuations from the inflationary era, as modified by gravitationally-driven growth. This would be an excellent assumption if the only matter in the universe were cold dark matter. Things become complicated, though, when we consider that 15 percent of the matter in the universe is baryonic. As far as gravity is concerned, a kilogram of hydrogen is the same as a kilogram of WIMPs; however, baryonic matter has the additional ability to interact with photons through the electromagnetic force. This ability significantly changes the amplitude of mass fluctuations in the universe. On relatively small mass scales, the ability of baryonic matter to cool by emitting photons permits the formation of high-density galaxies on a mass scale $M < 10^{12}\,\text{M}_\odot$ and the formation of ultra-high-density stars on a mass scale $M < 100\,\text{M}_\odot$. On mass scales bigger than the largest supercluster, $M > 10^{17}\,\text{M}_\odot$, the ability of baryons to interact with photons has also placed its mark on the power spectrum of density perturbations. It has done so through the mechanism of *baryon acoustic oscillations*.

To understand the origin of baryon acoustic oscillations, go back in time to the era when baryons decoupled from photons, at a redshift $z_{\text{dec}} \approx z_{\text{ls}} \approx 1090$. As we noted in Section 8.5, the densest regions in the photon–baryon fluid at the time of last scattering were regions that had just managed to compress themselves to maximum density before rebounding under their own pressure. The size of these high density regions was comparable to the sound horizon distance at the time of last scattering. From Equation 8.65, the sound horizon distance had a physical length[6]

$$d_s(t_{\rm ls}) = 0.145\,{\rm Mpc}. \tag{11.84}$$

After the photons and baryons parted ways, the overdensity of the photons in these high density regions reveals itself as hot spots on the last scattering surface, with a characteristic angular size $\theta \approx \theta_s \approx 0.7^\circ$.

However, just as the overdense photons left an observable mark on the universe, the overdense baryons left their mark as well. A physical size of $d_s = 0.145$ Mpc at a redshift $z_{\rm ls} = 1090$ corresponds to a comoving length

$$r_s = d_s(t_{\rm ls})(1 + z_{\rm ls}) \approx 160\,{\rm Mpc}. \tag{11.85}$$

This comoving length, called the *acoustic scale*, in turn corresponds to a mass scale

$$M_s = \frac{4\pi}{3} r_s^3 \rho_{m,0} \approx 7 \times 10^{17}\,{\rm M}_\odot. \tag{11.86}$$

The density fluctuations that are present on a comoving scale $r_s \approx 160$ Mpc are called "baryon acoustic oscillations" since they are the baryonic manifestation of the acoustic oscillations that were present in the photon–baryon fluid just before decoupling. Immediately after decoupling, the baryon overdensity on the scale r_s was small in amplitude. However, the Matthew effect has caused the initially low-amplitude density fluctuations to grow to detectable levels at the present day. (Despite this growth, the density fluctuations on a length scale $r_s \approx 160$ Mpc are still not high enough in amplitude to destroy our assumption that the universe is homogeneous and isotropic on scales larger than 100 Mpc.)

The best way to detect low-amplitude overdensities on a scale as large as 160 Mpc is to look at the *correlation function* of galaxies in space. Suppose that the average number density of galaxies at the present day is $n_{\rm gal}$. Choose a galaxy at random, then look at a small volume dV at a comoving distance r from the chosen galaxy. The correlation function $\xi(r)$ is a way of telling you how many galaxies are likely to be found in that small volume dV. Expressed mathematically, the expected number dN is given by the relation

$$dN = n_{\rm gal}[1 + \xi(r)]dV. \tag{11.87}$$

If the galaxies have a Poisson distribution, then the correlation function is $\xi(r) = 0$; the number of galaxies you expect to find is simply the number density $n_{\rm gal}$ times the volume dV. However, galaxies emphatically do not have a Poisson distribution in space, as Figure 11.1 illustrates; instead, they tend to cluster, giving a correlation function ξ that is positive at small separations.

[6] Given the existence of an inflationary epoch, we must slightly redefine the sound horizon distance as the distance traveled by sound since the universe was refilled with particles at $t \sim t_f \sim 10^{-34}$ s, during the reheating period at the end of inflation.

When discussing the correlation function $C(\theta)$ of temperature fluctuations in the cosmic microwave background, we found it useful to break it down, with the use of spherical harmonics, into different multipole moments l. Similarly, it is useful to break down the correlation function $\xi(r)$ for galaxies, with the use of Fourier transforms, into different wavenumbers $\vec{k}$. In particular, we can express the correlation function $\xi(r)$ in terms of the statistical properties of the initial mass fluctuations in the universe. On sufficiently large scales, the galaxy distribution is a mapping of the mass density distribution $\delta(\vec{r})$; that is, the number density of galaxies is highest in regions where the initial overdensity δ was largest. For a Gaussian density field $\delta(\vec{r})$, the correlation function $\xi(r)$ is simply the Fourier transform of the power spectrum $P(k)$:

$$\xi(r) = \frac{V}{(2\pi)^3} \int P(k) e^{-i\vec{k}\cdot\vec{r}} d^3k. \tag{11.88}$$

For an isotropic power spectrum $P(k)$, this can be written as

$$\xi(r) = \frac{V}{2\pi^2} \int P(k) j_0(kr) k^2 dk, \tag{11.89}$$

where $j_0(x) = \sin x/x$ is yet another spherical Bessel function. Since the power spectrum $P(k)$ and the correlation function $\xi(r)$ are Fourier transforms of each other, they contain the same information. Which one you chose in a given situation depends on convenience. If you are doing a redshift survey of galaxies, it is generally more convenient to compute the correlation function $\xi(r)$ of the galaxies' distribution in space.

The acoustic scale is $r_s \approx 160\,\text{Mpc} \approx 0.04c/H_0$. Thus, a redshift survey of galaxies that goes to redshifts significantly greater than $z \approx 0.04$ should have the ability, if it contains enough galaxies, to observe an extra bump in the correlation function $\xi(r)$ at a comoving scale r_s, reflecting the presence of baryonic acoustic oscillations in the distribution of matter. Figure 11.6 shows the correlation function $\xi(r)$ for a sample of nearly a million luminous galaxies in a redshift range $0.4 < z < 0.7$, near the end of the matter-dominated era. Notice the enhancement of $\xi(r)$ at a comoving length scale $r \sim r_s \sim 160\,\text{Mpc}$. This "BAO bump" is a modern-day relic of the fluctuations that were present in the photon–baryon fluid at the time of last scattering, $t_{\text{ls}} \approx 0.37\,\text{Myr}$.

Baryon acoustic oscillations provide a useful tool for computing cosmological parameters. Since the acoustic scale is a standard yardstick, whose length is determined by well-understood physical principles, it can be used, for instance, to probe the properties of dark energy. By seeing how the angular size of the BAO bump varies with redshift during the recent dark-energy-dominated era, we can determine (once we have enough data) whether the dark energy is a simple cosmological constant with $w = -1$, or whether it has a different value of w. In addition (once we have enough data), we can determine whether w for dark energy is changing with time.

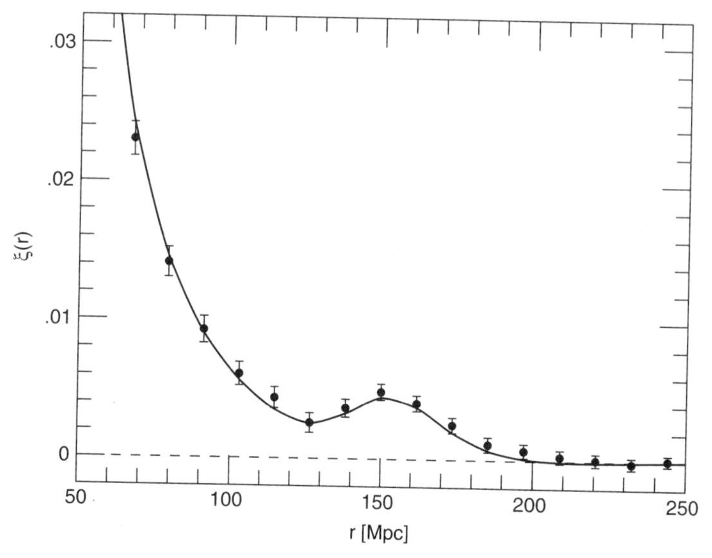

Figure 11.6 The correlation function as a function of comoving radius r. Points with error bars are the data for a large sample of luminous galaxies from the Sloan Digital Sky Survey. The solid line is a model including the effects of baryon acoustic oscillations. [data from Anderson *et al.* 2014, *MNRAS*, **441**, 24]

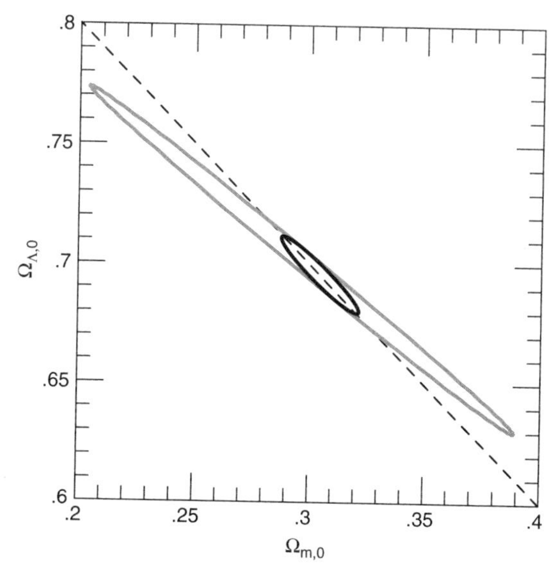

Figure 11.7 The gray elliptical contour gives the 95% confidence interval for $\Omega_{m,0}$ and $\Omega_{\Lambda,0}$ from the joint supernova and CMB data (compare with Figure 8.7). The black elliptical contour is the 95% confidence interval when the baryon acoustic oscillation observations are added. [Anže Slosar & José Alberto Vázquez, BNL]

If we assume, for the moment, that the ΛCDM model is correct, and that $w = -1$ for the dark energy, the location and amplitude of the BAO bump in Figure 11.6 enable us to place further constraint on the values of $\Omega_{m,0}$ and $\Omega_{\Lambda,0}$. For instance, in Figure 11.7, the gray contour gives the values of $\Omega_{m,0}$ and $\Omega_{\Lambda,0}$ that are consistent, at the 95% confidence level, with the type Ia supernova results and the CMB temperature fluctuations observed by *Planck*. The smaller black ellipse gives the 95% confidence interval once the BAO measurements are added to the supernova and CMB measurements. The tiny size of the black ellipse (centered on $\Omega_{m,0} = 0.305$ and $\Omega_{\Lambda,0} = 0.696$ for this particular combination of data) is the result of a happy concurrence among the three sources of information, and is the basis for the "Benchmark Model" that we have been using in this text. The fact that the black ellipse clings so closely to the $\kappa = 0$ line is also the source of the $|\Omega_0 - 1| \leq 0.005$ limit that we used when discussing the flatness problem in Section 10.1.

Exercises

11.1 Consider a spatially flat, matter-dominated universe ($\Omega = \Omega_m = 1$) that is *contracting* with time. What is the functional form of $\delta(t)$ in such a universe?

11.2 Consider an empty, negatively curved, expanding universe, as described in Section 5.2. If a dynamically insignificant amount of matter ($\Omega_m \ll 1$) is present in such a universe, how do density fluctuations in the matter evolve with time? That is, what is the functional form of $\delta(t)$?

11.3 A volume containing a photon–baryon fluid is adiabatically expanded or compressed. The energy density of the fluid is $\varepsilon = \varepsilon_\gamma + \varepsilon_{\rm bary}$, and the pressure is $P = P_\gamma = \varepsilon_\gamma/3$. What is $dP/d\varepsilon$ for the photon–baryon fluid? What is the sound speed, c_s? In Equation 11.26, how large an error did we make in our estimate of λ_J(before) by ignoring the effect of the baryons on the sound speed of the photon–baryon fluid?

11.4 Suppose that the stars in a disk galaxy have a constant orbital speed v out to the edge of its spherical dark halo, at a distance $R_{\rm halo}$ from the galaxy's center. If a bound structure, such as a galaxy, forms by gravitational collapse of an initially small density perturbation, the minimum time for collapse is $t_{\rm min} \approx t_{\rm dyn} \approx 1/\sqrt{G\bar{\rho}}$. Show that $t_{\rm min} \approx R_{\rm halo}/v$ for a disk galaxy. What is $t_{\rm min}$ for our own galaxy? What is the maximum possible redshift at which you would expect to see galaxies comparable in v and $R_{\rm halo}$ to our own galaxy? (Assume the Benchmark Model is correct.)

11.5 Within the Coma cluster, as discussed in Section 7.3, galaxies have a root mean square velocity of $\langle v^2 \rangle^{1/2} \approx 1520\,{\rm km\,s^{-1}}$ relative to the center of mass of the cluster; the half-mass radius of the Coma cluster is $r_h \approx 1.5\,{\rm Mpc}$. Using arguments similar to those of the previous problem, compute the minimum time $t_{\min}$ required for the Coma cluster to form by gravitational collapse.

11.6 Suppose that the density fluctuations $\delta(\vec{r})$ in the early universe constitute a Gaussian field with a power spectrum $P(k)$ that equals zero above some maximum wavenumber $k_{\max}$. This maximum wavenumber corresponds to a minimum length scale $r_{\min} = 2\pi/k_{\max}$ and a minimum mass scale $M_{\min} = (4\pi/3)\rho_m r_{\min}^3$. Show that for $M < M_{\min}$, the mean square mass fluctuation, $\langle(\delta M/M)^2\rangle$, is equal to σ_δ^2 for the density field.

12

Structure Formation: Baryons and Photons

If the only matter in the universe were cold dark matter, then structure formation would be driven solely by gravity. Adding baryonic matter, capable of absorbing, emitting, and scattering light, complicates the process of structure formation. As we saw in Section 11.2, before the time of decoupling at $z_{\rm dec} = 1090$, the baryonic matter was coupled to the photons, thanks to the ability of photons and electrons to scatter from each other. At $z > 1090$, therefore, the interaction of baryonic matter with photons *prevented* dense baryonic lumps from forming. At lower redshifts, however, the interaction of baryonic matter with photons *encouraged* dense baryonic lumps to form, since the ability to radiate away excess thermal energy is necessary to make objects with high mass density.

The average density of baryonic matter today, at $t_0 \approx 13.7\,\text{Gyr}$, is

$$\rho_{\rm bary,0} = 4.2 \times 10^{-28}\,\text{kg m}^{-3} = 6.2 \times 10^{9}\,\text{M}_\odot\,\text{Mpc}^{-3}. \tag{12.1}$$

However, some parts of the universe are far above average when it comes to baryonic density. Let's look at a typical suburban location in a luminous galaxy: the region within a few hundred parsecs of the Sun. In the solar neighborhood, we find that the density of stars and interstellar gas is $\rho_{\rm sn} \approx 0.095\,\text{M}_\odot\,\text{pc}^{-3} \approx 6.4 \times 10^{-21}\,\text{kg m}^{-3}$. This represents an overdensity

$$\delta_{\rm sn} = \frac{\rho_{\rm sn} - \rho_{\rm bary,0}}{\rho_{\rm bary,0}} \sim 2 \times 10^{7} \tag{12.2}$$

relative to the average baryonic density of the universe today. Now let's look at an individual main sequence star: the Sun itself. The Sun's average internal density is $\rho_\odot \approx 1400\,\text{kg m}^{-3}$, representing an overdensity $\delta_\odot \sim 3 \times 10^{30}$ relative to the average baryonic density today.[1] However, although baryons are capable of forming very dense objects, the majority of baryonic matter today is still in the form of low density intergalactic gas. To understand why some of the

[1] You are slightly less dense than the Sun, so $\delta_{\rm you} \sim 2 \times 10^{30}$.

baryonic matter in the universe forms condensed knots such as stars, while most remains low in density, we start by making a census of the baryonic matter in the universe today.

12.1 Baryonic Matter Today

Although we know the average baryon density $\rho_{\rm bary,0}$ quite well, the task of making a more detailed census of neutrons and protons (and their electron sidekicks) is frustratingly difficult. For example, in Section 7.1, we attempted to find the mass density of stars today, $\rho_{\star,0}$. Since stars glow at wavelengths that astronomers are highly experienced at detecting, stars should be the easiest baryonic component to detect. Nevertheless, we made only the rough estimate (Equation 7.4)

$$\rho_{\star,0} \approx 4 \times 10^8\,{\rm M_\odot\,Mpc^{-3}} \approx 3 \times 10^{-29}\,{\rm kg\,m^{-3}}, \tag{12.3}$$

representing about 7 percent of the baryonic mass. Estimates of the other contributions to the baryonic matter are equally rough, if not more so. With that caveat in mind, Figure 12.1 shows how the baryonic component of the universe is divided up. The wedge labeled "Stars, etc." represents the 7% of the baryonic matter that is in the form of stars, stellar remnants, brown dwarfs, and planets. Proceeding clockwise around the pie chart of Figure 12.1, about 1% of the baryonic mass is in the *interstellar gas* that fills the space between stars within the luminous portion of a galaxy. About 3% is in the *circumgalactic gas* that is gravitationally bound within the dark halo of a galaxy, but lying outside the main distribution of the galaxy's stars. (Since there is no clean dividing line between the interstellar gas and the circumgalactic gas, you can think of them together as gas associated with individual galaxies, providing about 4% of the baryonic matter in the universe.) Approximately 4% of the baryonic matter is in the *intracluster gas* that is gravitationally bound within a cluster of galaxies, but is not bound to any particular galaxy within the cluster; we have already encountered intracluster gas (Figure 7.3) in the form of the hot, X-ray emitting gas of the Coma cluster. Thus, we conclude that baryonic matter associated with gravitationally bound systems (galaxies and clusters of galaxies) contributes just 15 percent of the total baryonic matter in the universe.

Where are the rest of the baryons? In the 1990s, astronomers began talking of a "missing baryon problem," when they began to realize that the baryonic matter in galaxies and clusters falls far short of $\rho_{\rm bary,0}$. The missing baryons, however, aren't truly missing; they are simply in a low-density and inconspicuous intergalactic medium. As shown in Figure 12.1, about 40% of the baryonic matter is in the *diffuse intergalactic gas*, which consists of gas widely distributed outside galaxies and clusters, at a temperature $T < 10^5$ K. The remaining $\sim$ 45% of the

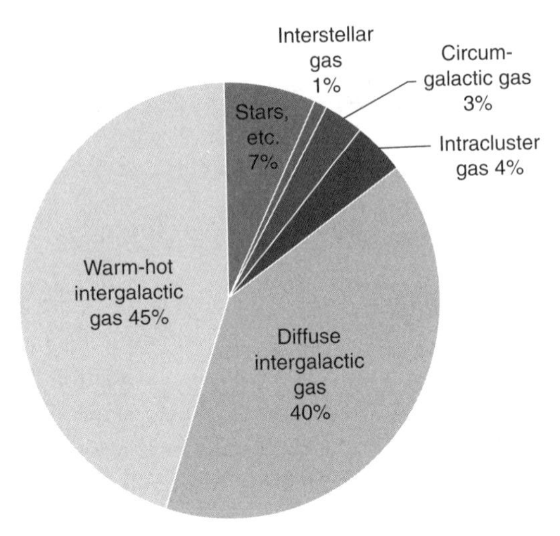

Figure 12.1 An approximate division of the current baryonic mass of the universe into its various components.

baryonic matter is in the *warm-hot intergalactic gas*, which is hotter than the diffuse intergalactic gas, typically having $10^5\,\mathrm{K} < T < 10^7\,\mathrm{K}$.[2] The warm-hot intergalactic gas is found in long filaments between clusters, as compared to the more smoothly distributed diffuse intergalactic gas.

One striking characteristic of intergalactic gas is its low density. Most of the volume in the universe today is filled with diffuse intergalactic gas with $\delta \leq 0$. Even the warm-hot intergalactic gas, which tends to be higher in density, typically lies in the overdensity range $3 < \delta < 300$. Another striking characteristic of intergalactic gas today is its high degree of ionization. In most of the intergalactic gas, the fractional ionization of hydrogen, X, is very close to one. However, when we looked at the physics of recombination in Section 8.3, we noted that in the early universe, between $z = 1480$ and $z = 1260$, the fractional ionization of hydrogen plummeted from $X = 0.9$ to $X = 0.1$. Obviously, something happened after the epoch of recombination that *reionized* the hydrogen in the universe.

How was hydrogen reionized? To answer that question, it helps to ask the preliminary question, "*When* was hydrogen reionized?" Hydrogen was mostly neutral just after recombination ($z \approx 1380$); it is mostly ionized today ($z \approx 0$). The time that elapses between $z \approx 1380$ and $z \approx 0$ is $t \approx 0.99998t_0 \approx 13.7\,\mathrm{Gyr}$. If we could pin down the time of reionization just a little more closely than that, we would have a useful clue about what physical mechanisms might be ionizing the hydrogen. (I am focusing on the reionization of hydrogen, and ignoring

[2] The fact that gas at 10^7 K is called "warm-hot" rather than "hot-hot" may seem like laughable understatement. However, it is being contrasted with the X-ray-emitting intracluster gas, which can reach $T \sim 10^8$ K or more.

helium, for the same reason I focused on hydrogen while discussing recombination in Section 8.2. Adding helium, and other heavier elements, makes the analysis more complicated mathematically, without changing the basic physical conclusions.)

12.2 Reionization of Hydrogen

Although we mainly think of the cosmic microwave background as telling us about the epoch of last scattering ($z_{\rm ls} \approx 1090$), it also contains useful information about the universe at lower redshifts. The reionized intergalactic gas provides an obstacle course of free electrons that the photons of the CMB must pass through to reach our microwave antennas. Each free electron has a cross-section $\sigma_e = 6.65 \times 10^{-29}\,{\rm m}^2$ for scattering with a photon. The rate at which a CMB photon scatters from free electrons in the reionized gas is (compare to Equation 8.14)

$$\Gamma = n_e \sigma_e c, \tag{12.4}$$

where n_e is the number density of free electrons. If the baryonic matter is reionized starting at a time t_*, then the optical depth for scattering from the reionized gas is (compare to Equation 8.43)

$$\tau_* = \int_{t_*}^{t_0} \Gamma(t)dt = c\sigma_e \int_{t_*}^{t_0} n_e(t)dt. \tag{12.5}$$

If the optical depth of the reionized gas were $\tau_* \gg 1$, then each CMB photon would be scattered many times in passing through the reionized gas, losing all information about its original direction of motion. Our view of the cosmic microwave background would then be completely smeared out, as if we were looking through a translucent screen. The fact that we can see temperature fluctuations down to small angular scales in the CMB, as shown in Figure 8.3, tells us that the optical depth of the reionized gas must be $\tau_* < 1$.

Looking out at the distant last scattering surface through the nearby reionized gas is like looking out through a slightly frosted window rather than one made of perfectly transparent glass. As a consequence, the CMB shows a small amount of smearing on small angular scales, corresponding to large values of the multipole l. The temperature fluctuations measured by the *Planck* satellite show slight suppression at high l compared to what you would expect in the absence of reionization. The amount of suppression is consistent with an optical depth $\tau_* = 0.066 \pm 0.016$ for the reionized gas. That is, about one CMB photon in 15 scatters from a free electron at low redshift.

We can use the optical depth τ_* to estimate the time t_* of reionization, if we make some assumptions. First, let's assume the baryonic portion of the universe is pure hydrogen, either in the form of neutral atoms, with number density $n_{\rm H}$, or in the form of free protons, with number density n_p. With this assumption,

$$n_{\rm H} + n_p = n_{\rm bary} = \frac{n_{\rm bary,0}}{a^3}. \tag{12.6}$$

Next, let's assume that the hydrogen undergoes complete, instantaneous reionization at the time t_*. In that case, the number density of free electrons before reionization is $n_e = 0$, and the number density after reionization ($t > t_*$) is

$$n_e = n_p = \frac{n_{\rm bary,0}}{a^3}. \tag{12.7}$$

Plugging this estimate for the number density of free electrons into the relation for optical depth (Equation 12.5), we find

$$\tau_* = \Gamma_0 \int_{t_*}^{t_0} \frac{dt}{a(t)^3}, \tag{12.8}$$

where

$$\Gamma_0 = c\sigma_e n_{\rm bary,0} = 1.58 \times 10^{-4}\,{\rm Gyr}^{-1} \approx 0.0023 H_0 \tag{12.9}$$

is the rate at which photons would scatter from free electrons today, if the baryonic matter were a perfectly uniform distribution of fully ionized hydrogen.

Changing the variable of integration from t to a, equation 12.8 becomes

$$\tau_* = \Gamma_0 \int_{a(t_*)}^{1} \frac{da}{\dot{a}a^3} = \Gamma_0 \int_{a(t_*)}^{1} \frac{da}{H(a)a^4}, \tag{12.10}$$

using the fact that $H = \dot{a}/a$. Alternatively, we can find the redshift of reionization, z_*, by making the substitution $1 + z = 1/a$:

$$\tau_* = \Gamma_0 \int_0^{z_*} \frac{(1+z)^2 dz}{H(z)}. \tag{12.11}$$

During recent times, when the matter-dominated universe has been giving way to the lambda-dominated universe, the Hubble parameter has been, from Equation 5.96,

$$H(z) = H_0 \left[\Omega_{m,0}(1+z)^3 + \Omega_{\Lambda,0}\right]^{1/2}. \tag{12.12}$$

Inserting this functional form for $H(z)$ into Equation 12.11 gives an integral with the analytic solution

$$\tau_* = \frac{2}{3\Omega_{m,0}} \frac{\Gamma_0}{H_0} \left([\Omega_{m,0}(1+z_*)^3 + \Omega_{\Lambda,0}]^{1/2} - 1 \right). \tag{12.13}$$

Using the relevant parameters from the Benchmark Model, the optical depth for scattering in the reionized universe is

$$\tau_* = 0.00485 \left([0.31(1+z_*)^3 + 0.69]^{1/2} - 1 \right). \tag{12.14}$$

Given the observed optical depth, $\tau_* = 0.066 \pm 0.016$, we find that the redshift of reionization was $z_* = 7.8 \pm 1.3$, corresponding to an age for the universe $t_* \sim 650\,\text{Myr}$. Thus, the "era of neutrality," when the baryonic matter consisted mainly of neutral atoms, was a relatively brief interlude in the history of the universe, with $t_* - t_{\text{rec}} \sim 0.05 t_0$.

One way to ionize hydrogen is to bombard it with photons of energy $hf > Q = 13.6\,\text{eV}$. The obvious place to look for sources of ionizing photons at $z \geq 8$ is in galaxies. The highest redshift galaxies that have been discovered (so far) are at $z \sim 10$, so we know that galaxies were present at the time of reionization. One source of ionizing photons in galaxies is the hot, luminous O stars that contribute much of the luminosity in star-forming galaxies. Only the most massive O stars, those with $M \geq 30\,\text{M}_\odot$, are hot enough to contribute significantly to the background of ionizing ultraviolet radiation. As an example, an O star with $M \approx 30\,\text{M}_\odot$ produces ionizing photons at a rate $\dot{N}_* \approx 5 \times 10^{48}\,\text{s}^{-1}$. Thus, during its entire lifetime $t \approx 6\,\text{Myr} \approx 2 \times 10^{14}\,\text{s}$, the star will produce $N_* \approx \dot{N}_* t \approx 10^{63}$ ionizing photons.

Another source of ionizing photons in a galaxy is an *active galactic nucleus*, or AGN. An AGN is a compact central region of a galaxy that is luminous over a broad range of the spectrum, including the range $hf > 13.6\,\text{eV}$ required to ionize hydrogen. AGNs are compact because they are fueled by accretion of matter onto a black hole. Luminous galaxies usually have a supermassive black hole at their center. The central black hole of our own galaxy, for instance, has a mass $M_{\text{bh}} \approx 4 \times 10^6\,\text{M}_\odot$, and thus a Schwarzschild radius $2GM_{\text{bh}}/c^2 \approx 0.08\,\text{AU}$. The galaxy NGC 4889, one of the two brightest galaxies in the Coma cluster (Figure 7.2), has a central black hole with mass $M_{\text{bh}} \approx 2 \times 10^{10}\,\text{M}_\odot$ and Schwarzschild radius $2GM_{\text{bh}}/c^2 \approx 400\,\text{AU}$. If a supermassive black hole is accreting gas, the heated gas can emit light before it slips through the event horizon; the energy of the light can be as much as $0.1mc^2$, where m is the mass of the accreted gas. A significant fraction of this emitted light takes the form of ionizing ultraviolet photons with $hf > 13.6\,\text{eV}$, coming from a region not far outside the Schwarzschild radius of the black hole. For a luminous AGN, the rate of production of ionizing photons is approximately

$$\dot{N}_* \approx 3 \times 10^{56}\,\text{s}^{-1} \left(\frac{L_{\text{AGN}}}{10^{13}\,\text{L}_\odot} \right). \tag{12.15}$$

The most luminous active galactic nuclei, with $L > 10^{13}\text{L}_\odot$, are often referred to as *quasars*.[3] A quasar with $L \sim 10^{13}\,\text{L}_\odot$ can emit as many ionizing photons in a month as our example O star does in its entire 6 million year lifetime.

[3] The term "quasar" is short for "quasi-stellar object," referring to the fact that these distant compact light sources are, like stars, unresolved by our telescopes. In other respects, quasars are dissimilar to stars; they are not powered by nuclear fusion, and they are tremendously more luminous than any single star.

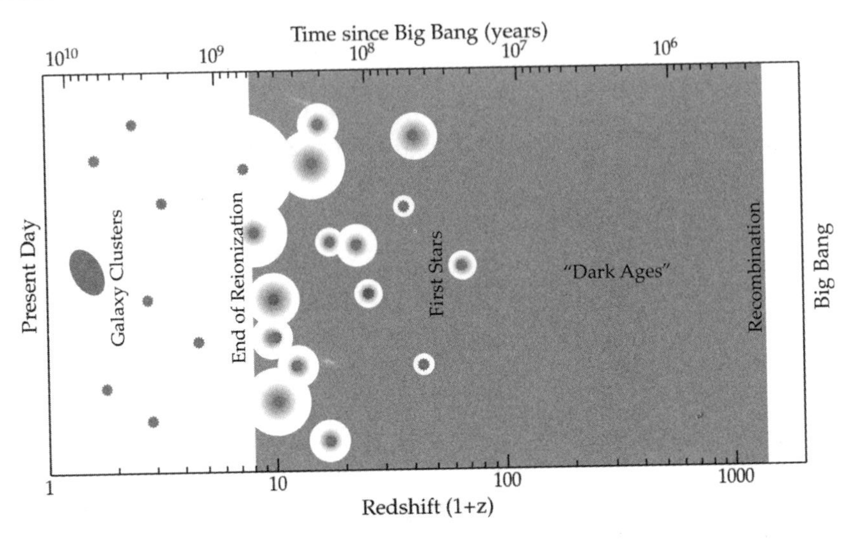

Figure 12.2 Schematic overview of the epoch of reionization; redshift decreases, and thus time increases, from right to left. The gray of the "Dark Ages" represents a gas of neutral atoms, while the white intergalactic spaces of the present day represent ionized gas.

12.3 The First Stars and Quasars

The time between recombination and the formation of the first stars and AGN is known to astronomers as the "Dark Ages" of the universe. The adjective "Dark," however, refers merely to the absence of starlight. The photons of the CMB were present during the Dark Ages. Indeed, immediately after the time of last scattering, at $z_{\rm ls} \approx 1090$, the temperature of the cosmic background radiation was $T \approx 2970\,{\rm K}$, about the temperature of an M star. If Heinrich Olbers had lived at the time of last scattering (admittedly an extremely large "if"), he would not have formulated Olbers' paradox. At $t \approx t_{\rm ls} \approx 0.37\,{\rm Myr}$, the entire sky *was* as bright as the surface of a star.

By the time the cosmic background radiation had cooled to a temperature $T \approx 140\,{\rm K}$, at $z \approx 50$, the universe was filled with a cosmic far infrared background, utterly ineffective at ionizing hydrogen. However, as shown in Figure 12.2, a redshift $z \approx 50$, corresponding to a cosmic time $t \approx 50\,{\rm Myr}$, is about the time when the very first stars began to emit light.[4] The Dark Ages ended as increasingly large numbers of stars and AGN poured out photons, some of them with $hf > 13.6\,{\rm eV}$. Figure 12.2 illustrates, in a schematic way, how regions of reionized gas began to grow around isolated galaxies, until the regions merged to form a single ionized intergalactic medium at $z \sim 8$.

[4] Exactly when the first stars formed is understandably conjectural, given our lack of direct observational evidence; most estimates fall in the range $z = 50 \rightarrow 20$, corresponding to $t = 50 \rightarrow 180\,{\rm Myr}$.

One complicating factor with using light from O stars and AGN to reionize intergalactic gas is that much of the ionizing radiation never escapes into intergalactic space. Instead, it is absorbed by the gas and dust within the galaxy in which the O star or AGN exists. The escape fraction, f_{esc}, represents the fraction of ionizing photons that actually leak out into the intergalactic gas. The escape fraction is poorly determined; as an optimistic first guess, we can adopt $f_{esc} = 0.2$, both from hot stars and from AGN. Now, let's consider a comoving cubic megaparsec of space, expanding along with the universal expansion, and compute how many ionizing photons we need to reionize its hydrogen. The comoving number density of baryons in intergalactic space is

$$n_{bary} = 0.25\,m^{-3} = 7.3 \times 10^{66}\,Mpc^{-3}. \tag{12.16}$$

In our pure hydrogen approximation, ionizing this many neutral hydrogen atoms requires a comoving number density of ionizing photons equal to

$$n_* = \frac{n_{bary}}{f_{esc}} = 3.7 \times 10^{67}\,Mpc^{-3}\left(\frac{0.2}{f_{esc}}\right). \tag{12.17}$$

Thus, assuming an escape fraction $f_{esc} \approx 0.2$, ionizing a comoving cubic megaparsec of hydrogen requires $N_* \sim 4 \times 10^{67}$ ionizing photons; this is the number created by $\sim 40\,000$ O stars, or by a $10^{13} L_\odot$ quasar shining for 4000 yr.

Given the immense ultraviolet luminosity of a quasar compared to even the hottest, brightest stars, it is tempting to assume that quasars perform most of the reionization. However, quasars, in addition to being breathtakingly luminous, are breathtakingly rare. Surveys like the 2dF survey and the Sloan Digital Sky Survey permit us to make quantitative statements about the scarcity of quasars as a function of time (or equivalently, of redshift). Figure 12.3 summarizes how the number of luminous quasars per comoving cubic megaparsec has changed over time. The comoving number density of quasars was greatest at a redshift $z \approx 2.5$, corresponding to a cosmic time $t \approx 2.6$ Gyr, long after reionization was complete. The comoving number density of quasars at $z_* \approx 8$ is more uncertain. Even the optimistic extrapolation shown in Figure 12.3 suggests that the number density was one luminous quasar for every $\sim 10^{10}$ comoving cubic megaparsecs. If reionizing one comoving cubic megaparsec of hydrogen requires 4000 years' worth of a quasar's luminosity, then reionizing $\sim 10^{10}$ comoving cubic megaparsecs would require $\sim 40\,000$ Gyr. Since the age of the universe at the time of reionization was only $t \sim 0.65$ Gyr, luminous quasars could have reionized only a small fraction of the intergalactic hydrogen.

Adding the ionizing radiation from lower-luminosity AGN to that from luminous quasars will help to boost the number of ionizing photons created. We know that in the present universe, there are more AGN with low luminosity than there are quasars with high luminosity. However, if the luminosity function of AGN at $z_* \sim 8$ is at all similar to that at $z \sim 0$, then even adding the efforts of the

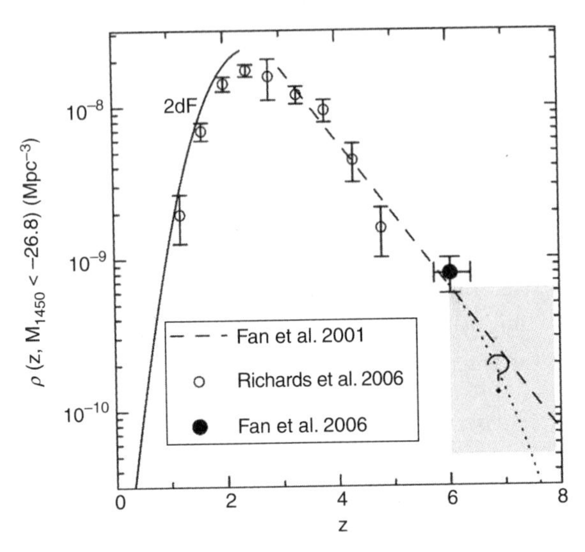

Figure 12.3 Evolution of the comoving number density of luminous quasars, based on data from the 2dF survey and the Sloan Digital Sky Survey. Redshift decreases, and thus time increases, from right to left. [Fan 2012, *RAA*, **12**, (8), 865]

lower-luminosity AGN will be inadequate to reionize the universe using the light from active galactic nuclei alone. It seems that most of the photons that reionized the universe came from hot, luminous stars.

At the present time, some galaxies are actively forming stars, while others are quiescent. However, if we average over a large volume containing a representative sample of galaxies, we find that the star formation rate today is $\dot{\rho}_{\star,0} \approx 20\,000\,\mathrm{M}_\odot\,\mathrm{Myr}^{-1}\,\mathrm{Mpc}^{-3}$. Since the baryon density today is $\rho_{\mathrm{bary},0} \approx 6 \times 10^9\,\mathrm{M}_\odot\,\mathrm{Mpc}^{-3}$, this means that 3 parts per million of the universe's baryons are being converted into stars every million years. However, studies of star formation at higher redshifts reveal that the star formation rate has varied with time. Figure 12.4, for instance, shows an observationally based estimate of the star formation rate per comoving cubic megaparsec as a function of redshift. The rate at which stars form today (within a comoving volume) is down by a factor of 10 from its maximum in the redshift range $z = 4 \to 1$, corresponding to cosmic times $t = 1.5 \to 6\,\mathrm{Gyr}$.[5]

At the time of reionization, $z_* \approx 8$, the star formation rate per comoving volume was about the same as it is today:

$$\dot{\rho}_\star(z = 8) \approx 20\,000\,\mathrm{M}_\odot\,\mathrm{Myr}^{-1}\,\mathrm{Mpc}^{-3}. \tag{12.18}$$

However, not all stars are equally useful for reionizing the intergalactic gas. In a comoving cubic megaparsec where stars are created at a rate of $20\,000\,\mathrm{M}_\odot\,\mathrm{Myr}^{-1}$,

[5] The intense star-forming era from $z = 4$ to $z = 1$ is when all the action was, baryonically speaking. We live today in a boring era, with fewer new stars and only feeble AGN.

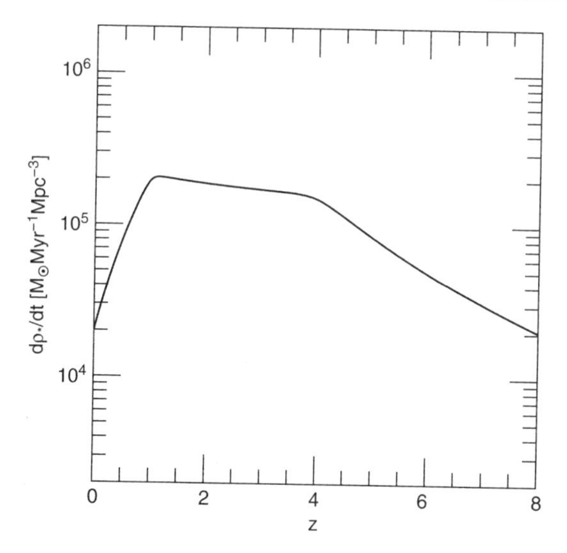

Figure 12.4 Evolution of the comoving density of star formation in the universe. Redshift decreases, and thus time increases, from right to left. [adapted from Yüksel *et al.* 2008, *ApJ*, **683**, 5]

only $\sim 2000\,\mathrm{M_\odot\,Myr^{-1}}$, or about 10 percent of the total, will take the form of O stars with $M \geq 30\,\mathrm{M_\odot}$, hot enough to emit a significant number of ionizing photons. If we make the simplifying assumption that all the O stars have $M \approx 30\,\mathrm{M_\odot}$, this implies the production of ~ 67 new O stars per comoving cubic megaparsec every million years. Since the lifetime of a $30\,\mathrm{M_\odot}$ star is $\sim 6\,\mathrm{Myr}$, this means there will be about 400 O stars present per comoving cubic megaparsec at any time, as long as the star formation rate remains at the level given by Equation 12.18. With 400 O stars each emitting ionizing photons at a rate $\dot{N}_* \approx 5 \times 10^{48}\,\mathrm{s^{-1}}$, this means that the total rate of ionizing photon production per comoving cubic megaparsec is

$$\begin{aligned}\dot{n}_* &\approx (5 \times 10^{48}\,\mathrm{s^{-1}})(400\,\mathrm{Mpc^{-3}}) \approx 2 \times 10^{51}\,\mathrm{s^{-1}\,Mpc^{-3}} \\ &\approx 6 \times 10^{64}\,\mathrm{Myr^{-1}\,Mpc^{-3}}.\end{aligned} \tag{12.19}$$

If we compare this rate to the total number n_* of ionizing photons required to reionize the hydrogen in the same comoving volume (Equation 12.17), we find that the star formation has to continue for a time

$$t = \frac{n_*}{\dot{n}_*} \approx \frac{3.7 \times 10^{67}\,\mathrm{Mpc^{-3}}}{6 \times 10^{64}\,\mathrm{Myr^{-1}\,Mpc^{-3}}}\left(\frac{0.2}{f_{\mathrm{esc}}}\right) \approx 600\,\mathrm{Myr}\left(\frac{0.2}{f_{\mathrm{esc}}}\right) \tag{12.20}$$

in order to reionize the intergalactic gas. As long as $f_{\mathrm{esc}} \geq 0.2$, this time is not impossibly long compared to the age of the universe at the time it was reionized, $t_* \approx 650\,\mathrm{Myr}$. Although many of the details of reionization remain to be worked out, this back-of-envelope accounting indicates that massive stars are capable of emitting most of the photons required to reionize the intergalactic gas.

Reionizing the baryonic universe with starlight requires forming stars at a comoving rate $\dot{\rho}_\star \approx 20\,000\,\mathrm{M_\odot\,Myr^{-1}\,Mpc^{-3}}$ for a time $t \approx 600\,\mathrm{Myr}$. By the time reionization is complete, the comoving number density of stars that have been formed is $\rho_\star = \dot{\rho}_\star t \approx 1.2 \times 10^7\,\mathrm{M_\odot\,Mpc^{-3}}$, or about 0.2% of the total baryon density. Thus, converting one part in 500 of the baryonic mass into stars has the side effect of reionizing the remaining baryonic gas. At times $t > t_*$, the ongoing star formation sketched out in Figure 12.4 drives the stellar mass density up to $\rho_{\star,0} \approx 4 \times 10^8\,\mathrm{M_\odot\,Mpc^{-3}}$ today. The accompanying production of ionizing photons from O stars (with help from AGN) enables the intergalactic medium to remain ionized despite the existence of radiative recombination.

12.4 Making Galaxies

The current mass density of stars, $\rho_\star \approx 4 \times 10^8\,\mathrm{M_\odot\,Mpc^{-3}}$, produces a luminosity density $\Psi_V = 1.1 \times 10^8\,\mathrm{L_{\odot,V}\,Mpc^{-3}}$. In addition, this mass density implies a stellar number density $n_\star \sim 10^9\,\mathrm{Mpc^{-3}}$ (the number we used back in Section 2.1, discussing Olbers' paradox). These stars are not uniformly distributed in space; instead, they tend to be contained within galaxies, which consist of a relatively small concentration of stars and interstellar gas in the midst of a larger halo, consisting mainly of dark matter with only a tenuous circumgalactic gas of baryons.

The observed luminosity function for galaxies, $\Phi(L)$, is defined so that $\Phi(L)dL$ is the number density of galaxies in the luminosity range $L \to L + dL$. It is found that the luminosity function is well fitted by the function[6]

$$\Phi(L)dL = \Phi^* \left(\frac{L}{L^*}\right)^\alpha \exp\left(-\frac{L}{L^*}\right)\frac{dL}{L^*}. \tag{12.21}$$

Surveys in the V band find a power-law slope $\alpha \approx -1$, a normalization $\Phi^* \approx 0.005\,\mathrm{Mpc^{-3}}$, and a characteristic luminosity $L_V^* \approx 2 \times 10^{10}\,\mathrm{L_{\odot,V}}$, comparable to the luminosity of our own galaxy. Figure 12.5 shows a characteristic luminosity function for galaxies.

Thanks to the exponential cutoff in the Schechter function, galaxies with $L > L^*$ are exponentially rare. Although a few galaxies with $L \approx 10L^*$ exist, such as NGC 4889 and NGC 4874 in the Coma cluster, they are very uncommon. They exist only in rich clusters of galaxies, where they have grown to vast size by cannibalizing other galaxies.[7] Our own galaxy, with $L_V \approx L_V^* \approx 2 \times 10^{10}\,\mathrm{L_{\odot,V}}$,

[6] This function is called the Schechter luminosity function, after the astronomer Paul Schechter, who pioneered its use.

[7] Although it sounds rather gruesome, "cannibalism" is the usual technical term for a merger between a small galaxy and a large one.

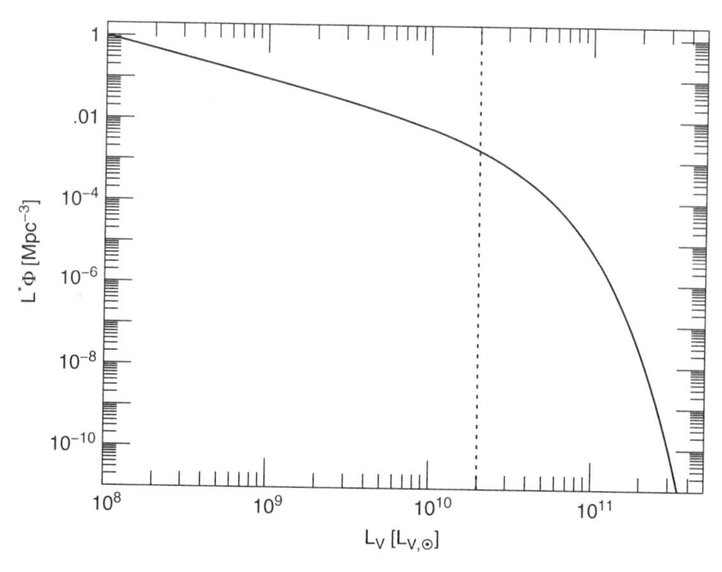

Figure 12.5 The luminosity function of galaxies today, observed in the V band. A Schechter function is assumed, with $\alpha = -1$ and $\Phi^* = 0.005\,\text{Mpc}^{-3}$. The vertical dotted line indicates the value of $L_V^* = 2 \times 10^{10}\, L_{\odot,V}$.

has a baryonic mass $M_{\text{bary}} \approx 1.2 \times 10^{11}\,\text{M}_\odot$, including all its interstellar and circumgalactic gas. Its total mass, provided mostly by the dark halo, is $M_{\text{tot}} \approx (1 \to 2) \times 10^{12}\,\text{M}_\odot$, depending on how far the halo extends. The exponential cutoff in the luminosity function at $L_V > L_V^*$ indicates that it is very difficult to make galaxies with a baryonic mass $M_{\text{bary}} > 10^{11}\,\text{M}_\odot$ and a total mass $M_{\text{tot}} > 10^{12}\,\text{M}_\odot$. It is definitely possible to make dark halos with a mass greater than $10^{12}\,\text{M}_\odot$; the dark halo of the Coma cluster, for instance, has $M \approx 2 \times 10^{15}\,\text{M}_\odot$. However, highly massive dark halos embrace multiple smaller galaxies rather than a single gargantuan galaxy. This hints that the difficulty in making jumbo galaxies, with $L_V > 2 \times 10^{10}\,\text{L}_{\odot,V}$ and $M_{\text{bary}} > 10^{11}\,\text{M}_\odot$, must have something to do with the properties of baryons rather than the properties of dark matter.

To see why there is an upper limit on the size of galaxies, we need a simplified model of how a galaxy forms. Consider a spherical overdense region at the time of radiation–matter equality ($t_{rm} \approx 0.050\,\text{Myr}$); this sphere will eventually become a luminous, star-filled galaxy and its surrounding dark halo. Initially, the relation among the mass M of the sphere, its radius R, and its overdensity δ is given by Equation 11.5:

$$M = \frac{4\pi}{3}\rho_m(t)[1 + \delta(t)]R(t)^3, \tag{12.22}$$

where $\rho_m(t) = \rho_{m,0}(1 + z)^3$ is the average mass density of the universe at time t and redshift $z = 1/a(t) - 1$.

At first, the overdensity is small: $\delta(t_{rm}) = \delta_{rm} \ll 1$. As long as $\delta(t) \ll 1$, the sphere's expansion is nearly indistinguishable from the Hubble expansion of the universe. However, the sphere reaches a maximum radius, and begins to collapse under its self-gravity, at the time $t_{\rm coll}$ when $\delta(t_{\rm coll}) \approx 1$. Given that $\delta \propto a \propto t^{2/3}$ during the matter-dominated epoch, we can make the approximation that $t_{\rm coll} \approx \delta_{rm}^{-3/2} t_{rm}$, or

$$1 + z_{\rm coll} \approx \delta_{rm}(1 + z_{rm}). \tag{12.23}$$

Thus, regions with a higher initial overdensity δ_{rm} will collapse at an earlier time, corresponding to a higher redshift.

At the moment when the sphere starts to collapse, its density $\bar{\rho}$ is

$$\bar{\rho}(t_{\rm coll}) \approx 2\rho_m(t_{\rm coll}) \approx 2\rho_{m,0}(1 + z_{\rm coll})^3. \tag{12.24}$$

After the sphere collapses, it oscillates in and out a few times before coming into an equilibrium state; it is now a gravitationally bound halo with a radius $R_{\rm halo} \approx R(t_{\rm coll})/2$. The process by which the collapsing structure comes into equilibrium is called "virialization," since the resulting halo obeys the virial theorem. Since the radius of the virialized halo is half the radius it had at $t_{\rm coll}$, the average density of the halo is now

$$\bar{\rho}_{\rm halo} \approx 8\bar{\rho}(t_{\rm coll}) \approx 16\rho_{m,0}(1 + z_{\rm coll})^3. \tag{12.25}$$

From the average density of a galaxy's virialized halo, you can deduce when it started its collapse. For instance, the mass of our own galaxy, contained mainly in its dark halo, is (from Equation 7.12)

$$M = 1.9 \times 10^{12}\,{\rm M}_\odot \left(\frac{R_{\rm halo}}{0.15\,{\rm Mpc}}\right), \tag{12.26}$$

giving an average density

$$\begin{aligned}\bar{\rho} = \frac{3M}{4\pi R_{\rm halo}^3} &= 1.4 \times 10^{14}\,{\rm M}_\odot\,{\rm Mpc}^{-3} \left(\frac{R_{\rm halo}}{0.15\,{\rm Mpc}}\right)^{-2} \\ &= 3400\rho_{m,0} \left(\frac{R_{\rm halo}}{0.15\,{\rm Mpc}}\right)^{-2}.\end{aligned} \tag{12.27}$$

Combined with Equation 12.25, this implies that our own galaxy started its collapse at a redshift given by the relation

$$1 + z_{\rm coll} \approx 6 \left(\frac{R_{\rm halo}}{0.15\,{\rm Mpc}}\right)^{-2/3}. \tag{12.28}$$

The process of virialization is not gentle; it involves subclumps of baryonic gas slamming into each other, creating shocks that heat the gas. A possible end state for the hot gas is a spherical distribution in hydrostatic equilibrium.

(As an example of such an end state, consider the hot intracluster gas in the Coma cluster, shown in Figure 7.3.) We can compute how hot the gas must be to remain in hydrostatic equilibrium. From Equation 7.41, a sphere of gas in hydrostatic equilibrium obeys the relation

$$M(r) = \frac{kT_{\rm gas}(r)r}{G\mu}\left[-\frac{d\ln\rho_{\rm gas}}{d\ln r} - \frac{d\ln T_{\rm gas}}{d\ln r}\right], \tag{12.29}$$

where $M(r)$ is the total mass within a radius r and μ is the mean mass per gas particle. For an ionized primordial mix of hydrogen and helium, with $Y = Y_p = 0.24$, the mean mass is $\mu = 0.59m_p$. For simplicity, assume a single temperature $T_{\rm gas}$ for all the gas in the halo, and a power law $\rho_{\rm gas} \propto r^{-\beta}$ for the gas profile. Evaluated at $r = R_{\rm halo}$, where $M(r) = M_{\rm tot}$, the equation of hydrostatic equilibrium yields

$$kT_{\rm gas} = \frac{GM_{\rm tot}\mu}{\beta R_{\rm halo}}. \tag{12.30}$$

This temperature is known as the *virial temperature* for gas in a virialized halo. Since the halo radius is

$$R_{\rm halo} = \left(\frac{3M_{\rm tot}}{4\pi\bar{\rho}_{\rm halo}}\right)^{1/3} = \left(\frac{3M_{\rm tot}}{64\pi\rho_{m,0}(1+z_{\rm coll})^3}\right)^{1/3}, \tag{12.31}$$

we can rewrite the virial temperature as a function of a halo's mass $M_{\rm tot}$ and the redshift $z_{\rm coll}$ at which its collapse began:

$$kT_{\rm gas} = \frac{4}{\beta}\left(\frac{\pi}{3}\right)^{1/3} G\mu\rho_{m,0}^{1/3}M_{\rm tot}^{2/3}(1+z_{\rm coll}). \tag{12.32}$$

Massive halos that collapsed early (and thus have high density) have the highest virial temperature.

If we assume the gas in the halo is hot enough to be mostly ionized, then $\mu \approx 0.6m_p$. Dark halos today have $\rho \propto r^{-2}$; if the hot gas started with a similar profile to the dark matter, we can take $\beta \approx 2$, and compute a numerical value for the temperature of the hot gas:

$$T_{\rm gas} \approx 1.0\times10^6\,{\rm K}\left(\frac{M_{\rm tot}}{10^{12}\,{\rm M}_\odot}\right)^{2/3}\left(\frac{1+z_{\rm coll}}{5}\right). \tag{12.33}$$

Here, I have scaled the value of $1+z_{\rm coll}$ to a collapse starting at redshift $z_{\rm coll} = 4$, at the start of the intense star-forming era, and consistent with the value for our own galaxy.

A universe with cold dark matter is a "bottom-up" universe, in which low-mass halos tend to collapse at earlier times than high-mass halos. However, the right panel of Figure 11.5 reminds us that the root mean square density fluctuation, $\delta M/M$, is not strongly dependent on mass; at $M \sim 10^{12}\,{\rm M}_\odot$, the dependence is $\delta M/M \propto M^{-0.14}$. Thus, the redshift of collapse should be only weakly dependent on halo mass. Consider, for example, a halo of mass $M_{\rm tot}$ that started as a modestly unusual 2σ density fluctuation; that is, at the time of radiation-matter equality, it had an overdensity δ_{rm} equal to $2 \times (\delta M/M)_{rm}$. For a halo mass $M_{\rm tot} = 10^{12}\,{\rm M}_\odot$, a 2σ fluctuation started its collapse at a redshift $z_{\rm coll} \approx 3.4$, and thus had a virial gas temperature $T_{\rm gas} \approx 9 \times 10^5$ K. For a much smaller halo mass, $M_{\rm tot} = 10^{10}\,{\rm M}_\odot$, a 2σ fluctuation started its collapse earlier, at $z_{\rm coll} \approx 6.7$, and had a virial temperature $T_{\rm gas} \approx 70\,000$ K. For a much larger halo mass, $M_{\rm tot} = 10^{14}\,{\rm M}_\odot$, a 2σ fluctuation started its collapse later, at $z_{\rm coll} \approx 1.0$, and had a virial temperature $T_{\rm gas} \approx 9 \times 10^6$ K.

This dependence of virial temperature on halo mass is the key to understanding why a low-mass halo can form a luminous, dense, star-filled galaxy and a high-mass halo cannot. To form a dense galaxy at the center of the dark halo, the baryonic gas must be able to cool by emitting light that escapes into intergalactic space. As it cools, the gas is no longer supported by pressure in a state of hydrostatic equilibrium, and falls to the halo's center. The scarcity of galaxies with mass $M_{\rm tot} > 10^{12}\,{\rm M}_\odot$ results from the fact that the hotter baryonic gas in higher mass halos is less efficient at radiating away its thermal energy.

Consider a $10^{10}\,{\rm M}_\odot$ halo, with a virial temperature $T_{\rm gas} \approx 70\,000$ K. At this temperature, although the hydrogen is ionized by collisions with other gas particles, the helium atoms are still able to retain one of their electrons. A $\rm He^+$ ion is able to radiate energy efficiently by line emission, as the remaining bound electron is excited to higher levels by collisions, then emits light as it falls back to the ground state. In general, halos with $T_{\rm gas} < 10^6$ K don't have their hydrogen and helium completely ionized, and can cool quickly by line emission from $\rm He^+$ or, at lower temperatures, from neutral He and H.

By contrast, halos with $T_{\rm gas} > 10^6$ K have hydrogen and helium that is almost completely ionized. In these halos, the ionized gas cools primarily by *bremsstrahlung*, also called free-free emission. Bremsstrahlung radiation is produced when a free electron is accelerated as it passes near a free proton or positively charged ion.[8] For a fully ionized primordial mix of hydrogen and helium, the luminosity density for bremsstrahlung emission is

$$\Psi = 5.3 \times 10^{-32}\,{\rm watts\,m^{-3}} \left(\frac{\rho_{\rm gas}}{10^{-24}\,{\rm kg\,m^{-3}}}\right)^2 \left(\frac{T}{10^6\,{\rm K}}\right)^{1/2}. \tag{12.34}$$

Since the low-density gas is highly transparent, all this luminosity is able to escape from the halo. The energy that is being radiated away is the thermal energy of the gas, which has energy density

[8] "Bremsstrahlung," in German, literally means "braking radiation." As the electron moves past the proton or ion, it emits a photon and loses kinetic energy. The alternative name of "free-free radiation" refers to the fact that the electron starts out free and ends free, without being captured by the proton or ion.

$$\varepsilon = \frac{3}{2}nkT = 2.1 \times 10^{-14}\,\mathrm{J\,m^{-3}} \left(\frac{\rho_{\rm gas}}{10^{-24}\,\mathrm{kg\,m^{-3}}}\right)\left(\frac{T}{10^6\,\mathrm{K}}\right). \tag{12.35}$$

The time it takes the ionized gas to cool by bremsstrahlung emission is then

$$t_{\rm cool} = \frac{\varepsilon}{\Psi} = 13\,\mathrm{Gyr}\left(\frac{\rho_{\rm gas}}{10^{-24}\,\mathrm{kg\,m^{-3}}}\right)^{-1}\left(\frac{T}{10^6\,\mathrm{K}}\right)^{1/2}. \tag{12.36}$$

Thus, if gas at $T > 10^6$ K is to cool in times less than the age of the universe, it must have a density $\rho_{\rm gas} > 10^{-24}\,\mathrm{kg\,m^{-3}}$. Is this a plausible density for gas in a virialized halo?

Suppose that the baryonic gas makes up a fraction f of the total mass of the virialized halo; if the baryon fraction in the halo is the same as that of the universe as a whole, we expect $f = 0.048/0.31 = 0.15$. The average mass density of the baryonic gas is then, making use of Equation 12.25,

$$\begin{aligned}\bar{\rho}_{\rm bary} &= f\bar{\rho}_{\rm halo} \approx 16f\rho_{m,0}(1+z_{\rm coll})^3 \\ &\approx 0.8 \times 10^{-24}\,\mathrm{kg\,m^{-3}}\left(\frac{f}{0.15}\right)\left(\frac{1+z_{\rm coll}}{5}\right)^3.\end{aligned} \tag{12.37}$$

We conclude that a virialized halo with $M_{\rm tot} \sim 10^{12}\,\mathrm{M_\odot}$ that starts its collapse at $z_{\rm coll} > 4$ will be hot enough to cool by bremsstrahlung (from Equation 12.33), and will be dense enough (from Equation 12.37) to cool in a time shorter than the age of the universe. However, we can also show that it is statistically unlikely for a halo with mass much larger than $10^{12}\,\mathrm{M_\odot}$ to collapse at a high enough redshift to be able to cool.

To begin our statistical analysis, consider the very first $10^{14}\,\mathrm{M_\odot}$ halo to have collapsed in the entire directly observable universe. (If the baryons in this halo could cool to form a single luminous galaxy, it would have ten times the baryonic mass of the huge cannibal galaxy NGC 4889.) Since the last scattering surface lies at a proper distance $d_p(t_0) \approx 14\,000$ Mpc, the total amount of mass inside the last scattering surface (and thus visible to our telescopes) is

$$M = \rho_{m,0}\frac{4\pi}{3}d_p(t_0)^3 \approx 4.3 \times 10^{23}\,\mathrm{M_\odot}. \tag{12.38}$$

We can divide this mass into 4.3×10^9 different regions, each of mass $10^{14}\,\mathrm{M_\odot}$. The very first $10^{14}\,\mathrm{M_\odot}$ halo to collapse is the one region out of 4.3 billion that had the highest overdensity at the time of radiation–matter equality. You can think of it as having won a "density lottery" with a probability $P = 1/4.3 \times 10^9 \approx 2.3 \times 10^{-10}$ of drawing the winning ticket. In a Gaussian distribution, this probability is equivalent to a 6.2σ deviation. From the cold dark matter $\delta M/M$ distribution (Figure 11.5), we can compute that a region of mass $M = 10^{14}\,\mathrm{M_\odot}$ with a 6.2σ overdensity begins its collapse at a redshift $z_{\rm coll} \approx 5.2$. After virialization, its gas has a virial temperature $T_{\rm gas} \approx 2.3 \times 10^7$ K (from Equation 12.33) and an average density $\bar{\rho}_{gas} \approx 1.5 \times 10^{-24}\,\mathrm{kg\,m^{-3}}$ (from Equation 12.37). The cooling time for the gas, from Equation 12.36, is therefore $t_{cool} \approx 42\,\mathrm{Gyr}$, longer than the present age of the universe. Moreover, the $10^{14}\,\mathrm{M_{\odot}}$ halos that collapse later will have slightly lower virial temperatures, but much lower densities. Therefore, the later-collapsing halos have even longer cooling times than the pioneering $10^{14}\,\mathrm{M_{\odot}}$ halo.

To adapt the "spherical cow" joke common among physicists, I've been using a "spherical, isothermal, virialized cow" model for the formation of galaxies. The true picture, unsurprisingly, is more complex. In particular, computer simulations of galaxy formation indicate that in a more realistic non-spherical collapse, not all the baryonic gas is shock-heated to the virial temperature T_{gas}. The relatively cold gas that escapes being heated is able to flow to the center of the halo on time scales shorter than the cooling time of Equation 12.36. These "cold flows" of gas, as they are called, permit the formation of the first galaxies at higher redshifts, signaling the end of the Dark Ages and the beginning of reionization. The portion of the baryonic gas that *is* shock-heated, however, must obey the cooling time argument. Thus, the spherical, isothermal, virialized model explains why the hot intracluster gas in a massive halo (like that of the Coma cluster) fails to form a single gargantuan galaxy with baryonic mass $M_{bary} \gg 10^{11}\,\mathrm{M_{\odot}}$.

12.5 Making Stars

Suppose that a collapsed, virialized halo contains gas that is below the virial temperature T_{gas} of Equation 12.30. This can be either because the gas is part of a cold flow that was never shock-heated, or because it cooled rapidly by line emission or bremsstrahlung. The gas is out of hydrostatic equilibrium, and falls toward the center of the dark halo. What happens to the cool, infalling gas then? We know it isn't all swallowed by the central supermassive black hole. In the local universe, the mass of a galaxy's central black hole is less than 1% of the total baryonic mass of the galaxy. (Even the extraordinarily massive black hole in the galaxy NGC 4889, with $M_{bh} \sim 2 \times 10^{10}\,\mathrm{M_{\odot}}$, is small compared to the total baryonic mass of that bloated cannibal galaxy, $M_{bary} \sim 2 \times 10^{12}\,\mathrm{M_{\odot}}$.) If $10^{5}\,\mathrm{M_{\odot}}$ of gas cools and falls inward, we know it doesn't form into a single "megastar." We also know that it doesn't form a trillion "microstars," each of mass $\sim 10^{-7}\,\mathrm{M_{\odot}}$. Instead, it forms about a million stars and brown dwarfs, with a typical mass $M_{\star} \sim 0.1\,\mathrm{M_{\odot}}$ and a power-law tail to higher masses, as illustrated in the initial mass function of Figure 7.1.

In our own galaxy, stars are observed to form in the dense central cores of *molecular clouds*, regions of interstellar gas that are relatively cold and dense; in these regions hydrogen takes the form of molecules (H_2) rather than individual atoms. In the dense cores of molecular clouds, the mass density is as high as

$\rho_{\rm core} \sim 10^{-15}\,{\rm kg\,m^{-3}}$. This is more than 10^{12} times $\rho_{\rm bary,0}$, the mean density of baryons today. However, it is still less than 10^{-18} times $\rho_\odot$, the mean density of the Sun. In the interstellar gas of our galaxy, the helium mass fraction has been raised, by pollution from early generations of stars, from its primordial value $Y_p = 0.24$ to a current value $Y = 0.27$.[9] The helium mass fraction is $X = 0.72$, leaving a mass fraction $Z = 0.01$ in other elements, primarily oxygen and carbon. Part of the carbon and oxygen is in the form of molecules and radicals such as CO, CH, and OH; however, some condenses into tiny dust grains made of silicates (minerals containing oxygen and silicon) or graphite (pure carbon). A molecular cloud core is dusty enough to be opaque at visible wavelengths; cores seen against a background of stars are called "dark nebulae."[10] A typical temperature for a molecular cloud core is $T_{\rm core} \approx 20\,{\rm K}$. This temperature results from a balance between heating by cosmic rays (high-energy charged particles that can penetrate the opaque core) and cooling by far infrared radiation from dust grains.

The dynamical time in a molecular cloud core is (Equation 11.13)

$$t_{\rm dyn} = \frac{1}{(4\pi G\rho_{\rm core})^{1/2}} \approx 1.1\times 10^{12}\,{\rm s}\left(\frac{\rho_{\rm core}}{10^{-15}\,{\rm kg\,m^{-3}}}\right)^{-1/2}. \qquad (12.39)$$

The mean molecular mass in a molecular cloud, given its mixture of H_2 and He, is $\mu = 2.3m_p$. This results in an isothermal sound speed

$$c_s = \left(\frac{kT_{\rm core}}{\mu}\right)^{1/2} \approx 270\,{\rm m\,s^{-1}}\left(\frac{T_{\rm core}}{20\,{\rm K}}\right)^{1/2}, \qquad (12.40)$$

only slightly slower than the sound speed in air at room temperature. The Jeans length in the molecular cloud core is then (Equation 11.20)

$$\lambda_J = 2\pi c_s t_{\rm dyn} \approx 1.9\times 10^{15}\,{\rm m}\left(\frac{\rho_{\rm core}}{10^{-15}\,{\rm kg\,m^{-3}}}\right)^{-1/2}\left(\frac{T_{\rm core}}{20\,{\rm K}}\right)^{1/2}. \qquad (12.41)$$

This means that the baryonic Jeans mass within a dense molecular cloud core is

$$\begin{aligned} M_J = \frac{4\pi}{3}\rho_{\rm core}\lambda_J^3 &\approx 3\times 10^{31}\,{\rm kg}\left(\frac{\rho_{\rm core}}{10^{-15}\,{\rm kg\,m^{-3}}}\right)^{-1/2}\left(\frac{T_{\rm core}}{20\,{\rm K}}\right)^{3/2} \\ &\approx 15\,{\rm M_\odot}\left(\frac{\rho_{\rm core}}{10^{-15}\,{\rm kg\,m^{-3}}}\right)^{-1/2}\left(\frac{T_{\rm core}}{20\,{\rm K}}\right)^{3/2}. \end{aligned} \qquad (12.42)$$

Since objects smaller than M_J are pressure-supported, this seems to indicate that regions of a molecular cloud that are less massive than $\sim 15\,{\rm M_\odot}$ cannot collapse to form stars. A look at the initial mass function of stars, plotted in Figure 7.1,

[9] O, the futility of stars! Their mass provides $< 10\%$ of the baryonic density today, and even after 13 billion years on the job, the helium they produce has increased the helium mass fraction Y by only $\sim 10\%$, just as the starlight they emit has increased the photon energy density ε_γ by only $\sim 10\%$.

[10] A dark nebula isn't called "dark" because it fails to interact with light (the way dark matter is dark). It's called "dark" because it absorbs light (the way dark chocolate is dark).

reveals that this conclusion is nonsense. Stars with $M_\star > 15\,\mathrm{M}_\odot$ are O stars, the rarest type of star; the preferred mass for stars is actually $M_\star \sim 0.1\,\mathrm{M}_\odot$.

The reason why molecular clouds can make stars smaller than the Jeans mass is that collapsing cores can cool. Consider a molecular cloud core of mass $M_{\rm core} \approx 15\,\mathrm{M}_\odot$, just above the Jeans mass. It contains a total of $N = M_{\rm core}/\mu = 7.8 \times 10^{57}$ gas particles and has a thermal energy

$$E_{\rm core} = N\left(\frac{3}{2}kT_{\rm core}\right) = 3.2 \times 10^{36}\,\mathrm{J}\left(\frac{M_{\rm core}}{15\,\mathrm{M}_\odot}\right)\left(\frac{T_{\rm core}}{20\,\mathrm{K}}\right). \tag{12.43}$$

The core starts its collapse. If energy didn't flow out of (or into) the core, the first law of thermodynamics tells us that the thermal energy would increase at the rate

$$\frac{dE_{\rm core}}{dt} = -P_{\rm core}\frac{dV_{\rm core}}{dt} = -\frac{NkT_{\rm core}}{V_{\rm core}}\frac{dV_{\rm core}}{dt}. \tag{12.44}$$

Since the core's volume is $V_{\rm core} \propto R^3_{\rm core}$, this implies

$$\frac{dE_{\rm core}}{dt} = -3NkT_{\rm core}\left(\frac{1}{R_{\rm core}}\frac{dR_{\rm core}}{dt}\right) = -2E_{\rm core}\left(\frac{1}{R_{\rm core}}\frac{dR_{\rm core}}{dt}\right). \tag{12.45}$$

If the core is to remain at a constant temperature $T_{\rm core}$ during its collapse, it must radiate away the increased thermal energy at the same rate it is generated. The luminosity required to keep the core at a constant temperature (let's call it the "isothermal luminosity") is

$$L^{\rm iso}_{\rm core} = -\frac{dE_{\rm core}}{dt} = \frac{2E_{\rm core}}{R_{\rm core}}\frac{dR_{\rm core}}{dt}. \tag{12.46}$$

For a freely collapsing core, $dR_{\rm core}/dt \approx R_{\rm core}/t_{\rm dyn}$, and the isothermal luminosity at $T_{\rm core} = 20\,\mathrm{K}$ is

$$L^{\rm iso}_{\rm core} \approx \frac{2E_{\rm core}}{t_{\rm dyn}} \approx 0.015\,\mathrm{L}_\odot\left(\frac{M_{\rm core}}{15\,\mathrm{M}_\odot}\right)\left(\frac{\rho_{\rm core}}{10^{-15}\,\mathrm{kg\,m^{-3}}}\right)^{1/2}. \tag{12.47}$$

Thus, as long as the core radiates away energy at a minimum rate $L^{\rm iso} \sim 0.015\,\mathrm{L}_\odot$, it can maintain a constant temperature of $T_{\rm core} = 20\,\mathrm{K}$ as it collapses on a dynamical time. But what will be the actual luminosity L of a molecular cloud core?

The main source of emission from a dusty molecular cloud core is the far infrared light from its dust grains. Looking at a dusty dark nebula, as in the left panel of Figure 12.6, our first guess might be that since it's an opaque object with a well-defined temperature $T_{\rm core}$, its luminosity is that of a blackbody. For a spherical core, this would be $L_{\rm bb} = 4\pi R^2_{\rm core}\sigma_{\rm sb}T^4_{\rm core}$, where $\sigma_{\rm sb} = 5.67 \times 10^{-8}\,\mathrm{watts\,m^{-2}\,K^{-4}}$ is the Stefan–Boltzmann constant for blackbody radiation. However, a molecular cloud core that is opaque at visible wavelengths is not necessarily opaque at infrared wavelengths. Interstellar dust grains are tiny, with a radius of 100 nanometers or less; for such small grains, the cross-section for absorbing visible and infrared light is a decreasing function of the wavelength λ.

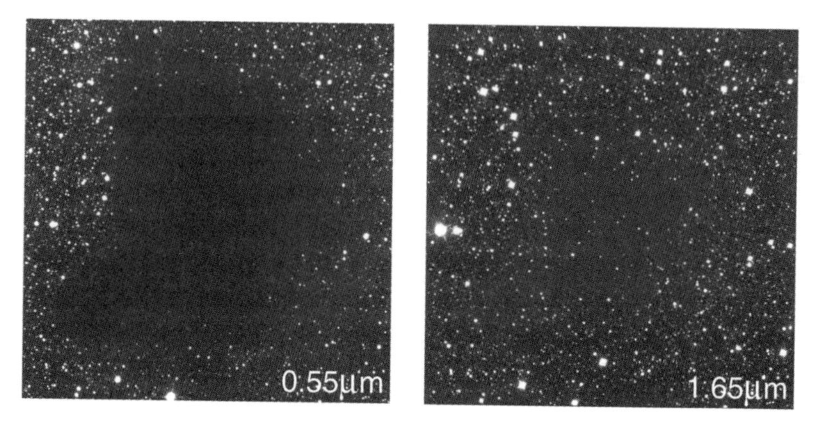

Figure 12.6 The dark nebula Barnard 68, which is at a distance $d \approx 125\,\text{pc}$ from the Earth and has a radius $R_{\text{core}} \approx 15\,000\,\text{AU} \approx 3 \times 10^6\,\text{R}_\odot$. Left: V-band image, at a wavelength $\lambda_{\text{vis}} \approx 550\,\text{nm}$. Right: Near infrared image, at a wavelength $\lambda_{\text{nir}} \approx 1650\,\text{nm}$. [European Southern Observatory]

Figure 12.6 shows a nearby dark nebula that is highly opaque at visible wavelengths, with $\tau \approx 23$ through its center at $\lambda_{\text{vis}} = 550\,\text{nm}$. However, at near infrared wavelengths it is much less opaque: $\tau \approx 4$ at $\lambda_{\text{nir}} = 1650\,\text{nm}$, so bright background stars can be glimpsed through the dusty core at this wavelength. At far infrared wavelengths, the core is largely transparent, with $\tau \approx 0.03$ at $\lambda_{\text{fir}} \approx 10^5\,\text{nm} \approx 0.1\,\text{mm}$. The optical depth of the core will increase as it is compressed to smaller radii; if n_{dust} is the average number density of dust grains in the core, then $n_{\text{dust}} \propto R_{\text{core}}^{-3}$ and therefore $\tau \propto n_{\text{dust}} R_{\text{core}} \propto R_{\text{core}}^{-2}$. However, different cores have different dust properties, so we can't say in the general case at what degree of compression a core will become opaque at far infrared wavelengths.

Mathematically, we can take into account the possible non-opacity of the core by writing its luminosity as

$$L_{\text{core}} = 4\pi R_{\text{core}}^2 \cdot f_e \sigma_{\text{sb}} T_{\text{core}}^4, \tag{12.48}$$

where the *efficiency factor* is $f_e \leq 1$. If the core is highly opaque at the wavelength of emission, with $\tau \gg 1$, then we expect $f_e \approx 1$. However, if it is largely transparent, with $\tau < 1$, then we expect $f_e < 1$. Scaled to the properties of our standard molecular cloud core, the core's radius is

$$R_{\text{core}} = \left(\frac{3M_{\text{core}}}{4\pi\rho_{\text{core}}}\right)^{1/3} \approx 10^4\,\text{AU}\left(\frac{M_{\text{core}}}{15\,\text{M}_\odot}\right)^{1/3}\left(\frac{\rho_{\text{core}}}{10^{-15}\,\text{kg}\,\text{m}^{-3}}\right)^{-1/3}, \tag{12.49}$$

and the luminosity actually emitted from the core at a temperature $T_{\text{core}} = 20\,\text{K}$ is

$$L_{\text{core}} \approx 1100\,\text{L}_\odot f_e \left(\frac{M_{\text{core}}}{15\,\text{M}_\odot}\right)^{2/3}\left(\frac{\rho_{\text{core}}}{10^{-15}\,\text{kg}\,\text{m}^{-3}}\right)^{-2/3}. \tag{12.50}$$

Any efficiency factor greater than the extremely modest value $f_e \sim 10^{-5}$ will enable the actual luminosity (Equation 12.50) to be larger than the isothermal luminosity required to maintain a constant temperature (Equation 12.47).

Let's see what happens as our $15\,M_\odot$ core collapses at a constant temperature $T_{\rm core} = 20\,{\rm K}$. Since the Jeans mass, from Equation 12.42, has the dependence $M_J \propto \rho^{-1/2}T^{3/2}$, then if the temperature is constant, the Jeans mass *decreases* as the core becomes smaller and its density thus becomes larger. Consider what happens when the radius of the core decreases from $R_{\rm core}$ to $R_{\rm core}/4^{1/3} \approx 0.63R_{\rm core}$, and the density thus increases from $\rho_{\rm core}$ to $4\rho_{\rm core}$. The Jeans mass has now fallen by a factor of 1/2, and the $15\,M_\odot$ core is now unstable, splitting to form a pair of $7.5\,M_\odot$ fragments. These fragments continue to collapse until their density increases by another factor of 4, then split into a total of four $3.75\,M_\odot$ fragments; and so on, and so forth.

Hierarchical fragmentation, as this process of repeated subdivision is called, naturally produces a power-law distribution of stellar masses, as long as the fragmentation process is slightly inefficient. Suppose, for example, that fragmentation has a failure rate of 1%. That is, if you start with 100 cores, each of mass $M_{\rm core}$, 99 of them will split, forming 198 fragments each of mass $M_{\rm core}/2$, but one will collapse directly to form a star of mass $M_\star = M_{\rm core}$. If the failure rate is the same at the next step, then of the 198 fragments of mass $M_{\rm core}/2$, we expect that 196 will split, forming 392 subfragments, each of mass $M_{\rm core}/4$, but two will collapse directly to form stars of mass $M_\star = M_{\rm core}/2$. As this process proceeds, you will expect one star of mass $M_{\rm core}$, two of mass $M_{\rm core}/2$, four of mass $M_{\rm core}/4$, and so forth. In general, after the nth round of fragmentation, there will be 2^n stars that each have mass $2^{-n}M_{\rm core}$. Thus, the number of stars produced per logarithmic interval of mass will be

$$\frac{dN}{d\log M} \propto \frac{1}{M}, \tag{12.51}$$

which is equivalent to an initial mass function $\chi(M) \propto M^{-2}$. (The actual initial mass function has the steeper dependence $\chi \propto M^{-2.3}$. This is a sign that the simple hierarchical fragmentation model doesn't capture all the physics involved.)

Starting with a core mass $M_{\rm core}$, after n rounds of fragmentation, the mass of each fragment will be $M_{\rm frag} = 2^{-n}M_{\rm core}$ and the density of each fragment will be $\rho_{\rm frag} = 2^{2n}\rho_{\rm core}$. What halts the fragmentation process after a finite number of rounds, and prevents it from proceeding *ad infinitum*, producing an infinite number of infinitesimal fragments? Notice that the isothermal luminosity required to keep the fragments at a constant temperature $T_{\rm core} = 20\,{\rm K}$ has the value (Equation 12.47)

$$L_{\rm frag}^{\rm iso} \approx 0.015\,L_\odot \left(\frac{M_{\rm frag}}{15\,M_\odot}\right)\left(\frac{\rho_{\rm frag}}{10^{-15}\ {\rm kg\,m^{-3}}}\right)^{1/2}. \tag{12.52}$$

However, since $M_{\rm frag} \propto 2^{-n}$ and $\rho_{\rm frag}^{1/2} \propto 2^n$, the isothermal luminosity is the same for every value of n. That is, every fragment, regardless of size, must radiate at the same luminosity to maintain a temperature of 20 K. However, since the actual luminosity is proportional to the square of the fragment's radius, smaller fragments will have lower luminosity. After some number of rounds of fragmentation, the small fragments will find it impossible to radiate away enough energy to keep cool.

The ratio of the actual luminosity of a fragment (Equation 12.50) to its required isothermal luminosity (Equation 12.47) is

$$\frac{L_{\rm frag}}{L_{\rm frag}^{\rm iso}} \approx 73\,000 f_e \left(\frac{M_{\rm frag}}{15 {\rm M}_\odot}\right)^{-1/3} \left(\frac{\rho_{\rm frag}}{10^{-15}\,{\rm kg\,m^{-3}}}\right)^{-7/6}. \tag{12.53}$$

After n rounds of fragmentation, the characteristic fragment mass will be $M_{\rm frag} = 2^{-n} M_{\rm core} \approx 2^{-n}(15\,{\rm M}_\odot)$, and the characteristic fragment density will be $\rho_{\rm frag} = 2^{2n}\rho_{\rm core} \approx 2^{2n}(10^{-15}\,{\rm kg\,m^{-3}})$. The four-fold increase of density with each round of fragmentation means that the ratio of the actual luminosity to the required isothermal luminosity will *decrease* with each round of fragmentation:

$$\frac{L_{\rm frag}}{L_{\rm frag}^{\rm iso}} \approx 73\,000 f_e 2^{n/3} 2^{-7n/3} \approx 73\,000 f_e 2^{-2n}. \tag{12.54}$$

Fragmentation stops when this ratio drops to one, and the temperature of the collapsing fragments starts to rise; this causes the Jeans mass to increase instead of decrease as the fragments become denser. From Equation 12.54, we find that cooling the fragments that result from $n = 9$ rounds of fragmentation would require an unphysically high efficiency $f_e \approx 3$. Thus, the maximum possible number of fragmentation rounds is $n = 9$, producing a minimum fragment mass $M_{\rm frag} \sim 2^{-9} M_{\rm core} \sim 0.03\,{\rm M}_\odot$, within the broad maximum of the initial mass function (Figure 7.1). Once fragments can no longer keep cool, they become *protostars*, objects that are almost, but not quite, in hydrostatic equilibrium. Protostars gradually contract inward as they radiate away energy, but only on time scales much longer than the dynamical time. A protostar with $M > 0.08\,{\rm M}_\odot$ becomes a star when the density and temperature at its center become large enough for nuclear fusion to begin.[11]

Hierarchical fragmentation requires that collapsing molecular cloud cores are able to cool efficiently through multiple rounds of fragmentation. If molecular cloud cores are very low in dust (which happens if they are deficient in heavy elements such as carbon, oxygen, and silicon), this drives down f_e, and halts fragmentation at an earlier stage. Thus, star formation in regions with low abundances of heavy elements has an initial mass function that peaks at a higher mass. When the very first stars formed at the end of the Dark Ages, there was no carbon,

[11] A protostar with $M < 0.08\,{\rm M}_\odot$ should really be called a "proto-brown-dwarf."

oxygen, and silicon; thus, there were no dust grains at all. The first stars could cool only through much less efficient processes involving molecular hydrogen. Thus, it is likely that the very first stars that formed in the universe were not the result of hierarchical fragmentation, but instead were all extremely massive stars, comparable in size to the Jeans mass of the gas from which they formed. These ultramassive stars would have lived fast and died young, ejecting the carbon, oxygen, and other heavy elements that they formed into the surrounding gas.

The hierarchical fragmentation scenario obviously represents a highly simplified "spherical isothermal cow that reproduces by fission." Realistic models of star formation must take into account, among other physical processes, the effects of magnetic fields and of turbulence within molecular clouds. In addition, the effects of angular momentum must be acknowledged. An initially slowly spinning fragment, thanks to conservation of angular momentum, will rotate more and more rapidly as it contracts adiabatically. The net result will be a protostar, made from the material with lower angular momentum, surrounded by a rotationally supported protoplanetary disk. Protoplanetary disks, it is found observationally, are unstable to the formation of the objects that we call "planets." Planets are a common, but minor, side effect of star formation. In the Solar System, for instance, the total mass of all the planets is $M_{\rm pl} = 2.67 \times 10^{27}$ kg, with more than two-thirds of this mass provided by Jupiter. The total planet mass thus represents a small fraction of the Sun's mass: $M_{\rm pl} = 0.0013\,{\rm M}_\odot$. If the Sun is not unusual in having a planetary system with a mass $\sim 0.13\%$ of its own mass, then we can estimate the density parameter of planets:

$$\Omega_{\rm pl,0} \sim 0.0013\Omega_{\star,0} \sim (0.0013)(0.003) \sim 4 \times 10^{-6}. \qquad (12.55)$$

From one point of view, planets are unimportant, because they provide just a few parts per million of the mass-energy density of the universe. From another point of view, however, planets are vitally important; planets of the right size, at the right distance from their parent star, provide a hospitable environment for the evolution of beings who ask the questions "Where do we come from?", "What are we?", and "Where are we going?"

Exercises

12.1 For the Schechter luminosity function of galaxies (Equation 12.21), find the number density of galaxies more luminous than L, as a function of L^*, Φ^*, and α. In the limit $L \to 0$, show why $\alpha = -1$ leads to problems, mathematically speaking. What is a plausible physical solution to this mathematical problem? [Hint: an acquaintance with incomplete gamma functions will be useful.]

12.2 For the Schechter luminosity function of galaxies, find the total luminosity density Ψ as a function of L^*, Φ^*, and α. What is the numerical value of the luminosity density Ψ_V in the V band, given $L_V^* = 2 \times 10^{10} L_{\odot,V}$, $\Phi^* = 0.005\,\text{Mpc}^{-3}$, and $\alpha = -1$?

12.3 On a mass scale $M = 10^{17}\,\text{M}_\odot$ the root mean square mass fluctuation is $\delta M/M = 0.12$ today (see Figure 11.5). Do you expect to see any gravitationally collapsed structures with a mass $M = 10^{17}\,\text{M}_\odot$ in the directly visible universe today? Explain why or why not.

12.4 The universe will end in a Big Rip if the dark energy takes the form of phantom energy with $w < -1$ (see Exercise 5.5). Since energy density has the dependence $\varepsilon \propto a^{-3(1+w)}$, the energy density ε_p of phantom energy increases as the universe expands; when ε_p/c^2 becomes larger than the mass density of a bound object, the object will be ripped apart. Suppose that the universe contains both matter and phantom energy with equation-of-state parameter $w_p = -1.1$. If the density parameters of the two components are $\Omega_{m,0} = 0.3$ and $\Omega_{p,0} = 0.7$, at what scale factor a_{gal} will a galaxy comparable to our Milky Way Galaxy be ripped apart? At what scale factor $a_\star$ will a star comparable to our Sun be ripped apart? If $H_0 = 68\,\text{km}\,\text{s}^{-1}\,\text{Mpc}^{-1}$, how many years before the Big Rip will a galaxy be ripped apart? How many years before the Big Rip will a sunlike star be ripped apart? [Hint: the Big Rip is defined as the time t_{rip} when $a \to \infty$. The result of Exercise 5.5 will be useful.]

Epilogue

A book dealing with an active field like cosmology can't really have a neat, tidy ending. Our understanding of the universe is still growing and evolving. During the twentieth century, the growing weight of evidence pointed toward the Hot Big Bang model, in which the universe started in a hot, dense state, but gradually cooled as it expanded. At the end of the twentieth century and the beginning of the twenty-first, cosmological evidence was gathered at an increasing rate, refining our knowledge of the universe. As I write this epilogue, on a clear spring day in the year 2016, the available evidence is explained by a Benchmark Model that is spatially flat and that has an expansion, which is currently accelerating. It seems that 69% of the energy density of the universe is contributed by a cosmological constant (or other form of "dark energy" with negative pressure). Only 31% of the energy density is contributed by matter (and only 4.8% is contributed by the familiar baryonic matter of which you and I are made).

However, many questions about the cosmos remain unanswered. Here are a few of the questions that currently nag at cosmologists:

- *What are the precise values of cosmological parameters such as H_0, $\Omega_{m,0}$, and $\Omega_{\Lambda,0}$?* Much effort has been invested in determining these parameters, but still they are not pinned down absolutely.
- *What is the dark matter?* It can't be made entirely of baryons. It can't be made entirely of neutrinos. Most of the dark matter must be in the form of some exotic stuff that has not yet been detected in laboratories.
- *What is the dark energy?* Is it vacuum energy that plays the role of a cosmological constant, or is it some other component of the universe with $-1 < w < -1/3$? If it is vacuum energy, is it provided by a false vacuum, driving a temporary inflationary stage, or are we finally seeing the true vacuum energy?
- *What drove inflation during the early universe?* Our knowledge of the particle physics behind inflation is still sadly incomplete. Indeed, some

cosmologists pose the questions, "Did inflation take place at all during the early universe? Is there another way to resolve the flatness, horizon, and monopole problems?"

- *Why is the universe expanding?* At one level, this question is easily answered. The universe is expanding today because it was expanding yesterday. It was expanding yesterday because it was expanding the day before yesterday... However, when we extrapolate back to the Planck time, we find that the universe was expanding then with a Hubble parameter $H \sim 1/t_P$. What determined this set of initial conditions? In other words, "What put the Bang in the Big Bang?"

The most interesting questions, however, are those that we are still too ignorant to pose correctly. For instance, in ancient Egypt, a list of unanswered questions in cosmology might have included "How high is the dome that makes up the sky?" and "What's the dome made of?" Severely erroneous models of the universe obviously give rise to irrelevant questions. The exciting, unsettling possibility exists that future observations will render the now-promising Benchmark Model obsolete. I hope, patient reader, that learning about cosmology from this book has encouraged you to become a cosmologist yourself, and to join the scientists who are laboring to make my book a quaint, out-of-date relic from a time when the universe was poorly understood.

Table of Useful Constants

Fundamental constants	
Gravitational constant	$G = 6.673 \times 10^{-11}\,\mathrm{m^3\,kg^{-1}\,s^{-2}}$
Speed of light	$c = 2.998 \times 10^{8}\,\mathrm{m\,s^{-1}}$
Reduced Planck constant	$\hbar = 1.055 \times 10^{-34}\,\mathrm{J\,s} = 6.582 \times 10^{-16}\,\mathrm{eV\,s}$
Boltzmann constant	$k = 1.381 \times 10^{-23}\,\mathrm{J\,K^{-1}} = 8.617 \times 10^{-5}\,\mathrm{eV\,K^{-1}}$
Electron rest energy	$m_e c^2 = 0.5110\,\mathrm{MeV}$
Proton rest energy	$m_p c^2 = 938.272\,\mathrm{MeV}$
Neutron rest energy	$m_n c^2 = 939.566\,\mathrm{MeV}$
Planck units	
Planck length	$\ell_P = (G\hbar/c^3)^{1/2} = 1.616 \times 10^{-35}\,\mathrm{m}$
Planck mass	$M_P = (\hbar c/G)^{1/2} = 2.177 \times 10^{-8}\,\mathrm{kg}$
Planck time	$t_P = (G\hbar/c^5)^{1/2} = 5.391 \times 10^{-44}\,\mathrm{s}$
Planck energy	$E_P = (\hbar c^5/G)^{1/2} = 1.956 \times 10^{9}\,\mathrm{J} = 1.221 \times 10^{28}\,\mathrm{eV}$
Planck temperature	$T_P = E_P/k = 1.417 \times 10^{32}\,\mathrm{K}$
Conversion of units	
Astronomical unit	$1\,\mathrm{AU} = 1.496 \times 10^{11}\,\mathrm{m}$
Megaparsec	$1\,\mathrm{Mpc} = 3.086 \times 10^{22}\,\mathrm{m}$
Solar mass	$1\,\mathrm{M_\odot} = 1.989 \times 10^{30}\,\mathrm{kg}$
Solar luminosity	$1\,\mathrm{L_\odot} = 3.828 \times 10^{26}\,\mathrm{J\,s^{-1}}$
Gigayear	$1\,\mathrm{Gyr} = 3.156 \times 10^{16}\,\mathrm{s}$
Electron volt	$1\,\mathrm{eV} = 1.602 \times 10^{-19}\,\mathrm{J}$
Cosmological parameters	
Hubble constant	$H_0 = 68 \pm 2\,\mathrm{km\,s^{-1}\,Mpc^{-1}}$
Hubble time	$H_0^{-1} = (4.54 \pm 0.13) \times 10^{17}\,\mathrm{s} = 14.4 \pm 0.4\,\mathrm{Gyr}$
Hubble distance	$c/H_0 = (1.35 \pm 0.04) \times 10^{26}\,\mathrm{m} = 4380 \pm 130\,\mathrm{Mpc}$
Critical energy density	$\varepsilon_{c,0} = 4870 \pm 290\,\mathrm{MeV\,m^{-3}}$
Critical mass density	$\rho_{c,0} = \varepsilon_{c,0}/c^2 = (8.7 \pm 0.5) \times 10^{-27}\,\mathrm{kg\,m^{-3}}$

Index

accelerating universe, 66, 119
acceleration equation, 60, 61, 104, 192, 214
  with cosmological constant, 64
acoustic oscillations, 162
active galactic nucleus (AGN), 237
age of universe (t_0)
  Benchmark Model, 92
  empty universe, 75
  flat universe
    matter + lambda, 92
    matter only, 80
    radiation only, 81
    single-component, 78
angular-diameter distance (d_A), 110–112, 157
  maximum value, 113
  vs. luminosity distance, 112
astronomical unit (AU), 2, 106
axions, 222

baryon, 19
baryon acoustic oscillations, 227
baryon–antibaryon asymmetry, 182
  in early universe, 183
baryon-to-photon ratio (η), 143, 151, 163, 176, 179, 183
  smallness of, 182
baryonic Jeans mass (M_J), 212
  after decoupling, 212
  before decoupling, 212
baryonic matter, 19
  circumgalactic gas, 233
  diffuse intergalactic gas, 233
  interstellar gas, 233
  intracluster gas, 233
  warm-hot intergalactic gas, 234
baryons
  energy density, 142
  number density, 142
bear
  polar, 107, 121
  teddy, 34, 43
Benchmark Model, 73, 85–86, 96–99, 187, 188, 230
beryllium (Be), 167, 178
Big Bang, 5, 16–18
Big Bang nucleosynthesis, 169–180
Big Bounce, 93, 94
  evidence against, 121
Big Chill, 86, 88, 90, 93
Big Crunch, 87, 88, 91, 93
  evidence against, 119
  time of (t_{crunch}), 88
Big Rip, 100
blackbody radiation, 21
blueshift, 13
bolometric flux (f), 107, 109
  Euclidean, 108
bottom-up scenario, 225
bremsstrahlung, 135, 246
  cooling time, 247
brown dwarf, 125, 126

Cepheid variable stars, 114
  period–luminosity relation, 114
chemical potential, 152
cluster of galaxies
  Abell 2218, 139
  Coma, 126, 130, 133–135
    mass of, 134
  Local Group, 10
  Virgo, 2, 115, 146

comoving coordinates, 43, 108, 111, 217, 239
Copernican principle, 11
correlation function, 227
  and power spectrum, 228
Cosmic Background Explorer (COBE), 144
cosmic microwave background, 23, 142
  amplitude, 146
  and baryon density, 163
  and flatness of universe, 163
  blackbody spectrum, 23, 144
  compared to starlight, 71–72
  cooling of, 25
  correlation function, 158, 159
  dipole distortion, 144
  discovery of, 143
  energy density, 23, 142
  number density, 24, 142
  origin of, 160–163
  temperature fluctuations, 146, 157
cosmic neutrino background, 72, 139
  energy density, 72
  number density, 139
cosmic time (t), 43
cosmological constant (Λ), 63–66, 194
  energy density, 64
  pressure, 65
cosmological principle, 12
  perfect, 17
cosmological proper time, *see* cosmic time
critical density (ε_c), 57
  current value, 57
curvature
  negative, 39
  positive, 38
curvature constant (κ), 40, 43

Dark Ages, 238
dark energy, 62, 69, 85, 226, 228
dark halo, 130
dark matter, 23, 131
  axions, 139
  cold, 221–222, 224–226
  hot, 221–223
  neutrinos, 139
  nonbaryonic, 128
  primordial black holes, 139
de Sitter universe, 83
deceleration parameter (q_0), 103
  Benchmark Model, 104
  sign convention, 103
density fluctuations
  lambda-dominated era, 215
  matter-dominated era, 216
  power spectrum, 219
  radiation-dominated era, 215
density parameter (Ω), 57
  baryons ($\Omega_{\rm bary}$), 128
  clusters of galaxies ($\Omega_{\rm clus}$), 135
  cosmological constant (Ω_Λ), 73
  matter (Ω_m), 73, 123
  planets ($\Omega_{\rm pl}$), 254
  radiation (Ω_r), 72
  stars ($\Omega_\star$), 126
deuterium (D), 167
deuterium abundance, 181
  determination of, 181
deuterium synthesis, 169, 174
  compared to recombination, 175
  temperature of, 176
deuteron-to-neutron ratio, 176
diproton, 173
distance modulus ($m - M$), 119
dust
  interstellar, 249

Einstein radius (θ_E), 137, 138
Einstein tensor, 51
Einstein's static universe, 65, 93
  instability, 65
  radius of curvature, 65
Einstein–de Sitter universe, 80
electron (e^-), 19
electron volt (eV), 3
empty universe, 74
  expanding, 75
  static, 75
energy density (ε)
  additive, 69
  baryons, 142
  CMB, 142
  flat universe, single-component, 79
  matter, 70
  radiation, 70
entropy, 59
equation of state, 61, 69
equilibrium
  chemical, 153, 175
  hydrostatic, 134, 209
  kinetic, 151, 170
  thermal, 21, 151
equivalence principle, 28
  and photons, 36
  and teddy bears, 34
expansion
  adiabatic, 58
  superluminal speed, 45

false vacuum, 200
driving inflation, 201
Fermat's principle, 36
first law of thermodynamics, 24, 58
flatness problem, 185–187
resolved by inflation, 194–195
fluid equation, 59, 61, 69
Fourier transform, 218
fractional ionization (X), 148, 154
free streaming, 222
freezeout, 172
Friedmann equation, 52, 61, 69, 186
with cosmological constant, 64
during inflation, 193
general relativistic, 55
Newtonian, 54
fundamental force, 190
electromagnetic, 27
electroweak, 190
Grand Unified Force, 190
gravitational, 27, 28
strong nuclear, 27
weak nuclear, 27, 172

galaxy
Andromeda (M31), 2, 10, 13, 129
distance, 115
Large Magellanic Cloud, 114, 115
distance to, 115
Milky Way, 2, 10, 129
central black hole, 237
luminosity of, 2
mass of, 130
mass-to-light ratio, 130
time of formation, 244
NGC 4874, 126
NGC 4889, 126
central black hole, 237
Small Magellanic Cloud, 114
galaxy formation, 243–247
cold flows, 248
galaxy luminosity function, 242
Gaussian field, 219
general relativity, 34–37
field equation, 37, 50, 51
geodesic, 36
geometry
Euclidean, 28
non-Euclidean, 49
gigayear (Gyr), 3
Grand Unified Theory (GUT), 189
phase transition, 191
gravitational instability, 206–209
gravitational lens, 136
cluster of galaxies, 138
MACHO, 136
gravitational potential energy, 131
gravity, Newton vs. Einstein, 36

helium (He), 167, 178
helium fraction (Y), 169
primordial, 173–174
homogeneity, 9, 11
horizon distance (d_{hor}), 9, 79, 195
Benchmark Model, 98
flat universe
matter only, 80
radiation only, 81
single-component, 79
at last scattering, 189
horizon problem, 185, 187–189
and isotropy of CMB, 189
resolved by inflation, 195–197
Hot Big Bang, 5, 25
and production of CMB, 147
Hubble constant (H_0), 13–15, 56, 103
Hubble distance (c/H), 17, 45
Hubble friction, 199, 201, 214
Hubble parameter (H), 56
Hubble time (H_0^{-1}), 16, 66
and age of universe, 16, 78
Hubble volume, 221
Hubble's law, 13
consequence of expansion, 15
hydrogen (H), 167

inflation, 192–202
and density perturbations, 202
and flatness problem, 194–195
and horizon problem, 195–197
and monopole problem, 197
inflaton field, 198, 201
ionization energy (Q), 148
isotropy, 9–11

Jeans length (λ_J), 210
radiation, 211

Keplerian rotation, 129
kinetic energy, 131

lambda, *see* cosmological constant
lambda CDM, 226
large scale structure, 204

last scattering, 147
time of (t_{ls}), 156
last scattering surface, 147
angular-diameter distance to, 157, 158, 189
thickness of, 157
Lemaître universe, 93
lepton, 19
lithium (Li), 167, 178
loitering universe, 93, 94
lookback time, 98
Benchmark Model, 99
luminosity distance (d_L), 107, 110
flat universe, 109
vs. angular-diameter distance, 112

MACHO, 136–138
magnetic monopole, 191
energy density, 192
number density, 192
magnitude, 117
absolute, 118
apparent, 118
mass
gravitational (m_g), 28
inertial (m_i), 28
mass-to-light ratio
cluster, 134
galaxy, 130
star, 124
matter-dominated universe
fate of, 89
negative curvature, 87
positive curvature, 87
matter–lambda equality, 73, 91, 92
Matthew effect, 206
Maxwell's equations, 31
mean free path, 149
megaparsec (Mpc), 2
megayear (Myr), 3
metric, 40
curved space, 40, 41
flat space, 40
homogeneous & isotropic, 41
Minkowski, 42
Robertson–Walker, 42
Milne universe, 75
molecular clouds, 248
moment of inertia, 132
monopole problem, 185, 189–192
resolved by inflation, 197

neutrino (ν), 20, 139, 222
flavors, 20
massive, 21
neutron (n), 19, 169
decay, 19, 170, 176
neutron-to-proton ratio, 171
after freezeout, 172
nuclear binding energy (B), 168
of deuterium, 168, 174
of helium, 168
nucleons, 167
null geodesic, 42
Minkowski space, 42
Robertson–Walker metric, 46

Olbers' paradox, 6–9, 17, 142
optical depth (τ), 156

parallax distance, 106
parsec (pc), 2
particle horizon distance, *see* horizon distance
peculiar motion, 55
perfect gas law, 61
phase transition, 191
loss of symmetry, 191
photodissociation, 21, 174
photoionization, 21, 148
photon (γ), 21
photon decoupling, 147, 150, 155
time (t_{dec}), 156
photon–baryon fluid, 162, 211
Planck (satellite), 144
Planck units
energy (E_P), 3
length (ℓ_P), 3
mass (M_P), 3
temperature (T_P), 3
time (t_P), 3, 82
Poisson distribution, 205, 220, 227
Poisson's equation, 29, 63, 161
compared to field equation, 50
with cosmological constant, 64
power spectrum, 219, 223
and correlation function, 228
Harrison–Zel'dovich, 219, 221
scale invariant, 221
pressure (P)
additive, 70
negative, 60

proper distance (d_p), 43
flat universe
lambda only, 83
matter only, 80
radiation only, 81
single-component, 79
Robertson–Walker metric, 44
time of emission, 77
time of observation, 76, 105
proton (p), 19, 169
protostar, 253
Pythagorean theorem, 37

quantum gravity, 82
quark, 19
quasar, 237
number density, 239

radiation–matter equality, 73, 95
radiative recombination, 148
radius of curvature (R), 40, 43
recombination, 147
compared to deuterium synthesis, 175
time of ($t_{\rm rec}$), 154
recombination temperature
crude approximation, 151
refined calculation, 154
redshift (z), 12
related to scale factor, 45–47
related to time, 74
redshift survey, 204
reference frame, 30
inertial, 30
reheating, 201
reionization, 234
and cosmic microwave background, 235
redshift of (z_*), 237
time of (t_*), 237
Robertson–Walker metric, 42

Sachs–Wolfe effect, 162
Saha equation, 153, 175
nucleosynthetic analog, 175
scale factor (a), 42, 43
empty universe, 75
flat universe
lambda only, 83
matter + lambda, 90, 91
matter only, 80
radiation + matter, 95
radiation only, 81
single-component, 78
negative curvature
matter only, 88
positive curvature
matter only, 88
Taylor expansion, 102, 103
solar luminosity ($L_\odot$), 2
solar mass ($M_\odot$), 2
sound speed (c_s), 210
baryonic gas, 212
photon gas, 212
special relativity, 30–34
first postulate, 31
second postulate, 31
spherical harmonics, 158
standard candle, 107
determining H_0, 114
standard yardstick, 110
angular resolution, 113
star
initial mass function, 124, 248
main sequence, 124
spectral type, 124
star formation, 250–253
first stars, 253
rate ($\dot\rho$), 240
statistical weight, 152
Steady State, 17–18
stress-energy tensor, 51
supercluster, 11, 205
supersymmetry, 140

Theory of Everything (TOE), 190
Thomson cross-section (σ_e), 149, 172
Thomson scattering, 149
top-down scenario, 224
topological defect, 191
cosmic string, 191
domain wall, 191
magnetic monopole, 191
transformation
Galilean, 31
Lorentz, 32
triangle
in curved space, 38, 39
in flat space, 37
tritium (^{3}He), 177
type Ia supernova, 116, 117
luminosity, 117

vacuum energy density, 66, 67
  vs. Planck energy density, 67
Virgocentric flow, 116
virial temperature, 245
virial theorem, 133, 244
  steady-state, 133
virialization, 244
visible universe, 80
void, 11, 205

Wilkinson Microwave Anisotropy Probe (WMAP), 144
WIMP, 23, 140, 221

year (yr), 2